D0208029

SOCIOLOGICAL METHODOLOGY
❧ 2016 ☙

AMERICAN SOCIOLOGICAL ASSOCIATION

SOCIOLOGICAL METHODOLOGY

2016

VOLUME 46

EDITOR

Duane F. Alwin

DEPUTY EDITOR

Jason R. Thomas

MANAGING EDITOR

Lisa Savage

EDITORIAL BOARD

Paul D. Allison

Carolyn J. Anderson

Jake Bowers

Elizabeth Bruch

Ronald S. Burt

Roberto P. Franzosi

Dana Garbarski

Guang Guo

Melissa Hardy

Guillermina Jasso

Burt L. Monroe

Stephen L. Morgan

Robert M. O'Brien

Raffaella Piccaretta

Lyn Spillman

Tanya Stivers

Katherine Stovel

Jeroen Vermunt

Kazuo Yamaguchi

EDITORIAL ASSISTANT: Judy L. Bowes

An official publication by SAGE Publishing for

THE AMERICAN SOCIOLOGICAL ASSOCIATION
SALLY T. HILLSMAN, *Executive Officer*

Sociological Methodology (SM) is the only American Sociological Association periodical publication devoted entirely to research methods. It is a compendium of new and sometimes controversial advances in social science methodology. Contributions come from diverse areas and have something new and useful—and sometimes surprising—to say about a wide range of methodological topics. *SM* seeks qualitative and quantitative contributions that address the full range of methodological problems confronted by empirical research in the social sciences, including conceptualization, data analysis, data collection, measurement, modeling, and research design. The journal provides a forum for engaging the philosophical issues that underpin sociological research. Papers published in *SM* are original methodological contributions including new methodological developments, reviews or illustrations of recent developments that provide new methodological insights, and critical evaluative discussions of research practices and traditions. *SM* encourages the inclusion of applications to real-world sociological data. *SM* is published annually as an edited, hardbound book. Manuscripts should be submitted electronically to http://mc.manuscriptcentral.com/smx.

Sociological Methodology is published annually—in August—by SAGE Publications, Inc., 2455 Teller Road, Thousand Oaks, CA 91320, on behalf of the American Sociological Association, 1430 K Street NW, Suite 600, Washington, DC 20005. Send address changes to *Sociological Methodology* c/o SAGE Publications, Inc., 2455 Teller Road, Thousand Oaks, CA 91320.

Non-Member Subscription Information: All non-member subscription inquiries, orders, back issues, claims, and renewals should be addressed to SAGE Publications, 2455 Teller Road, Thousand Oaks, CA 91320; telephone: (800) 818-SAGE (7243) and (805) 499-0721; fax: (805) 375-1700; e-mail: journals@sagepub.com; http://www.sagepublications.com. **Subscription Price:** Institutions: $446 (online/print), $402 (online only). Individual subscribers are required to hold ASA membership. For all customers outside the Americas, please visit http://www.sagepub.co.uk/customerCare.nav for information. **Claims:** Claims for undelivered copies must be made no later than six months following month of publication. The publisher will supply replacement issues when losses have been sustained in transit and when the reserve stock will permit.

Member Subscription Information: American Sociological Association member inquiries, change of address, back issues, claims, and membership renewal requests should be addressed to the Executive Office, American Sociological Association, 1430 K Street NW, Suite 600, Washington, DC 20005; Web site: http://www.asanet.org; e-mail: customer@asanet.org. Requests for replacement issues should be made within six months of the missing or damaged issue. Beyond six months and at the request of the American Sociological Association the publisher will supply replacement issues when losses have been sustained in transit and when the reserve stock permits.

Abstracting and Indexing: Please visit http://sm.sagepub.com and, under the "More about this journal" menu on the right-hand side, click on the Abstracting/Indexing link to view a full list of databases in which this journal is indexed.

Copyright Permission: Permission requests to photocopy or otherwise reproduce material published in this journal should be submitted by accessing the article online on the journal's Web site at http://sm.sagepub.com and selecting the "Request Permission" link. Permission may also be requested by contacting the Copyright Clearance Center via their Web site at http://www.copyright.com, or via e-mail at info@copyright.com.

Advertising and Reprints: Current advertising rates and specifications may be obtained by contacting the advertising coordinator in the Thousand Oaks office at (805) 410-7772 or by sending an e-mail to advertising@sagepub.com. To order reprints, please e-mail reprint@sagepub.com. Acceptance of advertising in this journal in no way implies endorsement of the advertised product or service by SAGE, the American Sociological Association, or the journal editor(s). No endorsement is intended or implied. SAGE reserves the right to reject any advertising it deems as inappropriate for this journal.

Change of Address for Non-Members: Six weeks' advance notice must be given when notifying of change of address. Please send the old address label along with the new address to the SAGE office address above to ensure proper identification. Please specify the name of the journal.

International Standard Serial Number ISSN 0081-1750
International Standard Book Number ISBN 978-1-5063-7756-8 (Vol. 46, 2016, hardcover)
Manufactured in the United States of America. First printing, August 2016.

Printed on acid-free paper

Reviewers

Paul Allison
Salvatore Babones
Michael Baumgartner
Andrew Bell
Robert Belli
Henning Best
Jaak Billiet
Ulf Bockenholt
Daniel M. Bolt
Jake Bowers
Elizabeth Bruch
Shawn Dorius
Glenn Firebaugh
Peer Fiss
Roberto Franzosi
Dana Garbarski
Jacques-Antoine
  Gauthier
Adam Glynn
Amir Goldberg
Melissa A. Hardy
Andrew Hayes

Douglas Heckathorn
Guillermina Jasso
Gary King
Kenneth C. Land
Liying Luo
Jennifer Madans
Jay Magidson
Aaron Maitland
Peter Marsden
Nancy Mathiowetz
Peter Miller
Burt L. Monroe
Stephen L. Morgan
Thomas Brendan
  Murphy
Robert M. O'Brien
Wendy Olsen
Kristen Olson
Raffaella Piccarreta
Jeremy Porter
Charles Ragin
Margaret Roberts

Nicolas Robette
Thomas Rotolo
Matthew Salganik
Nora Cate Schaeffer
Rainer Schnell
Jason Seawright
Herbert L. Smith
Jeffrey Smith
Lyn Spillman
Brandon Stewart
Volker Stocké
Luca Tardella
Roger Tourangeau
Stephen Vaisey
Ashton Verdery
Barbara Vis
Edward Walker
Christopher Winship
Yu Xie
Kazuo Yamaguchi
Yang Yang

Past Editors
Sociological Methodology

Volumes	Years	Editors
1	1969	Edgar F. Borgatta, editor
2	1970	Edgar F. Borgatta and George W. Bohrnstedt, coeditors
3–5	1971–1974	Herbert L. Costner, editor
6–8	1975–1977	David R. Heise, editor
9–11	1978–1980	Karl F. Schuessler, editor
12–14	1981–1984	Samuel Leinhardt, editor
15–16	1985–1986	Nancy Brandon Tuma, editor
17–20	1987–1990	Clifford C. Clogg, editor
21–25	1991–1995	Peter V. Marsden, editor
26–28	1996–1998	Adrian E. Raftery, editor
29–31	1999–2001	Michael E. Sobel and Mark P. Becker, coeditors
32–36	2002–2006	Ross M. Stolzenberg, editor
37–39	2007–2009	Yu Xie, editor
40–45	2010–2015	Tim Futing Liao, editor
46–48	2016–2018	Duane F. Alwin, editor

Volume 46 August 2016

SOCIOLOGICAL METHODOLOGY

Contents

Symposium on Interviewing

Commentaries

Rejoinder

Survey Measurement

Sociological Methodology
2016, Vol. 46(1) ix–xvi
© American Sociological Association 2016
DOI: 10.1177/0081175016664881
http://sm.sagepub.com

DEDICATION

IN THE SHADOW OF A GIANT[1]

This volume of *Sociological Methodology* (*SM*) is dedicated to the journal's founder and first editor, Edgar F. "Ed" Borgatta, who died in Seattle, Washington, on February 20, 2016, at age 91. Some people referred to him as "Big Ed," as in physical height he was over six feet, but he was a "giant" in more than one sense. Ed was a beloved leader, scholar, teacher, and friend to many in the sociological community. He mentored many postdocs and graduate students who went on to become well known in their own right. He had a special relationship with a number of Italian graduate students, who in turn significantly influenced the Italian social science professional community. I was one of several graduate students taught by Ed Borgatta at the University of Wisconsin–Madison during the late 1960s. In fact, I was a member of Ed's first cohort of students in the National Institutes of Health (NIH) training program in methodology at Wisconsin in 1966, and he supervised the completion of my master's thesis. To members of my generation, he was an influential figure in sociology who did important things. Many of those in more recent cohorts of sociologists do not know very much about him, nor about his part in shaping the role of *SM* and other journals in the discipline of sociology. Hence, in addition to dedicating this volume to Ed Borgatta, I wanted to provide a short summary of his biography, which played out during an important formative period in American sociology. His life and his work are worth knowing about, and in this tribute I present some of his background and briefly discuss his role in developing scholarly journals dealing with methodology in

Edgar F. Borgatta, 1924–2016 (photo credit © Marie L. Borgatta)

sociology. I try to express the essence of my generation's appreciation of his life and work.

Ed Borgatta was born in Milan in 1924, of Mexican and Italian heritage. His father was Mexican. His mother was Italian. His parents emigrated from Italy to New York in 1929, in part to get away from fascism but also related to his father's pursuit of a career in communications. Ed arrived in New York City as a kindergartner who spoke no English, but he soon caught up, graduating from high school at age 16. He attended Queens College in New York City the following year. He entered the military at the age of 18 (serving from 1942 to 1946). During World War II, according to his personal narrative, he spent a lot of time on KP duty, and the closest he came to danger was at the invasion of Okinawa.

After the war, Ed returned to New York, obtaining a master's from New York University in 1949 and completing a PhD in sociology in 1951. After receiving his doctorate, he took a position as lecturer and research associate in social relations at Harvard University for three years (1951–1953),[2] and from there his professional career (among many other memberships, appointments, and consultancies) encompassed employment by the Russell Sage Foundation (1954–1959), New York University (1954–1959), Cornell University (1959–1961), the University of Wisconsin–Madison (1961–1972), the Graduate Center of the City University of New York (1972–1981), and the University of Washington (1980–1993), where he was also the first director of the Institute on Aging (1980–1986). At the time of his retirement from the University of Washington, Otto Larsen, former American Sociological Association (ASA) executive officer and an emeritus professor at the University of Washington, said that Ed's curriculum vitae reflected "one of the most vital and productive careers in the history of American sociology. The scope and quality of the scholarly contents are unprecedented. Gathered under one cover, they would constitute an encyclopedia" (Bohrnstedt and Montgomery 2016).

Borgatta's scholarly research interests were in quantitative methodology, social psychology, factorial ecology and demography, and, later, social gerontology. He had a rich set of professional relationships over his career. The extent of his ties to important figures in the history of social science is one of the most illuminating things about his background. For instance, during his graduate studies and prior to receiving his PhD in sociology from New York University, he was supported as a research assistant and instructor with the Sociometric Institute in New York, working with Jacob L. Moreno, the founder of the institute and the founder (in 1937) of the journal *Sociometry*, who is also considered the founder of early social network analysis. At the institute, Borgatta became involved in Moreno's research agenda, which included sociometric analysis and a variety of other activities. Borgatta essentially worked as Moreno's research assistant for several years.

Later, after he moved to Harvard, Borgatta continued working with Moreno on *Sociometry* and became an associate editor of the journal in the early 1950s. Eventually this turned into the position of (acting) editor for four years, from 1952 through 1955 (volumes 15–18).[3] Then, in 1956, Moreno worked out an arrangement to transfer ownership of the journal, as a gift without any conditions, to the American Sociological Society (ASS), as the ASA was known at the time. With volume 19, under the editorship of Leonard S. Cottrell Jr., it officially became *Sociometry: A Journal of Research in Social Psychology*, published by the ASS. Ed played a role in that transition. It was the second periodical that was sponsored by the national association, the first being the *American Sociological Review*. The name of the journal was later changed to *Social Psychology Quarterly*, as it is known today. Borgatta retained a position on the editorial board for many years thereafter, and through his support of the journal, he was an important force within the sociological social psychology research community.

At Harvard, Borgatta initially became an apprentice of Samuel A. Stouffer, the legendary social psychologist, who had been the lead author on the famous publication *The American Soldier*, a postwar study of the military and the impact of World War II on the lives of Americans. Stouffer originated the concept of "relative deprivation," an essential idea in modern social psychology, popularized by Robert Merton and others. Stouffer also specialized in the measurement of attitudes, a topic that attracted Borgatta. Ed often spoke endearingly of the things he had learned from "Sam" Stouffer. In addition to his work with Stouffer, and to his work editing *Sociometry* for Moreno, he established his own program of research at Harvard. He was considered the "go to" person on research methods and data analysis in the department. In addition to Stouffer, he had contact at Harvard with an interdisciplinary array of established scholars, including Talcott Parsons, George Homans, Robert Freed Bales, Frederick Mosteller, and Clyde Kluckhohn, among others. Borgatta particularly associated himself with Freed Bales and coedited a volume on small groups with Bales and A. Paul Hare (who was at Haverford). Their reader, *Small Groups: Studies in Social Interaction* (Hare, Borgatta, and Bales 1966), was one of the standard textbooks in the field of group dynamics in the 1960s—I still have a marked-up copy in my library.

After several career moves, Borgatta was recruited by the University of Wisconsin–Madison, where he chaired the Department of Sociology during the turbulent 1960s. At that time, Borgatta undertook a number of far-reaching initiatives and, with the support of others, helped propel the Wisconsin sociology department into one of the top graduate programs in the United States (some people would say *the* top program in quantitative sociology), a position it held for many years. One of these was his leadership in promoting the idea that support for graduate training was in the national interest and should be subsidized with federal funding. Wisconsin pioneered in establishing federally funded (mostly NIH-funded) training programs—in methodology, demography, economic development, social psychology, and mental health—a model other universities followed. In 1966, Borgatta established and led the Wisconsin methodology training program, assisted also by his postdocs David R. Heise and George W. Bohrnstedt. This training program continued to receive federal funding for at least a decade (thanks to the efforts of Robert M. Hauser, Jerry Marwell, and others), and over the years, this was a program through which literally scores of Wisconsin graduate students passed. Bob Hauser kept the methodology seminar going for many years, long after the NIH funding ended. Another of Borgatta's initiatives, which he undertook in collaboration with William H. Sewell, among others, was to convince the dean of arts and sciences at Wisconsin to strengthen the Department of Sociology through the recruitment of young research-oriented scholars. This resulted in the recruitment of sizable cohorts of junior faculty members to Wisconsin in the 1960s, among whom were Larry Bumpass, Geraldine Clausen, John DeLamater, David Featherman, Elaine Hatfield, Robert M. Hauser, Gerald Marwell, Andrew Michener, Jane Piliavin, Irving Piliavin, Shalom Schwartz, Jim Sweet, Donald Treiman, and Hal Winsborough.

Borgatta had broad research interests in the study of social behavior, and he was an early proponent of quantitative measurement, especially using techniques such as Guttman scaling (also known as scalogram analysis) and exploratory factor analysis. Ed was proud to be known as a "factor analyst." He was a prolific writer, the author of more than 30 books and literally hundreds of articles and book chapters, many dealing with methodology. He was also a founding editor of several professional journals. Not only did Borgatta found *SM*, he also founded (with George Bohrnstedt) *Sociological Methods and Research* and, later, *Research on Aging*, all now published by SAGE Publications. His last major publishing effort was the massive five-volume *Encyclopedia of Sociology*, which he coedited with Marie Borgatta in its first edition (Borgatta and Borgatta 1992) and later with Rhonda Montgomery in its second edition (Borgatta and Montgomery 2000). He was a major international figure in American (and also Italian) sociology and during the 1970s and 1980s was active in the International Sociological Association's Research Committee on Logic and Methodology (RC 33). Over his lifetime, Ed received many accolades for his intellectual and scholarly accomplishments, but perhaps he will be best remembered for his work in promoting high standards for methodological training and his establishment of journals devoted entirely to the study of sociological methods.

Academia was not Ed's only pursuit—my impression was that to him it was more of a job than a lifestyle. For example, he was interested in human rights, especially the rights of children, working and publishing on these matters. And anyone who knew Ed was aware of the importance he placed on his family. His enduring loyalty to and passion for his wife, Marie Borgatta (a scholar in her own right), with whom he enjoyed 69 years of marriage, was well known, as well as the affection he had for his three daughters, Lynn, Kim, and Lee (all intentionally given androgynous names). Ed lived life fully, adventuring into fields completely unrelated to his professional life in sociology. For example, few people know that Ed was a "rock hound," a hobby he pursued for many years. After becoming involved in a stone polishing school project with one of his daughters, he developed a keen interest in gemology and became a Gemology Institute of America certified gemologist. He enjoyed many years of working with the stones himself, and he donated rough specimens and finished gems to a number of museums. Perhaps the most important of his donations was to the Smithsonian Institution in Washington, which "received the donation by Marie and Edgar Borgatta of the American Golden Topaz, 22,000 + carats, faceted by Leon Agee of Washington, one of the largest faceted gemstones in the world" ("American Golden Topaz" 2016).

Ed was also, after a fashion, a farmer. He loved Vermont, where he and Marie had a place in Bennington County in the southern part of the state for many years. Starting with the house by the covered bridge in Arlington, they expanded their territory to include considerable forest acreage near Rupert, Vermont. When we could not locate Borgatta when he was teaching at Wisconsin, the message that frequently came back was that he was in Rupert. Ed enjoyed this lifestyle, and he reportedly became one of the first tree farmers recognized by the state of Vermont, selling off some of the timber to support other activities. Upon his move to Seattle in 1980, leaving the farm in Vermont behind, Ed turned his attention to boating and sailing in Puget Sound and environs. He and Marie owned a house on Bainbridge Island, and their flower garden became the focus of a new hobby of flower photography. The cover art for this volume of *SM* is one of Ed's flower photos—an iris in their garden.

One of Borgatta's traits that many of us relished was his sense of humor, especially his use of humor in making arguments about theory and methods in sociology. There were several astonishingly satirical pieces he wrote that were favorites of his students. One of Ed's good-natured expositions, "My Student the Purist: A Lament" (Borgatta 1968), was ostensibly about meeting the measurement assumptions involved in the implementation of statistical tests, but it was really a plea for the improvement of our measures in sociology. The narrative in this essay was that "his student" objected to his use of statistical tests that assumed interval scaling when his measures were no more than ordinal. After all, Blalock's *Social Statistics* (Blalock 1960), the standard statistics textbook used in most sociology departments in those days, insisted that certain assumptions must be met in order to apply these statistical tools. Rather than view this issue in absolutist terms, Ed discussed the matter in terms of trade-offs, arguing that given the quality of our measures, there were many advantages to looking at one's data, regardless of how crude our empirical tests may be.

Ed Borgatta was an extraordinary, multifaceted person. This is one of those cases in which it is literally true that "when he was born, they threw away the mold." Whether as husband, father, friend, sociologist, social psychologist, gerontologist, methodologist, factor analyst, scholar, editor, mentor, teacher, researcher, gemologist, farmer, photographer, or visionary, Ed Borgatta will be remembered by all who knew him.[4] What is important in the present context is his role in the training of graduate students in methodology and establishing methodology journals for future generations of sociologists. In this sense, Ed was a visionary. In his own words, regarding his training activities in Madison, "possibly the most important thing in which [he] was involved was in maintaining a major cohort of graduate students who represented the future standards for methodologically well-trained sociologists." Soon after he established these training activities, he founded *SM*, with the support of the ASA, and a few years later, he and George Bohrnstedt founded *Sociological Methods and Research*, both of which continue to be thriving enterprises aimed at publishing innovations and applications of methodology. Ed's cohorts of "well-trained sociologists" have carried these activities forward and have preserved a culture that focuses on methodology, involving the most rigorous of methodological innovations but also an emphasis on their practical application in the study of social relations. It is thus with a sense of humility and respect that we dedicate this volume of *SM* to Ed Borgatta, a giant in whose shadow we stand.

—Duane F. Alwin
Editor

Notes

1. Much of the factual information presented here was obtained from *Edgar F. Borgatta: Freedom in Sociology*, edited by Alberto Gasparini and Bruno Tellia (2008), and from his obituary in the *Seattle Times* (March 17 to March 20, 2016). *Freedom in Sociology* includes a lengthy autobiographical narrative written by Borgatta, from which I have drawn in writing this essay. Also, there is a wonderful obituary for him in the American Sociological Association's *Footnotes*, written by George Bohrnstedt and Rhonda Montgomery (2016), two of his students. I acknowledge the input of Alan Sica, Dick Campbell, Jason Thomas, and Lowell Hargens in crafting this essay.
2. The Department of Social Relations at Harvard, begun in 1946, was an interdisciplinary collaboration of three social science departments (anthropology, psychology, and sociology), which lasted until 1972, when the program was disbanded and the departments resumed their separate status.
3. In other words, Ed edited *Social Psychology Quarterly* before it became *Social Psychology Quarterly*. There was also another journal involved, which he edited in connection with Moreno. His role with the journal *Group Psychotherapy and Psychodrama* is discussed in his narrative and was an important part of Borgatta's early career interests. It is baffling to imagine a newly minted PhD essentially editing two journals, while beginning a teaching career and starting a research program of his own.
4. I have taken pains to list as many of the roles Ed would have listed to describe himself if asked, an effort inspired by my master's thesis research, which he supervised, a study titled "A Factor Analysis of the 'Who Am I' Test."

References

"American Golden Topaz." 2016. Retrieved August 3, 2016 (https://en.wikipedia.org/wiki/American_Golden_Topaz).

Blalock, Hubert M. 1960. *Social Statistics*. New York, NY: McGraw-Hill.

Bohrnstedt, George, and Rhonda Montgomery. 2016. "Ed Borgatta: 1924–2016." *Footnotes*. Retrieved August 3, 2016 (http://www.asanet.org/news-events/footnotes/jul-aug-2016/announce/obituaries).

Borgatta, Edgar F. 1968. "My Student the Purist: A Lament." *Sociological Quarterly* 9(1):29–34.

Borgatta, Edgar F., and Marie Borgatta, eds. (1992). *Encyclopedia of Sociology*. New York: Macmillan.

Borgatta, Edgar F., and Rhonda Montgomery, eds. (2000). *Encyclopedia of Sociology*. 2nd ed. New York: Macmillan.

Gasparini, Alberto, and Bruno Tellia, eds. 2008. *Edgar F. Borgatta: Freedom in Sociology*. Gorizia, Italy: Institute of International Sociology.

Hare, A. Paul, Edgar F. Borgatta, and Robert Freed Bales. 1966. *Small Groups: Studies in Social Interaction*. New York: Knopf.

Sociological Methodology
2016, Vol. 46(1) xvii–xxxiv
© American Sociological Association 2016
DOI: 10.1177/0081175016664887
http://sm.sagepub.com

PROLOGUE

THE PAST

To borrow a line from Shakespeare, "what's past is prologue" (*The Tempest*, Act II, Scene 1), and in keeping with this thought, we begin this prologue by talking briefly about the *past*. We focus on some of the trends in methodology over the time *Sociological Methodology* (*SM*) has been in existence and briefly discuss the nature of improvements to methodology in sociology. We then summarize the chapters in the *present* volume, focusing on their unique contributions to sociological methodology. Finally, we discuss our expectations with respect to the kinds of methodological issues *future* volumes will bring.

When Borgatta introduced the first volume of *SM*, he cautioned,

> While many major improvements in methodology are occurring in sociology, it is clear that few are unique to sociology. Most of these developments have been borrowed from other fields and sometimes blindly applied to sociological problems. Each may be viewed as the "breakthrough" by the profession and adopted as though it is most certainly the methodological panacea all have been waiting for. These fads come and go; ultimately, however, these tools are only as good as the scientific intuition and imagination of the researcher using them. (Borgatta 1969:xiv)

True enough, historically, new developments in sociological methodology were formulated on the basis of earlier innovations in other fields. This is probably *not* as true today as it was then, because it is increasingly the case that methods have been developed within the discipline and reflect improvements over past solutions (see below). One example that may have been on Borgatta's mind was the publication of several articles on path analysis in the first few volumes of *SM* (Heise 1969; Land 1969; Duncan 1969, 1970). This is an example of what

Borgatta observed, in that path analysis was developed in the field of genetics by Sewall Wright as a deductive tool for inferring correlations among genetic traits across generations. At the time, it may have been considered a "fad" in sociology. Path analysis, based on Pearson correlations, was essentially an algebra for writing variances and covariances among variables, given a path or flow diagram. Although it would be accurate to say that "nobody uses path analysis anymore," which one often hears, it would be a mistake to conclude that path analysis was no more than a fad. In fact, with the work of Hauser and Goldberger (1971), and the influence of psychometrician Karl Jöreskog and his students on confirmatory factor analysis, including especially Bengt Muthén (1997), among others, topics such as path analysis and confirmatory factor analysis morphed into the field known as "covariance structure analysis" or "structural equation modeling" (SEM) in today's language. At this time, we might put all of the following in the SEM category: multilevel models, latent class factor models, item response theory models, growth models, and many other techniques, especially if they include both a measurement model and a structural model simultaneously. But it is true that in the modern world of counterfactuals, collider variables, fixed effects, random effects, and so on, "nobody does path analysis anymore."

We would argue, however, that the use of any model in sociology is based on a logic of causal relations that can be learned from path analysis (correlated causes, spuriousness, indirect effects, reciprocal effects, and measurement errors). The descendant models (i.e., SEM models) are alive and well within the field, and although not a panacea, they have been prominent in the development of sociological methodology. There is even an entire journal devoted to SEM (*Structural Equation Modeling: A Multidisciplinary Journal*, published by Taylor & Francis), begun in the early 1990s. One could argue that these developments provided an alternative to publishing articles on SEM in *SM*, but by our estimation, SEM models continue to be an important topic covered by publications in *SM*.

In part, the changes in the models attended to by sociological methodology are changes in the languages that surround the analysis of certain problems. For example, an early *SM* article on time-dependent probability processes, or "stochastic processes," by Thomas Fararo (1969) focused on a set of mathematical models for expressing human behavior in terms of a function of time. This type of application, from mathematics, found a home in sociology, among other fields, involving a set of models that eventually would be classified as "survival analysis" or "event-history analysis" (e.g., see Allison 1982; Barber, Murphy, and Verbitsky 2004; Flinn and Heckman 1982; Tuma 1982; Tuma and Hannan 1979), and it also serves as an important building block in the analysis of time-series data (Brockwell and Davis 2002). People who use event-history techniques often do not relate their applications to these earlier foundations, but much can be gained from applying the theory of stochastic processes to the analysis of waiting times (Aalen, Borgan, and Gjessing 2008).

Although some applications of sociological methodology have evolved from developments in other fields, what we are now seeing is many more innovations within sociology. For example, on the basis of the "chain-referral sampling designs" discussed early by James Coleman (1958), sociologists have developed the use of respondent-driven sampling (RDS), formerly referred to as snowball sampling (also known as link-tracing sampling and random-walk sampling) to estimate the parameters of hard-to-find populations (see Heckathorn 1997). The work of Salganik and Heckathorn (2004), published in *SM*, has pioneered an approach that relies on information from social networks to develop a sampling frame for finding members of hidden populations and then developing statistical models for estimating their characteristics. These methods have been used in the work of researchers in several disciplines (e.g., epidemiology and demography), but sociology continues to be a test bed for sampling hard-to-find populations. It is indicative of the perpetuation of this thread of methodological work that the present volume includes two chapters dealing with these issues (see below).

Perhaps the iconic example of methodology that has primarily matured through the efforts of sociologists consists of the techniques for analyzing cross-classified data (e.g., Clogg 1992; Raftery 2001; Sobel 1996). The development of log-linear and association models, among others, has been featured in *SM* over the past several decades, including several key contributions (e.g., Davis 1973; Goodman and Hout 1998; Hauser 1980; Hout, Duncan, and Sobel 1987) as well as more recent advances (Németh and Rudas 2013; Yamaguchi 2012). These developments draw attention to the intersection of sociology and statistics, which has grown over the past few decades. Principally through the mechanism of dual training, as well as the attraction to sociology of scholars trained initially in statistics, we now have a subset of sociological methodologists who are also trained in statistics. We caution, however, that there is a difference—that is, between _methodology_ and _statistics_. Borgatta taught that research methodology included statistics, but methodology was not the same as statistics. Adherence to Borgatta's principle is one of the things that has prevented _SM_ from devolving entirely into a journal on multivariate statistics.

At the same time, the intersection of statistics and sociology has been an important development, and this is evident in the pages of _SM_. There is no better description of this trend than Raftery's (2001) comprehensive review, "Statistics in Sociology," which argued that the inroads statistics has made into the field of sociology have "greatly improved the standard of scientific rigor in the discipline" (p. 1). Statistical methods were not always present in sociology until the developments of the past few decades. Writing in 1950, Patricia Kendall and Paul Lazarsfeld commented that, despite many advances in various aspects of survey research procedures, "there is very little discussion of the art of analyzing material once it has been collected" (Lazarsfeld and Merton [1950] 1974:133). At that time, the use of cross-tabulations was the mainstay of that early work, which yielded eventually to work (mentioned above) on categorical data analysis. In the short space of 50 or 60 years, we have come a long way. Raftery described several trends in statistical methodology in sociology that suggest a series of developments that have emerged since the 1950s, including categorical data analysis, SEM models, latent class models, Bayesian inference and model selection, event-history analysis, limited dependent variables, social networks, spatial analysis, multilevel models, approaches to missing data, and causality. We would emphasize that in addition to statistical methods, the pages of _SM_ have revealed a relatively continuous concern over the years with issues of measurement, either issues of design and process (e.g., Cannell, Miller, and Oksenberg 1981) or those concerned with data quality (e.g., Alwin 1992). Although all of these topics have received fairly consistent attention over time, others have made only brief or sporadic appearances on the pages of _SM_. It is interesting to ask, for example, if the two-sided logit model (Logan 1998), relative distribution methods (Handcock and Morris 1998), multilevel diffusion curves (Rossman, Chiu, and Mol 2008), multichannel sequence analysis (Gauthier et al. 2010), fuzzy sets (Montgomery 2000), or progressive supervised-learning approaches to text analysis (Nardulli, Althaus, and

Hayes 2015) are indeed fads, as cautioned by Borgatta, or simply methods ahead of their time.

THE PRESENT

The material in the *present* volume covers a range of issues—from data collection, to statistical models, to data analysis. We have organized the chapters into six major sections, each containing two chapters: (1) interviewing, (2) survey measurement, (3) inferences to hidden populations, (4) aggregation issues in statistical models, (5) methods for demographic and economic data, and (6) comments.

Interviewing

The interview is still considered by many to be the gold standard for obtaining information from individuals and households in social research, although meeting the costs of this mode of inquiry is becoming increasingly difficult. Many government surveys continue to rely on face-to-face interviews, in addition to popular surveys such as the General Social Survey, which has used this mode of interviewing throughout its history. Hence, interviewing remains an important focus of study by sociologists, and through transcribed interviews, the characteristics of interviewer-respondent interaction can be investigated in a serious manner.

This approach is revealed in the symposium on interviewing presented in this volume. There are two major articles, one from a standardized survey interviewing perspective, and one from a qualitative interviewing point of view. "Interviewing Practices, Conversational Practices, and Rapport: Responsiveness and Engagement in the Standardized Survey Interview," by Dana Garbarski, Nora Cate Schaeffer, and Jennifer Dykema, focuses specifically on the evaluation of the concept of rapport as an interactional phenomenon in the survey interview. The concept of rapport is at the heart of traditional standards for evaluating the quality of information gained in the survey interview, and yet it may not always lead to improved data quality and future participation in the study. Garbarski et al. established several criteria for defining rapport as a phenomenon in the survey interview and focus critical attention on what the concept of rapport means. They use the method of *behavior coding* to assess the behavior of both the interviewer and the respondent. They identify aspects of rapport (responsiveness and engagement) and describe their existence using data from questions about end-of-life planning in the Wisconsin Longitudinal Study. Garbarski et al.'s chapter is accompanied by commentaries by scholars from three areas of expertise: survey methods (Peter V. Miller), cognitive psychology (Michael F. Schober), and conversation analysis (Johanna Ruusuvuori). Each of these informative commentaries finds interesting implications of the work, and they are generally unified behind the importance of studying interviewer-respondent interaction.

The article on qualitative interviewing (also in the symposium), "Eliciting Frontstage and Backstage Talk with the Iterated Questioning Approach," by Laura Robinson and Jeremy Schulz, is also about data quality, and it is an interesting complement to the standardized survey interview discussed by Garbarski et al., because it is essentially a "nonstandardized" (or "in-depth") interviewing approach. This offers an interesting counterpoint to Garbarski et al.'s chapter, by underscoring the traditional assumption that the development of *rapport* is essential to obtaining useful data. This work is in the tradition of Robert Weiss's (1994) *Learning from Strangers*. Invoking an idea from Erving Goffman's (1959) dramaturgical framework of behavior, particularly the differences between frontstage versus backstage talk, Robinson and Schulz present the iterated questioning approach, an innovative approach to gathering data in qualitative research, by repeated interviewing. The authors' presentation advocates for a four-step process: (1) establishing the baseline iterated question, (2) eliciting frontstage talk, (3) going backstage, and (4) eliciting backstage talk. They rely on data from the application of this technique from studies of high school students and business professionals. The basic idea is that by "going backstage" to elicit information that complements what was gathered frontstage, the authors assert, through the use of these innovative strategies, the overall quality of the information and the replicability of findings are increased.

Survey Measurement

Survey measurement plays an extraordinarily important role in contemporary social science, and given the substantial resources invested each year in survey data collection, research into the best approaches to survey measurement is strongly warranted. Measurement error in sample surveys has serious consequences in the study of social behavior, regardless of the characteristics of the subject population. Without valid and reliable measurement, the quantitative analysis of data hardly makes sense; yet there is a general lack of empirical information about these problems. This section includes two chapters on survey measurement. "Assessing the Effectiveness of Anchoring Vignettes in Bias Reduction for Socioeconomic Disparities in Self-rated Health among Chinese Adults," by Hongwei Xu and Yu Xie, seeks to address the linkage between the nature of the measures and estimates of the nature of social processes. The authors focus on potential biases in the study of health disparities and the use of "vignettes methodology" for anchoring responses to questions about self-rated health. By adjusting self-ratings on the basis of the vignette data and thereby reducing measurement error, Xu and Xie suggest that they can improve estimates of relationships between socioeconomic variables and health self-ratings. Specifically, they take as problematic the relationship between socioeconomic factors and self-reported health, noting that a large literature from the United States and Europe consistently demonstrates a positive relationship, but studies in developing countries often show an inverse relationship. They propose that rather than suggesting cross-national differences, these findings may reflect differences in referential comparisons, and they suggest that using vignettes to anchor responses may improve estimates of health disparities. They argue that persons of different socioeconomic status backgrounds may be thinking of quite different things when they self-report their health, and the authors discover a standard principle in cognitive psychology (e.g., see Kahneman 2011) that anchoring heuristics affect response variability. Using a concept of "reporting heterogeneity," the authors explore the question using data from the People's Republic of China: the China Family Panel Studies. They conclude that the use of anchoring techniques may have more general applicability beyond questions of self-rated health.

In "The KISS Principle in Survey Design: Question Length and Data Quality," Duane F. Alwin and Brett A. Beattie emphasize the importance of simplicity in survey question design and report the results of a systematic study of the relationship between question length (i.e., the number of words in a question), and the reliability of measurement.[1] The literature in survey methods is mixed in its advice about this aspect of question writing, and little empirical evidence has been brought to bear in assessing question length in relation to data quality. Alwin and Beattie argue that longer questions may increase the complexity of the task and demands for comprehension on the part of the respondent, creating the potential for measurement error. Focusing on single survey questions, they estimate the effects of question length on reliability using data from a larger project dealing with the improvement of the quality of the data obtained from surveys of the U.S. population. Their results indicate that question length interacts with survey context—whether the question was a part of a topical series, or battery, versus being a stand-alone question—and that the effects of question length are robust with regard to controls for question content. The results reinforce conclusions from previous studies that excessive verbiage in survey questions, either in the question text or in the introduction to the question, has potentially negative consequences for the quality of measurement, thus supporting the KISS principle concerning simplicity and parsimony in survey questionnaire design.

Inferences to Hidden Populations

In many cases, standard survey sampling strategies do not work, because it is not possible to develop well-defined sampling frames for the target populations of interest, because they are either rare or "hidden" for some other reason (e.g., intravenous drug use). In this section we present two articles on sampling hard-to-find populations. "Generalizing the Network Scale-up Method: A New Estimator for the Size of Hidden Populations," by Dennis M. Feehan and Matthew J. Salganik, falls directly in the lineage of this early work and focuses on the network scale-up method that enables researchers to estimate the size of hidden populations. Their new generalized network scale-up estimator improves upon earlier scale-up estimators by relaxing some assumptions and accommodating incomplete sampling frames and complex sample designs. The second article, "The Graphical Structure of Respondent-driven Sampling," by Forrest W. Crawford, focuses on some of the fundamental problems with the recruitment process in RDS and the fact that hidden social networks are only partially observed. This chapter develops an alternative approach on the basis of simple and realistic assumptions that interprets the hidden network as an exponential random graph model (ERGM). The author validates the method using simulated data and applies the technique to an RDS study of drug users in Russia.

Aggregation Issues in Statistical Models

In this section we include two articles that deal in one way or another with aggregation issues prominent in sociology, including robust estimation of measures of inequality for small areas and goodness of fit in subgroups in multilevel latent class (MLLC) models. The first article, "Robust Estimation of Inequality from Binned Incomes," by Paul T. von Hippel, Samuel V. Scarpino, and Igor Holas, focuses on the estimation of income inequality from binned income data. Typically, information on income is gathered using bracketed income categories, otherwise known as "binned data" on income. The chapter was inspired by the lead author's interest in the measurement of inequality at the level of the school district in the United States. The authors argue that popular methods for estimating inequality from such data are not robust in small samples, such as neighborhoods, school districts, and counties. They present the largest evaluation of binned-data estimators conducted to date, improving our understanding of the problems of estimating inequality and making their computational approach available to others.

Latent class models are often used to develop a latent or unobserved classification of subjects on the basis of multiple observed categorical variables. The extension of the latent class model to the multilevel framework— MLLC models—allows the simultaneous classification of individuals and groups, but it increases the complexity of evaluating the appropriateness of the model. In "Goodness-of-fit of Multilevel Latent Class Models for Categorical Data," Erwin Nagelkerke, Daniel L. Oberski, and Jeroen K. Vermunt focus on fit statistics within the MLLC model. The lack of local fit statistics within the standard MLLC models and the reliance solely on global fit statistics hinder the search for model improvements where the assumptions of local independence may not hold. The authors propose a strategy that supplements standard global fit statistics, such as the Bayesian information criterion and Akaike information criterion, with local fit statistics that single out and test one particular area of the model. They apply their approach to data from a study of employees and work groups on the basis of task characteristics, plus a small simulation study, and conclude that their results provide proof of concepts that should stimulate further investigations.

Methods for Demographic and Economic Data

Two chapters deal with matters involving demographic and economic data and potential advances through improved measurement. In

"Modeling Incidence of Nuptiality," Juha Alho focuses on "the two-sex problem in demography," examining the incidence of cross-sex nuptiality from a new perspective. As is well known, in the examination of fertility, demographers focus exclusively on children born to women, and in the examination of mortality, the near-universal approach is to examine the sexes separately. The situation of the statistical description of the formation of marriages is that traditionally, the rates of marriage of one sex depend on the available supply of potential partners from the other. The present article proposes a solution that considers the sexes jointly, relying on the idea of generalized averages. The model is formulated in stochastic terms and is parameterized in terms of the overall level of nuptiality, its relative propensity by age, and the mutual relative attraction of spouses at different ages. The methods are illustrated by both annual and monthly nuptiality data from Finland.

On an unrelated demographic topic, "Addressing Measurement Error Bias in GDP with Nighttime Lights and an Application to Infant Mortality with Chinese County Data," by Xi Chen, focuses the application of a new measure of economic development—satellite-based nighttime lights. The article proposes the use of unconventional, innovative, nonsurvey information involving luminosity data as a proxy variable. Building upon recent popularity of the use of such data, this article focuses on the study of the relationship of economic development and infant mortality rates in the People's Republic of China. The author suggests that observed indicators of economic and demographic data in developing countries, especially for local levels, tend to have large systematic errors relative to more developed regions. The article focuses specifically on gross domestic product data for China, arguing that the true extent of measurement error at the county or local level is often unknown. The study raises an important issue about the nature of the bias in the assessed effects of economic development on demographic processes and includes the nighttime lights data to improve these estimates. The study concludes that nighttime lights data have great potential for dealing with flawed economic indicators in developing regions or small areas where high-quality measures of economic and demographic variables are lacking.

Comments

Given its limited space, *SM* does not have much history publishing commentary, although there is certainly justification for doing so in some cases. We view as positive the potential for commentary on *SM* articles to improve the level of discourse in sociological methodology. In this issue, we have included two comments inspired by articles that have appeared in past issues of *SM*. The comment written by Alrik Thiem and Michael Baumgartner, "Modeling Causal Irrelevance in Evaluations of Configurational Comparative Methods," focuses on the original article in a symposium on qualitative comparative analysis (QCA) by Lucas and Szatrowski (2014). That original article was critical of the QCA approach, and by including it in a symposium, there was considerable discussion (see volume 44 of *SM*). Thiem and Baumgartner's comment adds to this discussion, specifically their defense of QCA, which identifies a fundamental flaw in Lucas and Szatrowski's simulation study used to illustrate confirmation bias in QCA. On the basis of their own simulations, implemented using their updated package for the R environment, Thiem and Baumgartner conclude that, when properly assessed, QCA adequately recovers the data-generating process. They do, however, note that further scrutiny of QCA is needed. The second article in this section, "Model Misspecification When Eliminating a Factor in Age-period-cohort Multiple Classification Models," by Robert M. O'Brien, critiques aspects of an article by Yang, Fu, and Land (2004) published in *SM*. Specifically, O'Brien questions the utility of using incremental fit tests to compare two- and three-factor age-period-cohort multiple classification models, a strategy advised by Yang et al. O'Brien clearly explicates the problem with such a model selection approach and offers alternative strategies for grappling with the age-period-cohort linear confounding.

THE FUTURE

As the only American Sociological Association (ASA) periodical devoted entirely to research methods, *SM* plays an important role in the discipline, and it holds a coveted position among sociology journals. *SM*'s privileged status among ASA journals reflects not only its high visibility and quality of the work published there but also the widespread support it has received from the membership of ASA. Past editors have demonstrated a commitment to methodological diversity in the journal, and there have been some clear efforts taken in the past to include greater breadth in coverage of research methods in the discipline. In the ideal scenario, *SM* will continue to publish broadly representative articles on research design, measurement, data collection, modeling, and data analysis, across the spectrum of approaches—qualitative versus quantitative, nomothetic versus idiographic, and so on. Statistical methods have tended to dominate the pages of *SM*, but we all maintain a commitment to publishing the best possible methodological work across these scientific territories. The ability to do this of course depends on the reputation of the journal for attracting such excellent submissions. This is not always easy to achieve, given that *SM* is an annual submissions-based periodical, without a great deal of leverage in being able to solicit particular kinds of research. Previous editors have, to the extent possible, developed the option of providing discussion of special topics without taking too much away from the normal submission-based publication needs of the journal, and at the same time giving voice to diverse views on relevant issues that are important to ASA research constituencies. The previous editor, Tim Futing Liao, took great advantage of this option and included several symposia in *SM*, and we fully intend to continue to exploit the potential of symposia to create opportunities for dialogue.

In addition to emphasizing methodological diversity, we also encourage a vision of looking to the future in terms of the kinds of methodological issues that will confront sociologists in the next decades (see Alwin 2013). The field of sociology has become inundated with data. With the continuation of many existing sources of social monitoring data, along with the digitization of many archival records, new demands are being placed on how we confront large-scale data sets. New approaches that merge computer science, visualization tools, and traditional statistics will be required to accommodate these growing resources, all vital for the future of sociology. In addition to the growing development of "big data" resources, training in the use of mathematical and statistical tools in the social sciences is changing, and *SM* has helped lead the challenges to "frequentist" logics involving probabilities of inferential errors. Bayesian thinking is slowly penetrating these historical traditions, and especially as we use large data structures, new ways of thinking will be required as a routine matter.

Among these public data sources, longitudinal designs are increasingly part of our sociological arsenal and in future years will become even more important. This includes everything from repeated cross-sectional surveys and panel surveys to retrospective life history calendars, which go to great lengths in dating the timing of events and their duration, as well as life histories presented in the form of narratives. Longitudinal data have several advantages over one-off cross-sectional studies, and the potential importance of these designs cannot be underestimated. Sociologists need to better understand how to use such data, and *SM* can help in this process.

In addition, in these studies, individual cases are often the primary focus of sociologists, and yet individual lives are linked to one another. People inhabit multilayered environments, or nested structures, and a growing body of knowledge is developing that involves the study of dyads, triads, and larger interpersonal structures. Greater integration of social network science and sociology is needed, and innovative methodological approaches, especially with regard to the gathering and analysis of data, are necessary to advance the interplay of human development and social structures as mediated by social environments and cultural norms. *SM* has a tradition of publishing contributions from network science, involving graph theory, visualization tools, block modeling, ERGMs, and so on, and it can continue to assist sociologists to incorporate information on network structures in their theories and data.

Modern sociology has its roots in the comparisons of cultures and social systems. Comparative sociology is not new, and although the use of intercultural comparisons as a research strategy is a timeworn tradition in the social sciences, the use of multinational surveys is a relatively recent development over the past 30 years, especially those using a large number of countries. The vastness of this ever expanding international survey database, along with the relative success of these research strategies, means that the use of multinational surveys is no longer the exclusive domain of international or "area" specialists. There are many challenges involved in making comparisons across cultures and countries, and ultimately sociological methodology can make a contribution to understanding the nature of institutional and normative differences across systems. The recent focus on QCA (Ragin 1987) (see the *SM* symposium in 2014) reinforces the conclusion that comparisons of social systems at a macro-level are of vital interest to sociologists.

With its origins in nineteenth-century Europe and pre–World War II American society, survey research plays an extraordinarily important role in contemporary social sciences throughout the world (Converse 1987). Vast amounts of survey data are collected for many purposes, including governmental information, public opinion and election surveys, advertising and marketing research, as well as basic social scientific research. Some have even described survey research as the *via regia* for modern social science (Kaase 1999:253)—the ideal way of conducting empirical science. We disagree with the idea that surveys are the only way to do social science, but there would be hardly any dissent from the view that survey research has become a mainstay for governmental planning, the research of large numbers of academic social scientists, and the livelihoods of a growing numbers of pollsters, marketing and advertising researchers, not to mention sociologists. Social methodologists must be aware of all the methods of survey research, including methods of sampling, questionnaire design, modes of communication, nonresponse, and measurement error. In the area of sampling, several developments in RDS in recent years portend further development.

Finally, the linkage between theory and data involves a number of considerations typically discussed under the specific heading of measurement. Ultimately, our concerns with measurement involve issues of quality, as is seen in many of the chapters of this volume of *SM*. In the case of survey data, many aspects of the information-gathering process are worthy of serious scrutiny given the potential for producing measurement error: aspects of survey questions, the cognitive mechanisms of information processing and retrieval, the motivational context of the setting that produces information, and the response framework in which the information is transmitted. But the issues of "validity claims" go well beyond the use of survey data and tend to permeate all types of methods and require attention regardless of the particular "logic" used to make sense of the social world. *SM* very much continues to welcome the submission of articles dealing with measurement and measurement quality.

ACKNOWLEDGMENTS

There are a number of people whose contributions made this volume of *SM* possible, and we take the time here to acknowledge them. We especially owe a huge debt to Tim Futing Liao, professor of sociology and statistics at the University of Illinois at Urbana-Champaign, the previous editor of *SM*, whose work over the past few years helped prepare the present volume. Many of the articles included in this volume were accepted by Tim and/or had been in various stages of review and revision during his editorship, prior to our term as editor. Without Tim's hard work and foresight with respect to the need for advance work on this volume, we would not have been able to produce this volume on time.

Lisa Savage, our managing editor, made our jobs much easier. Lisa has been the managing editor of *SM* since 2009. Her experience, knowledge, and good judgment lend themselves to a smooth operation of the reviewing process. Her careful tracking and monitoring of submissions through the ScholarOne system has been an extraordinary asset. Most important, without Lisa, it would have been difficult to manage the transition of editorial offices this year.

We wish to acknowledge the expert assistance of Stephanie Magean in copyediting all the chapters and other text for this volume, and Sara Sarver, our production editor at SAGE, who has been extraordinarily helpful and continually demonstrates an exceptional level of professionalism in dealing with authors. We are grateful for her flexibility and openness to new ideas, as well as her emphasis on best practices in journal publication. Karen Edwards at ASA provided indispensable guidance and support.

We also add our thanks to *new* editorial board members—Paul Allison, Ron Burt, Melissa Hardy, Guillermina Jasso, Burt L. Monroe, and Robert O'Brien—for accepting our invitation to serve in this capacity. We want to especially acknowledge Paul Allison for his willingness to serve another term on the editorial board, and we are indebted to the support we have received from continuing members.

Susan Welch, dean of the College of the Liberal Arts, Pennsylvania State University, provided financial, course release, and other forms of support through our department for locating the editorial office of *SM* in University Park.

Last but not least, we wish to thank Marie Borgatta for permission to reprint the photograph of Ed Borgatta that appears in the Dedication, and to Marie and Larry Chomsky (son-in-law to Ed and Marie Borgatta) for access to more than 60 flower photographs taken by Ed Borgatta, and specifically for providing high-density scans of some that were chosen for the cover art for this and future volumes.

—Duane F. Alwin
Editor
—Jason R. Thomas
Deputy Editor

Note

1. This article was submitted and reviewed during the editorship of Tim Futing Liao. Decision making on this paper was made entirely during his editorship. The only decision made by the present editor was where to place the chapter in this volume.

References

Aalen, Odd, Ornulf Borgan, and Håkon Gjessing. 2008. *Survival and Event History Data: A Process Point of View.* New York: Springer-Verlag.

Allison, Paul D. 1982. "Discrete-time Methods for the Analysis of Event Histories." Pp. 61–98 in *Sociological Methodology*, vol. 13, edited by Samuel Leinhardt. San Francisco, CA: Jossey-Bass.

Alwin, Duane F. 1992. "Information Transmission in the Survey Interview: Number of Response Categories and the Reliability of Attitude Measurement." Pp. 83–118 in *Sociological Methodology*, vol. 22, edited by Peter V. Marsden. Washington, DC: American Sociological Association.

Alwin, Duane F. 2013. "Reflections on Thirty Years of Methodology and the Next Thirty." *Bulletin of Sociological Methodology/Bulletin de Méthodologie Sociologique* 120(1):28–37.

Barber, Jennifer S., Susan A. Murphy, and Natalya Verbitsky. 2004. "Adjusting for Time-varying Confounding in Survival Analysis." Pp. 163–92 in *Sociological Methodology*, edited by Ross M. Stolzenberg. Washington, DC: American Sociological Association.

Borgatta, Edgar F. 1969. "The Current Status of Methodology in Sociology." Pp. ix–xiv in *Sociological Methodology*, vol. 1, edited by Edgar F. Borgatta. San Francisco, CA: Jossey-Bass.

Brockwell, Peter J., and Richard A. Davis. 2002. *Introduction to Time-series and Forecasting.* 2nd ed. New York: Springer-Verlag.

Cannell, Charles F., Peter V. Miller, and Lois Oksenberg. 1981. "Research on Interviewing Techniques." Pp. 389–437 in *Sociological Methodology*, vol. 12, edited by Samuel Lienhardt. San Franciso, CA: Jossey-Bass.

Clogg, Clifford C. 1992. "The Impact of Sociological Methodology on Statistical Methodology." *Statistical Science* 7(2):183–207.

Coleman, James S. 1958. "Relational Analysis: The Study of Social Organization with Survey Methods." *Human Organization* 17(1):28–36.

Converse, J. M. 1987. *Survey Research in the United States: Roots and Emergence, 1890–1960.* Berkeley: University of California Press.

Davis, James A. 1973. "Hierarchical Models for Significance Tests in Multivariate Contingency Tables: An Exegesis of Goodman's Recent Papers." Pp. 189–231 in *Sociological Methodology*, vol. 5, edited by Herbert L Costner. San Francisco, CA: Jossey-Bass.

Duncan, Otis Dudley. 1969. "Contingencies in Constructing Causal Models." Pp. 74–112 in *Sociological Methodology*, vol. 1, edited by Edgar F. Borgatta. San Francisco, CA: Jossey-Bass.

Duncan, Otis Dudley. 1970. "Partials, Partitions, and Paths." Pp. 38–47 in *Sociological Methodology*, vol. 2, edited by Edgar F. Borgatta and George W. Bohrnstedt. San Francisco, CA: Jossey-Bass.

Fararo, Thomas. 1969. "Stochastic Processes." Pp. 245–56 in *Sociological Methodology*, vol. 1, edited by Edgar F. Borgatta. San Francisco, CA: Jossey-Bass.

Flinn, Christopher J., and James J. Heckman. 1982. "New Methods for Analyzing Individual Event Histories." Pp. 99–140 in *Sociological Methodology*, vol. 13, edited by Samuel Leinhardt. San Franciso, CA: Jossey-Bass.

Gauthier, Jacques-Antoine, Eric D. Widmer, Philipp Bucher, and Cédric Notredame. 2010. "Multichannel Sequence Analysis Applied to Social Science Data." Pp. 1–38 in *Sociological Methodology*, vol. 40, edited by Tim Futing Liao. Hoboken, NJ: Wiley-Blackwell.

Goffman, Erving. 1959. *The Presentation of Self in Everyday Life*. New York: Anchor.

Goodman, Leo A., and Michael Hout. 1998. "Statistical Methods and Graphical Displays for Analyzing How the Association between Two Qualitative Variables Differs among Countries, among Groups, or over Time: A Modified Regression-type Approach." Pp. 175–230 in *Sociological Methodology*, vol. 28, edited by Adrian E. Raftery. Washington, DC: American Sociological Association.

Handcock, Mark S., and Martina Morris. 1998. "Relative Distribution Methods." Pp. 53–97 in *Sociological Methodology*, vol. 28, edited by Adrian E. Raftery. Washington, DC: American Sociological Association.

Hauser, Robert M. 1980. "Some Exploratory Methods for Modeling Mobility Tables and Other Cross-classified Data." Pp. 413–58 in *Sociological Methodology*, vol. 11, edited by Karl F. Schuessler. San Francisco, CA: Jossey-Bass.

Hauser, Robert M., and Arthur S. Goldberger. 1971. "The Treatment of Unobserved Variables in Path Analysis." Pp. 81–117 in *Sociological Methodology*, vol. 3, edited by Herbert L. Costner. San Francisco, CA: Jossey-Bass.

Heckathorn, Douglas D. 1997. "Respondent-driven Sampling: A New Approach to the Study of Hidden Populations. *Social Problems* 44(2):174–99.

Heise, David R. 1969. "Problems in Path Analysis." Pp. 38–73 in *Sociological Methodology*, vol. 1, edited by Edgar F. Borgatta. San Francisco, CA: Jossey-Bass.

Hout, Michael, Otis D. Duncan, and Michael E. Sobel. 1987. "Association and Heterogeneity: Structural Models of Similarities and Differences." Pp. 145–84 in *Sociological Methodology*, vol. 17, edited by Clifford C. Clogg. Washington, DC: American Sociological Association.

Kaase, M. (Ed.). 1999. *Quality Criteria for Survey Research*. Berlin: Akademie Verlag.

Kahneman, Daniel. 2011. *Thinking Fast and Slow*. New York: Farrar, Straus.

Land, Kenneth C. 1969. "Principles of Path Analysis." Pp. 3–37 in *Sociological Methodology*, vol. 1, edited by Edgar F. Borgatta. San Francisco, CA: Jossey-Bass.

Lazarsfeld, Paul F., and Robert K. Merton. [1950] 1974. *Continuities in Social Research: Studies in the Scope and Method of "The American Soldier."* New York: Arno.

Logan, John A. 1998. "Estimating Two-sided Logit Models." Pp. 139–73 in *Sociological Methodology*, vol. 28, edited by Adrian E. Raftery. Washington, DC: American Sociological Association.

Lucas, Samuel R., and Alisa Szatrowski. 2014. "Qualitative Comparative Analysis in Critical Perspective." Pp. 1–79 in *Sociological Methodology*, vol. 44, edited by Tim Futing Liao. Thousand Oaks, CA: Sage.

Montgomery, James D. 2000. "The Self as a Fuzzy Set of Roles, Role Theory as a Fuzzy System." Pp. 261–314 in *Sociological Methodology*, vol. 30, edited by Michael E. Sobel and Mark P. Becker. Washington, DC: American Sociological Association.

Muthén, Bengt. 1997. "Latent Variable Modeling of Longitudinal and Multilevel Data." Pp. 453–80 in *Sociological Methodology*, vol. 27, edited by Adrian E. Raftery. Washington, DC: American Sociological Association.

Nardulli, Peter F., Scott L. Althaus, and Matthew Hayes. 2015. "A Progressive Supervised-learning Approach to Generating Rich Civil Strife Data." Pp. 148–58 in *Sociological Methodology*, vol. 45, edited by Tim Futing Liao. Thousand Oaks, CA: Sage.

Németh, Renáta, and Tomás Rudas. 2013. "On the Application of Discrete Marginal Graphical Models." Pp. 70–100 in *Sociological Methodology*, vol. 43, edited by Tim Futing Liao. Thousand Oaks, CA: Sage.

Raftery, Adrian. 2001. "Statistics in Sociology, 1950–2000: A Selective Review." Pp. 1–45 in *Sociological Methodology*, vol. 31, edited by Michael E. Sobel and Mark P. Becker. Washington, DC: American Sociological Association.

Ragin, Charles. 1987. *The Comparative Method: Moving beyond Qualitative and Quantitative Strategies*. Los Angeles: University of California Press.

Rossman, Gabriel, Ming Ming Chiu, and Moeri M. Mol. 2008. "Modeling Diffusion of Multiple Innovations via Multilevel Diffusion Curves: Payola in Pop Music Radio." Pp. 201–30 in *Sociological Methodology*, vol. 38, edited by Yu Xie. Hoboken, NJ: Wiley-Blackwell.

Salganik, Matthew J., and Douglas D. Heckathorn. 2004. "Sampling and Estimation in Hidden Populations Using Respondent-driven Sampling." Pp. 193–239 in *Sociological Methodology*, vol. 34, edited by Ross M. Stolzenberg. Boston: Blackwell.

Sobel, Michael E. 1996. "Clifford Collier Clogg, 1949–1995: A Tribute to His Life and Work." Pp. 1–38 in *Sociological Methodology*, vol. 26, edited by Adrian E. Raftery. Washington, DC: American Sociological Association.

Tuma, Nancy B. 1982. "Nonparametric and Partially Parametric Approaches to Event-history Analysis." Pp. 1–60 in *Sociological Methodology*, vol. 13, edited by Samuel Leinhardt. Washington, DC: American Sociological Association.

Tuma, Nancy B., and Michael T. Hannan. 1979. "Approaches to the Censoring Problem in Analysis of Event Histories." Pp. 209–40 in *Sociological Methodology*, vol. 10, edited by Karl F. Schuessler. San Francisco, CA: Jossey-Bass.

Weiss, Robert. 1994. *Learning from Strangers: The Art and Method of Qualitative Interviewing Studies*. New York: Free Press.

Yamaguchi, Kazuo. 2012. "Log-linear Causal Analysis of Cross-classified Categorical Data." Pp. 257–85 in *Sociological Methodology*, vol. 42, edited by Tim Futing Liao. Thousand Oaks, CA: Sage.

Yang, Yang, Wenjiang J. Fu, and Kenneth Land. 2004. "A Methodological Comparison of Age-period-cohort Models: The Intrinsic Estimator and Conventional Generalized Linear Models." Pp. 75–110 in *Sociological Methodology*, vol. 34, edited by Ross M. Stolzenberg. Washington, DC: American Sociological Association.

Sociological Methodology
2016, Vol. 46(1) 1–38
© American Sociological Association 2016
DOI: 10.1177/0081175016637890
http://sm.sagepub.com

ॐ 1 ॐ

⑤SAGE

INTERVIEWING PRACTICES, CONVERSATIONAL PRACTICES, AND RAPPORT: RESPONSIVENESS AND ENGAGEMENT IN THE STANDARDIZED SURVEY INTERVIEW

Dana Garbarski*
Nora Cate Schaeffer[†]
Jennifer Dykema[†]

Abstract

"Rapport" has been used to refer to a range of positive psychological features of an interaction, including a situated sense of connection or affiliation between interactional partners, comfort, willingness to disclose or share sensitive information, motivation to please, and empathy. Rapport could potentially benefit survey participation and response quality by increasing respondents' motivation to participate, disclose, or provide accurate information. Rapport could also harm data quality if motivation to ingratiate or affiliate causes respondents to suppress undesirable information. Some previous research suggests that motives elicited when rapport is high conflict

*Loyola University Chicago, Chicago, IL, USA
[†]University of Wisconsin–Madison, Madison, WI, USA

Corresponding Author:
Dana Garbarski, Loyola University Chicago, Department of Sociology, 1032 W. Sheridan Road, 440 Coffey Hall, Chicago, IL 60660, USA
Email: dgarbarski@luc.edu

with the goals of standardized interviewing. The authors examine rapport as an interactional phenomenon, attending to both the content and structure of talk. Using questions about end-of-life planning in the 2003–2005 wave of the Wisconsin Longitudinal Study, the authors observe that rapport consists of behaviors that can be characterized as dimensions of responsiveness by interviewers and engagement by respondents. The authors identify and describe types of responsiveness and engagement in selected question-answer sequences and then devise a coding scheme to examine their analytic potential with respect to the criterion of future study participation. The analysis suggests that responsive and engaged behaviors vary with respect to the goals of standardization; some behaviors conflict with these goals, whereas others complement them.

Keywords

interviewer-respondent interaction, rapport, responsiveness, engagement, survey interview, standardization

1. INTRODUCTION

Survey researchers have long invoked the concept of "rapport," either to describe qualities of the interaction between the interviewer and sample member (e.g., Dundon and Ryan 2010) or as an intervening mechanism that influences the respondent's decision to participate and motivated effort during the interview (e.g., Thornton, Freedman, and Camburn 1982). Rapport has been examined in interviews (e.g., Belli, Lepkowski, and Kabeto 2001; Lavin and Maynard 2001), but the concept, its dimensions, and the features of the interaction that might indicate its qualities have not received comprehensive analysis, contributing to the concept's "disrepute" (e.g., Goudy and Potter 1975; Schaeffer 1991). An important challenge to such an analysis is to identify what opportunities for creating and maintaining rapport are available within the constraints of a standardized survey interview, which is meant to produce high-quality data by ensuring that all respondents are asked the same questions in the same way to minimize interviewer variance in measurement.

In this study, we examine rapport as an observable feature of interaction rather than focusing on its (undoubtedly related) psychological aspects. We consider the various dimensions of rapport and their interactional expressions both for interviewers and respondents in standardized interviews. In particular, we recognize that a general concept such as "rapport" needs to be adapted to the specific task environment of the survey interview and the roles and goals of the actors. Thus, we identify two actor-specific concepts to be examined: the interviewer's responsiveness and the respondent's engagement. We characterize sets of utterances as interactionally responsive and engaged in various ways by considering their content and sequential placement in the interaction. Rather than viewing rapport as a violation of standardization, we examine whether the behaviors that constitute responsiveness and engagement complement or conflict with the practices of standardization to accomplish the task of obtaining codable answers to survey questions. We then translate our qualitative descriptions to a coding scheme to examine the analytic potential of the dimensions of responsiveness and engagement with respect to the criterion of future study participation.

2. BACKGROUND

2.1. *Standardization in Survey Interviews*

Despite the potential cost savings of self-administered modes, interviewer-administered surveys continue to be central to data collection (see overview in Schaeffer, Dykema, and Maynard 2010). Interviewers can be used with area, telephone, and list sampling frames, and response rates typically are higher when an interviewer recruits participants. The use of interviewers also facilitates the selection of respondents, administration of complex instruments, and collection of auxiliary measures (e.g., biomeasures, cognitive tests, linkages between survey data and sensitive external records such as Social Security data) that are increasingly incorporated into study designs (e.g., the Panel Study of Income Dynamics, the Health and Retirement Study). For both panel and cross-sectional studies, the motivation of sample members to participate, to work to provide accurate and honest answers, and to consent to providing sensitive information is critical to data quality, and interviewers play a key role in obtaining respondents' cooperation in fulfilling these requirements.

Many studies that use interviewers are standardized survey interviews. The practices of standardized interviewing aim to control interviewer variability (Hyman [1954] 1975; O'Muircheartaigh and Campanelli 1998; Schaeffer et al. 2010; Schnell and Kreuter 2005). Fowler and Mangione (1990) codified the rules of standardization: read questions as written, probe inadequate answers nondirectively, record answers without discretion, and be interpersonally nonjudgmental regarding the substance of answers.

If survey questions are clearly written and fit the target population, standardized interviews should consist of a series of "paradigmatic" question-answer sequences (Schaeffer and Maynard 1996, 2008), in which the interviewer reads the question as scripted and the respondent provides an answer that is codable, that is, one of the response options that the interviewer can code (e.g., "yes" for a yes/no question); optionally, the interviewer may acknowledge the respondent's answer before moving on to the next question. However, answers to survey questions are interactional accomplishments, and nonparadigmatic question-answer sequences arise for many reasons, including respondents' displays of the uncertainty, understandings, accommodations, and adjustments that may occur while respondents formulate answers to survey questions (Schaeffer and Maynard 1996, 2008). Some ways interviewers respond in these situations may go beyond the rules of standardization but still be consistent with the goals of measurement, although it may be difficult to demonstrate their impact on measurement.

Variants of strict standardization address three possible ways to relax its rigidities: (1) allow the interviewer more freedom in diagnosing and attending to the respondent's comprehension problems, with the goal of improving understanding of key concepts (Schober and Conrad 1997); (2) allow the interviewer to improvise during event history calendar interviews in ways that are predicted to improve the respondent's recall (Belli, Bilgen, and Al Baghal 2013); and (3) motivate the respondent with utterances that recognize or acknowledge emotions such as sadness, frustration, or irritation in a prior utterance, as in Dijkstra's (1987) "personal style" of interviewing, which augmented standardization with restricted types of person-oriented feedback. Notwithstanding these studies, much large-scale production interviewing relies on traditional standardization using large corps of interviewers of varying backgrounds, and we still know little about the implications of traditional standardization for the engagement or motivation of the respondent.

2.2. *Studies of Rapport*

Rapport is a concept often invoked as instrumental in motivating respondents to participate and provide complete and accurate information in the interview. However, conceptual and operational definitions of rapport vary (Goudy and Potter 1975; Weiss 1970). It has been described by referring to perceptions of the actors or an observer, such as a sense of affiliation, friendliness, comfort, or empathy between interactional partners; a willingness to disclose; or a motivation to please (e.g., see Cannell and Axelrod 1956; Weiss 1968). Or rapport can be used to refer to the content of utterances, such as acts of disclosure, expressions of empathy, or positive evaluations (Dijkstra 1987; Houtkoop-Steenstra 2000), or to specific verbal or nonverbal behaviors such as laughter, smiling, eye gaze, head nods, and acknowledgment (e.g., Cassell and Miller 2008; Foucault Welles 2013; Weiss 1970). Any of these—perceptions, content of utterances, or behaviors—might be used as an indicator of a latent construct of rapport (e.g., Belli et al. 2013). Finally, rapport can be defined structurally, as a property of the interaction itself, for example, as synchronization or coordination in talk or laughter (e.g., Cappella 1990; Lavin and Maynard 2001). One influential conceptualization of rapport includes several of these elements: mutual attentiveness, positivity, and coordination (Tickle-Degnen and Rosenthal 1990).

Behaviors used to indicate rapport have sometimes been operationalized as present or absent, without attending to the context within which the behavior occurs. For example, investigations of verbal behavior during the interview have examined rapport by noting whether laughter or digressions from the interview script occurred (e.g., Belli et al. 2001, 2013). Failing to consider the context within which behaviors such as laughter occur within and across question-answer sequences risks misunderstanding the actions performed by the behaviors. For example, previous research indicates that respondents initiate laughter more frequently than do interviewers and that whether and how interviewers reciprocate the laughter depends on how they are trained or on features of the instrument (Cannell, Fowler, and Marquis 1968; Lavin and Maynard 2001). It is probable that context is similarly important for other behaviors, so repairing this omission in current research is critical.

Conversation analysis is a research method that examines both the content and structure of talk in interaction. The interactional work of "rapport" and of related concepts such as "empathy" and "affiliation" has been investigated by conversation analysts, although only in a limited way for the survey interview. In talk outside the survey interview (e.g., when a salesperson reciprocates a customer's evaluation of something), the sequence of talk may create the opportunity for subsequent affiliative turns (Clark, Drew, and Pinch 2003). Lavin and Maynard (2001:454) defined rapport narrowly, as the occurrence of reciprocal laughter. Or rapport may involve accounts of personal experiences that create empathic moments with which the recipient is invited to align, with alignment depending on the recipient's experience with respect to the event in question (Heritage 2011; Ruusuvuori 2013). The related concept of "affiliation" has been invoked to describe heterogeneous interactional practices (Lindström and Sorjonen 2013): social solidarity, preference organization, affective stance, alignment, and responsiveness, among others. These conversation analytic observations can inform the study of rapport.

2.3. *Interviewer-respondent Interaction in Survey Interviews*

Data obtained through the survey interview are a collaborative achievement accomplished through talk: the actual interactional practices include some features of conversation but also a series of paradigmatic question-answer sequences that serve institutionalized purposes (Drew and Heritage 1992; Schaeffer and Maynard 2008; Suchman and Jordan 1990). Thus, an additional consideration for the analysis of interaction in the survey interview is the various, and potentially conflicting, rules of talk that govern both conversation more generally and standardization in the survey interview.

In the standardized survey interview, the interviewer-respondent interaction has two interrelated dynamics (see also Clark 1996; Haan, Ongena, and Huiskes 2013; Maynard and Marlaire 1992). The interview-management dynamic is oriented, under the leadership of the interviewer, to the task, its requirements, and the rules of standardization. The interviewer must correct the course of the interview and sometimes train the respondent to keep the interview within the constraints of standardization. The interviewer has more experience than the respondent, having been trained as an interviewer and familiarized with the set of items. Yet the respondent also affects the course of the interview: the respondent's actions and inactions determine how often the interviewer must correct the course of the interview to steer it back to the confines of standardization.

The conversation-management dynamic is oriented to the maintenance and organization of the interaction. Conversational practices learned through everyday conversation and interactions may be deployed automatically, because communicating in the survey interview is similar to communicating in other talk. Yet conversational practices may conflict with the practices of a standardized interview; for example, Suchman and Jordan (1990) went so far as to propose that the standardized survey interview suppresses conversational resources that could clarify meaning and interpretation and so undermines the validity of the data obtained (see also Schober and Conrad 1997). The interview-management and conversation-management dynamics operate in tandem over the course of the interview, and the participants' behaviors may be oriented to the goals and structure of the task, to other conversational goals, or to both.

2.4. *The Present Study*

As noted earlier, a global concept such as "rapport" needs to be modified to recognize the different roles of interviewers and respondents in the task environment of the survey interview (Maynard, Freese, and Schaeffer 2010; Schaeffer et al. 2013; Weiss 1970). In this study, we focus on the interviewer's responsiveness and the respondent's engagement.

We define the interviewer's responsiveness as the interviewer's fitting a response to the respondent's preceding talk. This "fit" or "alignment" can occur with respect to the survey task, rules of standardization, content of the talk, or conversational practices. This definition extends previous work that has focused on responsiveness as talk outside the practices of standardization (Dijkstra 1987; Schober and Conrad 1997) or on how interviewers respond to sample members' concerns (e.g., see Broome 2015; Schaeffer et al. 2013). The respondent's engagement is defined as behaviors consistent with motivation to perform the task (Cannell, Miller, and Oksenberg 1981; Dijkstra 1987), that is, behaviors that display attention to the survey task, interest in the survey task, or both. Because we observe few opportunities for respondents to ingratiate themselves with interviewers, we focus on engagement with the survey task, although we note possible instances in which ingratiation might occur.

We describe interviewers' responsiveness and respondents' engagement by considering the content and placement of utterances in the interaction. This approach allows us to describe where the opportunities arise, or are missed, for interviewers to display responsiveness and respondents to display engagement. We define *behaviors* as what is said and *actions* as what the behaviors accomplish, or the "main job" the turn is performing (Levinson 2013). Thus, the dimensions of responsiveness and engagement are actions, accomplished by behaviors characterized in terms of content and placement in the interaction, not affective states or conscious processes that interviewers and respondents engage in (which may be present but are not directly observable).

We characterize dimensions of the interviewer's responsiveness and the respondent's engagement and examine whether responsiveness and engagement complement or conflict with the practices of standardization for obtaining a codable answer to each question. In contrast to previous work, we (1) use actor-specific concepts (the interviewer's responsiveness and the respondent's engagement) and their subdimensions to examine "rapport" within the standardized survey interview; (2) reconceptualize "rapport" as something that can be described in terms of displayed engagement or interest in the survey task and the responsiveness or fit of utterances with respect to the survey task, rules of standardization, content, or conversational practices; and (3) highlight ways in which dimensions of rapport may conflict with but also complement the goals of standardization.

We study questions about end-of-life planning and treatment preferences in a survey of older adults, the Wisconsin Longitudinal Study (WLS). To identify and characterize actions (i.e., other than the paradigmatic question-answer sequences) as dimensions of responsiveness or engagement, we examine a portion of a survey with questions of varying task difficulty[1] and sensitivity,[2] to observe a range of behaviors.

3. METHODS

3.1. *Data*

We use digitally recorded telephone interviews conducted in the 2003–2005 wave of the WLS, a one-third random sample of the Wisconsin high school class of 1957 that has been interviewed periodically in the intervening decades. The investigation presented in this article is part of a larger study that uses future study participation as a criterion to assess the consequences of responsiveness and engagement. We use future participation as a criterion for the larger study from which this study is drawn because it is an outcome of interest to survey researchers and because it is probably associated with rapport, although we cannot assess whether dimensions of rapport lead to future participation or if dimensions of rapport and future survey participation are indicators of an underlying propensity to cooperate. We use providing a saliva sample when requested as our criterion because the request is challenging enough that we expect only the most motivated respondents to comply. Only 55 percent of respondents in the graduate sample provided the saliva sample (for field efforts initiated in 2007) compared with 81 percent who answered a follow-up self-administered instrument mailed a few months after their phone interviews (conducted between 2004 and 2005). If propensity to participate is associated in varying ways with responsiveness and engagement displayed during a survey interview, then our method of selecting cases should at least ensure a range of behaviors from both respondents and interviewers.

The complete analytic sample consists of 105 matched pairs of transcribed cases (210 cases) selected from the sample of original high school graduates who were interviewed from 2003 to 2005.[3] These cases were selected to simulate a case-control design: the cases form pairs in which one respondent provided a saliva sample and one did not. The pairs are matched on past participation in the WLS, gender, and estimated propensity to provide a saliva sample.[4] The sample is stratified by three levels of propensity to participate to ensure a sufficient number of pairs at each level. The sample is organized into random replicates to facilitate developing and applying coding systems independently in the larger study. The study presented here uses the first replicate of this sample, composed of 15 pairs or 30 cases.

3.2. *Analytic Strategy*

We use the 30 cases (15 pairs) to conceptualize and describe dimensions of responsiveness and engagement displayed in the interaction. In our transcripts, a "line" is a turn of talk, although a turn by the interviewer may be broken up into two lines if the interviewer begins the next question within the turn. We use "(F)" to denote a first name, with an F for each syllable. Our analytic strategy combines features of conversation analysis and content analysis, although we do not use either method in a strict way.

We drew on conversation analysis to uncover the practices used to accomplish social actions through talk and to characterize behaviors as interactionally responsive or engaged. Central concepts that inform the analysis include the description of an "action" (see above), as well as the sequencing of actions in adjacency pairs (Stivers 2013), the preference structure of initiating and responding actions (Pomerantz and Heritage 2013), repair (Kitzinger 2013), and overlapping talk. We used an inductive methodology, going through multiple iterations to describe (and revise) the analysis of each turn, question-answer sequence, and case, and to identify features that persisted across contexts. We also drew on our experience examining interviewer-respondent interaction to identify and search for keywords. These include apologetic utterances ("sorry"), mitigators ("just"), and acknowledgments ("okay").

We examine the interviewer-respondent interaction for the 13 to 29 questions (depending on skip patterns) in each of the 30 cases. Each author listened to a set of cases and went through the transcripts to identify interactional dimensions of responsiveness and engagement, and we developed group consensus about the dimensions and their alignment with standardization. We note that although we characterize utterances on the basis of their content and placement, nuances that were not captured in these transcripts could lead to different interpretations, such as a sarcastic tone for "sorry," smile voice (Tartter and Braun 1994), pauses within and between turns, and other forms of nonverbal communication. Thus, our descriptions are plausible but subject to measurement error. Furthermore, these characterizations of interactional responsiveness and engagement do not imply a psychological assessment of the actors' interpretations of the interaction.

In the "Results" section, we present conceptual descriptions of the interviewer's responsiveness and the respondent's engagement, then illustrate these dimensions and how they complement or conflict with the rules of standardization using selected question-answer sequences. These sequences are selected to describe how responsiveness and engagement are displayed in the interaction and to highlight their complexity rather than to summarize the frequency with which these behaviors occur. Finally, we present an analysis using a coding scheme based on the qualitative descriptions and applied to the set of 30 cases. We do this to illustrate the analytic potential of our concepts and their operationalizations.

4. RESULTS

4.1. *The Interviewer's Responsiveness*

As noted above, our definition of the interviewer's responsiveness is the interviewer's fitting a response to the respondent's talk in a way that considers the survey task, rules of standardization, content, or conversational practices. We identified various dimensions of the interviewer's responsiveness:

1. In its most basic form, responsiveness is displayed when the interviewer's acknowledgments display *listening*, such as saying "okay" after the respondent answers the survey question.

2. Another set of responsive practices displays how the interviewer *understands* what the respondent has said and how he or she will treat that content in formulating an answer; these recapitulations by the interviewer also *allow choice or correction* by the respondent, underscoring the collaboration involved in answering survey questions. For example, by repeating the respondent's utterance, the interviewer displays attentiveness similar to the simplest form of "reflective listening" in motivational interviewing: showing an interest in and understanding of what the participant has to say by repeating or rephrasing what has been said (Miller and Rollnick 2002); an example of this is the interviewer saying "your children" after the respondent answers "my children." Another practice that displays understanding and allows choice or correction is when the interviewer appends a "verification" to his or her initial reading of the survey question. Verifications occur in situations in which the respondent has provided the answer to the current question before the interviewer finishes reading the question or while answering a preceding question, such as "who is that first person, and you said your husband?" A related practice that often attends these practices is an explicit request for confirmation using the phrase "Is that correct?" or its variants.

3. Another dimension of the interviewer's responsiveness is *apologetic utterances*, such as "sorry," which acknowledge and align with potential interactional troubles, such as interruptions or expressions of irritation or frustration.

4. Behaviors that *address the respondent's uncertainty, difficulties, or other problems* in answering the survey questions respond to the substantive content of the respondent's answer and manage troubles, for example, by following up markers of uncertainty, mismatch, or difficulty. Although interviewer follow-up actions are responsive in terms of addressing the respondent's uncertainty, the actions are so to varying degrees and with varying implications for standardization.

5. Behaviors that *facilitate efficient progress through the interview* include reading questions exactly as scripted or offering to turn the respondent's utterance into an answer, for example, by "coding" an uncodable answer with a directive follow-up such as "Is that a 'no' then?"
6. Behaviors that reinforce the structure of the task are responsive by *aiding the respondent's cognitive processing* within the task, for example, by linking the current question to the previous line of questioning with a preface such as "and" or by naming a person the respondent previously mentioned ("now for Mary").
7. Behaviors that *downgrade* the force of the task or the request include phrases, such as "would you say" and its variants.
8. Finally, interviewers are responsive when they provide reasons for behaviors that deviate from the task or conversational practices to *address potential misunderstandings* of the interviewer's prior, current, or subsequent behavior. Examples of this include "I have to read all the response options" when the respondent provides an answer before the interviewer finishes question reading, or "we'll ask some more questions about those," which validates the ancillary information offered by the respondent as being relevant for future questions before returning to the interview script.

Clearly, responsive behaviors vary in consistency with the practices of standardization. Some behaviors are responsive in several ways at once, but a behavior might also be responsive in one way and unresponsive in another. In addition, one action could be accomplished by several different behaviors, and one behavior can perform several actions simultaneously. We attend to these possibilities in our descriptions of excerpts below, organized by the type of respondent talk that occasions these behaviors.

4.1.1. *When the Respondent's Answer Anticipates a Subsequent Question.*

Interviewers have an opportunity to display responsiveness when the respondent's answer anticipates a subsequent question. In excerpt 1, the respondent answers question 2, a yes/no question (see Appendix A in the online journal), by specifying the "anyone." This excerpt illustrates the influence of the question and the intrusion of conversational practices: the respondent's answer correctly predicts that a "yes" answer will be followed by a request to specify the person (this "filter + follow-up question" structure is common in surveys and other interactional contexts), and by providing that information, he implies

Excerpt 1, female interviewer, male respondent, case ID 30

Line	Question	Actor	Text
4	2	I	mkay have you discussed your health care plans and preferences with anyone?
5	2	R	with my brother
6	3	I	ok and let's see the first person would be your brother you said?
7	3	R	mhmm
8	3	I	ok

that his answer to the first question would be "yes." If the interviewer simply read question 3 as scripted, she could be perceived by the respondent as having not heard him. After the interviewer reads question 3, however, she appends the information the respondent just provided in a "verification." Applying our analysis of the types of responsiveness, we propose that a verification indicates that the respondent has been heard and understood and offers the interviewer's understanding for correction by the respondent. Verifications also facilitate efficient progress through the interview by incorporating information previously provided to suggest a codable answer at the current question. Although verification prevents the awkwardness that might result from ignoring this information (Schaeffer and Maynard 2008), this practice conflicts with strict standardization, because the interviewer does not read the question exactly as worded and instead offers a candidate answer in a way that might be considered directive.[5,6]

In excerpt 2, the interviewer formulates a verification at line 485 using information provided at line 463. The interviewer uses the daughter's name ("FF" at line 485) in her verification coupled with a request for confirmation ("correct?") and then announces the relationship that she is coding, "and she's your child" (line 487); "and" connects the current action to the discussion of (FF) at line 485 (Heritage and Sorjonen 1994). Although the verification and request for confirmation at lines 485 and 487 display listening by repeating information, they might also suggest that the respondent's earlier elaboration (at line 463) was not clear. The interviewer appears to orient to this possibility, addressing potential misunderstanding by providing a reason for her behavior (i.e., "just making sure double checking" in line 489). The mitigator "just" reduces the force of the action: the interviewer is only "making sure"

Excerpt 2, female interviewer, female respondent, case ID 27

Line	Question	Actor	Text
463	5	R	because I do not want to be put on life support I do not want tube feedings and my daughter and I disagree she says what if you come out of the comatose situation I says (FF) don't put me through that

. . .

Line	Question	Actor	Text
485	12	I	who has that authority you said (FF) correct?
486	12	R	(FF)
487	12	I	and she's your child
488	12	R	yeah
489	12	I	just making sure double checking

and not doing something else, such as implying that the respondent did not provide adequate information earlier.

In excerpt 3, the interviewer follows up the codable answer at question 6 by asking for the son's name (line 180); the mitigator ("just so . . .") modifies the purpose of the request ("I can"), but the request ends with the respondent's answer at line 181 before the proposed use is expressed. The interviewer confirms her hearing of the name (line 182). At line 183, the interviewer prefaces question 7 with "and," emphasizing the structure of the task by linking the current question to the previous question-answer pair (Heritage and Sorjonen 1994). This application of conversational practices within standardized interviews supports the respondent's cognitive processing by displaying connections between questions. Yet referring to the son as "this person" ignores information in the common ground (Clark and Brennan 1991). In particular, using "this person" violates "lexical entrainment," or referring to the same object with the same phrase in conversation, subverting the establishment of a "conceptual pact" by not using the term ("FF" or "your son") established in prior talk (Brennan and Clark 1996). Because the interviewer could have incorporated information that the respondent already provided, reading the question as scripted may display a lack of responsiveness.

Excerpt 3, male interviewer, female respondent, case ID 18

Line	Question	Actor	Text
178	6	I	and who would the next person be?
179	6	R	our son
180	6	I	ok and what is his name just so that I can
181	6	R	(FF)
182	6	I	(FF)?
183	7	I	and how well does this person understand your preferences and plans for future medical treatment?

Excerpt 4, male interviewer, male respondent, case ID 9

Line	Question	Actor	Text
1085	15	I	who if anyone have you given these written instructions to?
1086	15	R	ah my wife and daughter
1087	15	I	your wife and daughter? ok
1088	15	R	yeah

The interviewer can also be responsive in following up a codable answer. In excerpt 4, the interviewer repeats (in line 1087) the respondent's answer (given in line 1086) before providing an acknowledgment ("ok" in line 1087). Such repetitions are distinct from verifications in that they do not occur in the question-asking slot and do not use information the respondent provided in response to a previous question; like verifications, they display the interviewer's understanding and allow possible correction by the respondent. This follow-up of a codable answer is unnecessary from the point of view of strict standardization, although this practice may be useful with answers that are codable but complex, like that in excerpt 4, in which the respondent answers with multiple people. (Note that for this question, interviewers are supposed to record all persons mentioned by the respondent.)

4.1.2. When the Respondent Provides an Uncodable Answer. The interviewer can also display responsiveness after uncodable answers, particularly when the respondent displays a problem in mapping his or her experience onto the response options or marks an answer as uncertain. The interviewer then determines whether and how to pursue a codable answer; the interviewer's next turn can vary both in responsiveness and in its conformity with standardization.

A "report" is an utterance that provides information relevant to the question but is not formatted as an answer to the survey question (terminology adapted from Drew 1984). When there is a lack of fit between the question and the respondent's situation (Schaeffer and Maynard 2008), a report by the respondent displays a problem for possible repair by the interviewer. In excerpt 5, the respondent's report (line 1487) notes that she "did" have arrangements "a few years ago," but indicates that she "[doesn't] know if that would" count because "things have changed over the years." This report functions to delay a dispreferred action: the "no" answer offered at line 1489 (e.g., see Pomerantz and Heritage 2013). The interviewer's "mhmm" (line 1488) allows the respondent's story to continue to an upshot, "so probably na it's not up to date": a candidate answer of "no" preceded by a marker of uncertainty ("probably") and followed by a summary of the account for that answer (line 1489). The respondent's narrative identifies an ambiguity in the question: has the respondent ever made such arrangements, are arrangements currently in effect, or are arrangements in effect that express the respondent's current wishes? In the absence of specific instructions, the interviewer may not clarify which of these interpretations is intended. So the interviewer directs the respondent to map her report onto one of the response options ("so would you say yes or no then," line 1490). This follow-up is consistent with standardization's requirement that follow-up actions be nondirective (Fowler and Mangione 1990). Yet the follow-up formulates the respondent's choice as a conclusion from her story (the "so" and "then" formulation), allowing the respondent to code herself.

Moreover, in the face of the displayed ambiguity, the interviewer uses "would you say," a distancing phrase that downgrades the request; it suggests that a "best match" is sufficient and the answer need not perfectly describe the respondent's experience (Horgan 2005; Garbarski, Schaeffer, and Dykema 2011). Thus, the interviewer simplifies the task within the constraints of standardization, addressing the problem the respondent displays by downgrading the request while maintaining the respondent's role in confirming her answer. Standardization does not require this follow-up, but it may be useful with complex answers, for

Excerpt 5, female interviewer, female respondent, case ID 2

Line	Question	Actor	Text
1486	11	I	have you made any legal arrangements for someone to make decisions about your medical care if you become unable to make those decisions yourself?
1487	11	R	uh you know I did a few years ago but I don't know if that would it it you know things have changed
1488	11	I	mhmm
1489	11	R	over the years so probably na it's not up to date
1490	11	I	so would you say yes or no then
1491	11	R	um no
1492	11	I	ok

Excerpt 6, male interviewer, male respondent, case ID 12

Line	Question	Actor	Text
824	20	I	do you have assets or property that will go to someone through a joint ownership or beneficiary designation?
825	20	R	just those mentioned in the will but I don't think I can answer your question
826	20	I	ok so then would you say no then to that?
827	20	R	yeah
828	20	I	ok

example, when respondents display uncertainty or mapping difficulties along with their codable answer.

Like the report in excerpt 5, those in excerpts 6 and 7 display uncertainty about how the respondent's experience fits the question. In contrast to excerpt 5, however, the interviewers in excerpts 6 and 7 address this uncertainty with directive follow-up actions. Directive follow-up actions such as these may be most likely when the respondent's utterance implicates one of the response categories (Moore and Maynard 2002), and the action is similar to the "coding" interviewers do when they treat an answer such as "probably" as a synonym for "yes" (Hak 2002).

Excerpt 7, male interviewer, male respondent, case ID 10

Line	Question	Actor	Text
4	2	I	all right so all right we'll ask some more questions about those um have you discussed your healthcare plans and preferences with anyone?
5	2	R	well I I put my wife that died we had that old you know but I haven't discussed with my new wife so but uh
6	2	I	oh ok so you have at some point though
7	2	R	mm
8	2	I	you have discussed it with someone at some point?
9	2	R	yeah with my wife my uh first wife the one that died you know
10	2	I	yup ok I'll write that down

In excerpt 6, the interviewer's follow-up at line 826 is similar to the interviewer's follow-up in excerpt 5, except that the interviewer in excerpt 6 proposes just one response option as a candidate answer; it is not clear that the candidate answer is appropriate, but the respondent accepts it. In excerpt 7, the interviewer proposes a candidate answer at lines 6 and 8 in a follow-up that formulates for the respondent one definition of the scope of the question, "with someone at some point"; this interpretation in turn suggests that the respondent's information at line 5 implies "yes." The formulation displays the interviewer's understanding of the respondent's talk and allows possible correction by the respondent, yet these excerpts show that the interviewer's follow-up to a report may display varying degrees of listening and understanding; although excerpt 6 can be heard as "don't know," the interviewer's action selects an understanding that favors the codable answer "no." Although directive follow-up actions manage the respondent's uncertainty and facilitate efficient progress through the interview, they violate common rules of standardization and may lead to misrepresentations of the respondent's true answer.

The respondent's report may volunteer a hypothetical response option that falls along the continuum offered by the response dimension but is not included in the question (Garbarski et al. 2011), making a potential uncertainty or difficulty available for possible repair by the

Excerpt 8, male interviewer, male respondent, case ID 18

Line	Question	Actor	Text
183	7	I	and how well does this person understand your preferences and plans for future medical treatment?
184	7	R	very well
185	7	I	mkay so would you say extremely well or somewhat well?
186	7	R	extremely well he has a copy

interviewer. The WLS investigators made reading the response options optional when the "how well" questions were repeated the second and third time (see Appendix A in the online journal), but omitting the response categories may lead to nonparadigmatic question-answer sequences. In excerpt 8, the interviewer reads question 7, and the respondent answers with a hypothetical response option at line 184: "very well" lies on the response dimension ("how well"), but the response options are "extremely well, somewhat well, not very well, or not at all." The respondent's answer implicates one region of the response scale but does not differentiate between the two options on that side ("extremely" or "somewhat"). The interviewer follows up with these implicated categories in a practice that has been called "tuning" (van der Zouwen and Dijkstra 2002); although tuning was not standard practice when the data were collected, the investigators authorized interviewers to use it in the WLS to reduce the burden on the respondent. The interviewer's tuning follow-up displays listening and understanding in that it attends to the information the respondent provided initially and allows the respondent to choose an answer from a reduced set of choices. This procedure is efficient in that the interviewer is being concise, reading only the response options implied by the respondent's hypothetical answer. However, tuning may conflict with the rules of standardization to follow up nondirectively, given that the meaning of any response category depends on the entire set of categories being considered (Fowler and Mangione 1990; Schaeffer and Charng 1991; Smit, Dijkstra, and van der Zouwen 1997).

By attending to both the content and the structure of the interaction, we can also locate missed opportunities for an interviewer to address the respondent's uncertainty. In excerpt 9, the respondent repeats the

Excerpt 9, male interviewer, male respondent, case ID 14

Line	Question	Actor	Text
286	19	I	and do you have a revocable trust?
287	19	R	um revocable trust oh I I I don't um I don't know I forgot what that was
288	19	I	ok

question topic at line 287, says "don't know," and then announces that he "forgot" in a way that could be heard as a request for clarification. Rather than following up with the optional definition of revocable trust that appears on the interviewer's computer screen, the interviewer acknowledges the respondent's "don't know" answer and moves on to the next question. Moving to the next question rather than following up is unresponsive in that it leaves the respondent's uncertainty unaddressed and may reduce the respondent's motivation to display these ambiguities or uncertainties later in the interview. Yet moving on also facilitates efficient progress through the interview; as a procedural compromise, it is analogous to a "good" reason for "bad" records (Garfinkel 1967; Heath and Luff 1996). However, moving on in this instance also conflicts with standardization. Fowler and Mangione (1990) noted that interviewers should follow up a "don't know" response unless it is clear that the respondent is offering "don't know" as an answer. Because a potential request for clarification follows the "don't know," the interviewer should have responded to the request.

4.1.3. When the Respondent Interrupts. Respondents may talk at various points during the reading of the question, for example, after something that can be heard as a complete statement or question or after the first relevant response category. This speech may not constitute an "interruption" from the respondent's point of view, but it poses challenges for the interviewer. The interviewer's action in resuming the reading of the question may imply that she is correcting the respondent's proffered answer, so the interviewer must do this in a way that preserves the respondent's engagement.

For example, the respondent in excerpt 10, who has just heard a question similar to the current question that also has the same response options, says "strictly" at a point that can be heard as a complete question but before the interviewer has repeated the response options. The

Excerpt 10, male interviewer, male respondent, case ID 12

Line	Question	Actor	Text
882	29	I	and how strictly would your spouse want you to follow her wishes?
883	29	R	strictly
884	29	I	ok I I have to read the rest of the question
885	29	R	ok
886	29	I	I'm sorry um would she like you to strictly follow her wishes or do what you think is best even if your preferences are different from her own? and you said strictly follow is that correct?
887	29	R	mhmm

interviewer then explains that he "has" to read the rest of the question, thus providing a reason for not accepting the answer. He prefaces his resumed reading with an apologetic utterance ("I'm sorry") (line 886), a practice that interviewers sometimes use when managing interactional troubles. The interviewer's verification incorporates the answer proffered by the respondent and appends a request for confirmation (line 886), thus acknowledging and displaying understanding of the proffered answer and allowing possible correction. Although the rules of standardized interviewing require this interviewer to continue reading the question after the respondent provides an answer, which could appear unresponsive, the combination of an apologetic utterance, reason, verification, and request for confirmation manage the interactional troubles that could ensue from the interviewer's repair.

4.1.4. *When the Respondent Laughs or Jokes.* The context for and features of laughter and joking in the survey interview are distinctive. Given the constraints placed on the standardized interviewer, laughter and joking are more commonly initiated by the respondent than the interviewer (Cannell et al. 1968; Lavin and Maynard 2001). Furthermore, the interactional force behind the respondents' joking or laughter may be ambiguous: in addition to affiliating and inviting reciprocation (Lavin and Maynard 2001), laughter or jokes may display a "troubles resistance" that indicates uncertainty or difficulty with the question or the answer (Jefferson 1984), or they may mark a question or

Excerpt 11, male interviewer, male respondent, case ID 14

```
Line  Question  Actor  Text
295   23        I      and when you think about the last
                       few days or weeks of your life do
                       you hope to spend these days in
                       your home at an hospital with
                       hospice care or in a nursing home?
296   23        R      huh {L}
297   23        I      {L}
298   23        R      home
299   23        I      ok
```

answer as sensitive. Indeed, the jokes by respondents that we see in our data are either self-deprecating with respect to the task or jokes about death. These issues thus complicate interpretation of the interviewer's responsiveness to the respondent's laughter and jokes. Although reciprocating laughter signals a momentary alignment between parties laughing with each other, the potentially sensitive and difficult nature of the questions in this particular study may require that interviewers not reciprocate laughter so as not to imply laughing about the topic, questions, or the respondent. The upshot from the point of view of standardization is similarly complicated: laughter sometimes invites digression, and survey centers vary in their training of interviewers (Viterna and Maynard 2002).

The survey question in excerpt 11 is conceptually complicated, because where one spends the last few weeks of life probably depends on unknown future circumstances. The respondent displays rumination or hesitation about the question with a token ("huh") followed by laughter ("{L}"), which the interviewer reciprocates (lines 296–297). The respondent's laughter is "question oriented" (Lavin and Maynard 2001): it follows the survey question and occurs where the answer would usually appear. The location of the laughter after the token ("huh") could mark the topic as sensitive, the question as difficult, or both, rather than serving as a humorous comment about the question. The interviewer's reciprocation of the respondent's laughter is responsive with respect to the dimension of listening, showing that she has heard the laughter and is responding in kind. However, if the respondent's laughter marks sensitivity or difficulty, the interviewer's laughter could take the place of potentially more responsive actions, such as an apologetic utterance, that would acknowledge these for the respondent.

Excerpt 12, male interviewer, male respondent, case ID 10

Line	Question	Actor	Text
50	24	I	um now I am going to ask two questions about your end of life treatment preferences suppose you had a serious illness today with very low chances of survival uh first what if you were mentally intact but in severe and constant physical pain? would you want to continue all medical treatments or stop all life prolonging treatments?
51	24	R	well if I didn't have no chance I wanna stop everything yeah
52	24	I	uh ok just says a very low chances of survival?
53	24	R	yeah man
54	24	I	so
55	24	R	I always say throw me in the river you know
56	24	I	{L}
57	24	R	{L}
58	24	I	so for with very low chances you've wanna stop too is that correct?
59	24	R	yeah
60	24	I	ok

In excerpt 12, the respondent offers a report at line 51, and the interviewer initiates a repair in a way that might be characterized as "delicate" in that it does not confront the respondent's mischaracterization of the question directly: the question refers to "very low chances of survival," but the respondent refers to "no chance." The interviewer acknowledges the respondent's report with "ok" and then initiates a correction at line 52, not by repeating the problematic part of the respondent's utterance but by repeating the portion of the question that presents the correction. (The positive valence of this and most other acknowledgment tokens suggests a role for acknowledgments in mitigating subsequent repairs.) Furthermore, the "embedded correction" (Jefferson 1987) does not "expose" the respondent's mistake by turning the matter into an explicit error correction sequence but allows the correction to be accomplished in the course of other interactional business. The respondent answers at line 53, and he expands the answer with a humorous comment (line 55). In contrast to the laughter in excerpt 11, the humorous comment is not placed to express difficulty or sensitivity, and thus can be affiliative and invite reciprocation. (However, given the topic, any joke by a respondent could be a way to manage the sensitivity of the topic in this relatively impersonal setting.) The interviewer's laughter at line 56 is responsive by displaying listening, although again, the laughter could impede progress through the interview if laughter is itself a digression or invites that possibility.

4.2. *The Respondent's Engagement*

The discussion of the interviewer's responsiveness highlights the interdependence of actors: it is impossible to discuss the interviewer's responsiveness without examining the respondent's behavior. With the preceding examples in mind, we now characterize the dimensions of the respondent's engagement: attention to or interest in the survey task. Expressions of engagement vary in the challenges they pose for interviewers trying to maintain standardization. Behaviors that display the respondent's engagement include the following:

1. Like interviewers, respondents display *listening* with acknowledgments. There are several different subtopics within the end-of-life planning section; when a new topic is announced, the end of the preamble that introduces the topic before the target question provides a site that some respondents use to display that they are listening, often with "mhmm" or "okay."

2. The section about end-of-life planning comes at the end of a telephone interview that lasts just under 75 minutes on average. At this point in the interview, many respondents appear to be "well trained"; that is, they wait for interviewers to finish reading the question before answering and choose one of the offered or implied response options in their first answer turn. Such *training* or cooperation with the practices of the interview may display a kind of engagement with the survey task, because such respondents must have attended to the procedures of standardized survey interviews; thus, a series of question-answer sequences with complete codable answers (e.g., "yes" for a yes/no question) may display evidence of this training.

3. Two variations of codable answers, double answers and restatements, may signal engagement with the task by *emphasizing* the respondent's answer. Double answers include repeating the answer (e.g., "yes yes")

or expanding on the answer provided with a synonym (e.g., "nope no"). Restatements use part of the question wording to formulate an answer to the question (e.g., "no plans" or "yes I have" as answers to question 1). Although these emphasizing answers do not challenge the interviewer's ability to maintain standardization, another type of emphasizing answer, the anticipatory answer, does. In excerpt 1, the respondent's answer provides additional information that is directly relevant to the topic of the item and implies and emphasizes "yes" by answering who "anyone" is. However, as discussed, this additional information makes it difficult for the interviewer to simultaneously indicate that the respondent's answer has been heard and to maintain standardization.

4. Respondents can also show that they have *learned* the pattern of questions. For example, in excerpt 10, the respondent displayed having learned the pattern of repetition in questions 28 and 29 and gave a codable answer before the question reading was completed. Such learning is interactionally cooperative and facilitates progress through the task. As seen in excerpt 10, however, these behaviors also challenge the interviewer's ability to be simultaneously responsive and standardized.

5. Some of the behaviors by respondents that we described earlier—such as reports that display uncertainty, mapping difficulties, or providing potentially relevant information, as in excerpts 5 through 9—may also display engagement and be interactionally cooperative. Such behaviors indicate that the respondent is *grappling with the topic and question*; in excerpts 5 through 9, the respondents display a potential lack of fit between what the question is asking and their situation. Similarly, the beginning of the module is a site at which respondents who have experience with the topic may elaborate on their thoughts and experiences when the topic is first introduced. As discussed earlier, such displays are consistent with conversational practices but challenge the interviewer's ability to maintain standardization.

5. CODING THE DIMENSIONS OF RESPONSIVENESS AND ENGAGEMENT

Generating a description of interactional responsiveness and engagement contributes to our understanding of rapport in the survey interview. However, even more compelling would be a demonstration that the dimensions of responsiveness and engagement have analytic and predictive potential; as noted in the background information in section 2, prior research offers conflicting results about the influence of "rapport" (variably defined) on survey outcomes. Thus, we developed a coding scheme to quantify these dimensions. This coding highlights the tensions and trade-offs of moving from a qualitative description of particular interactions to applying a coding scheme, with rules for inclusion and exclusion, to quantify the dimensions of responsiveness and engagement. These findings are located in Appendix B in the online journal. We applied this coding scheme to the 30 cases and used this small data set to analyze differences in the behavioral expressions of responsiveness and engagement between the matched pairs of future participants and nonparticipants in a case-control design.

To illustrate the analytic potential of these codes with the larger set of data (105 pairs), we consider the interviewer's exact reading of the question, which is key to standardization and responsive in at least one way: it facilitates progress through the interview. Because it is efficient and professional, exact question reading could motivate respondents to continue to participate through a pleasant scientific experience that they later recall when asked to provide a saliva sample. The direction of causality is ambiguous, however: the respondent allows the interviewer to be standardized when the respondent answers immediately after the interviewer completes the question, and so the interviewer's exact reading could reflect the respondent's engagement. Although we cannot sort out these explanations, we are able to establish whether question reading is associated with future participation.

We coded whether the interviewer read all the questions exactly as scripted for five of the questions asked of all respondents (questions 1, 2, 11, 24, and 25 in Appendix A in the online journal). We find that reading these questions exactly (compared with reading at least one of the questions with any changes) is associated with increased odds of providing a saliva sample in the future (conditional logistic regression: odds ratio = 2.20, $p < .05$). Although several alternative operationalizations for question reading exist, and criteria other than future participation are also important, this analysis illustrates the analytic potential of the dimensions of responsiveness and engagement within the standardized survey interview.

6. DISCUSSION

This study makes several contributions to studies of survey interviewing. First, drawing on previous theory and our observations, we propose actor-specific concepts to describe what rapport might look like within standardized survey interviews: the responsiveness of the interviewer and the engagement of the respondent. Second, we describe specific behaviors of each actor, the sequential structure of the behaviors, and some actions these behaviors might accomplish as dimensions of responsiveness and engagement. By refining "rapport" in this way and considering the variable dimensions and expressions of responsiveness and engagement with respect to the survey task, standardization, or conversational practices, we see that responsiveness and engagement consist of behaviors that have varying implications for standardization. Finally, we illustrate how these qualitative observations can be the basis for a coding system (although some important nuances are lost in that transformation) and how the behaviors of respondents and interviewers might predict subsequent participation in a longitudinal study (Appendix B in the online journal).

Maintaining the respondent's engagement is one goal of survey researchers. Respondents who are motivated to complete the task are potentially more likely to provide accurate answers, as illustrated by Dijkstra's (1987) experiment about a personal style of interviewing; such motivation may also be an indicator of their underlying propensity to participate in future surveys. Respondents display engagement with the survey task in a variety of ways that show that they are listening, that they have absorbed training about how the interview works, and that they have learned the patterns the questions follow; respondents may also exert more than minimal effort by emphasizing answers and displaying that they are grappling with the topic. However, these displays of engagement vary in the extent to which they provide the interviewer with challenges in maintaining standardization.

Although opportunities for interviewers to display responsiveness are constrained in standardized survey interviews, we nevertheless identify several dimensions of responsive behavior by the interviewer: displays of listening, understanding and allowing choice or correction; apologetic utterances that acknowledge troubles; addressing uncertainty, difficulty, or other problems in answering questions; facilitating efficient progress through the interview; aiding the respondent's cognitive processing; downgrading the force of the task or the request; or addressing potential misunderstandings. Although some of these dimensions of responsiveness have been described previously, our approach further characterizes types of responsiveness from the interviewer that orient to task goals.

Such behaviors may be motivated by the reciprocity that underlies the interview: The respondent has agreed to do this long interview, and the interviewer has the tools and training to simplify the task, albeit with varying implications for standardization.

The various dimensions of responsiveness discussed above, with their behavioral expressions, fall under at least three categories with respect to standardization: (1) they may conflict with standardization, (2) they may be neutral from the point of view of standardization, or (3) they may align with the rules of standardization. Responsive practices that may conflict with standardization include verifications, linkages across questions (that modify question reading), and directive or tuning follow-up actions for uncodable answers. Acknowledgments that display listening may also conflict with standardization to the extent that "okay" or "right" are interpreted by respondents as a positive comment on the content of the answer; thus, some survey organizations train their interviewers not to provide such acknowledgments. Responsive practices that are probably neutral under the rules of standardization include repeating or paraphrasing (by reexpressing the answer, such as "sure," in terms of the response categories, "yes") the respondent's codable answer; following up codable answers that are complex or express uncertainty or difficulty; providing reasons for the interviewer's behavior; and apologetic utterances and mitigators embedded in other talk (although these would conflict with the rules of standardization if incorporated within question reading). One responsive practice that is part of standardization is a nondirective follow-up after an uncodable answer. Overall, by broadening our consideration of the interviewer's responsiveness, we see that the responsive behaviors by the interviewer do not always conflict with the rules of standardization and may even enhance the goals of standardization when used skillfully.

We have also shown the importance of considering sequencing in the interaction because the interviewer's opportunities for responsiveness are occasioned by the respondent's preceding behavior. In addition, examining the sequential nature of the interaction allows us to locate opportunities for responsiveness that were missed, for example, by failing to incorporate into question reading information that is in the common ground or failing to address the respondent's uncertainty, difficulties, or other problems in answering survey questions.

Intensive study of the dimensions of responsiveness and engagement during the survey interview is an integral first step in identifying practices that support reliable and valid measurement as well as motivation and willingness to continue participating in a study. Thus, the implications of the dimensions of responsiveness and engagement for data quality and future study participation will need to be addressed in future research. Important next steps include using the total survey error perspective (reviewed by Groves and Lyberg 2010) or the comprehension-retrieval-judgment-response model of respondents' cognitive processing (reviewed by Tourangeau, Rips, and Rasinski 2000) to guide future research on whether and when the dimensions of responsiveness and engagement contribute to data quality and survey participation. In addition, understanding how rapport can be developed in ways that do not conflict with the practices and thus the goals of standardization is increasingly important as new interviewing techniques must be developed to accompany new technologies and varied and demanding types of data collection, such as anthropometrics and biomarkers.

The present study highlights the multifaceted and context-specific nature of a seemingly straightforward sequence of question-answer-acknowledgment by describing situations in which interviewers must deal with the tension between practices of conversation and standardized interviewing. Our study also has some implications for question design and the training of interviewers. First, this study indirectly illustrates the important role of question design, because features of questions may elicit nonparadigmatic sequences that might be avoided with changes to the design of questions (Dykema, Schaeffer, and Garbarski 2012, 2013; Schaeffer and Dykema 2011a, 2011b; van der Zouwen and Dijkstra 2002). For example, researchers should minimize optional or conditional phrases in questions by integrating them into a single standard question (e.g., Dykema, Schaeffer, Garbarski, Nordheim, et al. 2013, forthcoming) or at least specify on the screen exactly when to read optional phrases so that this information is given to respondents in a standardized way.

Second, a contribution of our qualitative analysis is to identify situations that interviewers encounter during interviews for which training and monitoring should provide guidance but for which current research may not yet specify which behaviors by interviewers will have a negative effect on the quality of data or future participation. The needs of interviewers that may not be dealt with comprehensively in current standardized training regimes include the following:

- How to recognize a codable answer. For example, the design of the question needs to project for the respondent what detail is required to adequately answer a question about "who": a name, a relationship, or both, and whether multiple names or relationships should be recorded as "other" or dealt with individually. Although this requirement seems obvious, what constitutes a codable answer varies according to the type of question, or the codable answers that an interviewer hears may be embedded in conversational practices (e.g., mitigating words such as *probably*); in any case, the interviewer must respond instantaneously.
- Whether and when to follow up codable answers that are complex, possibly internally inconsistent, or hedged in uncertainty or difficulty.
- How to identify and respond to a variety of nonparadigmatic sequences.
- Methods to manage interactional troubles such as interruptions or silences. These include, for example, giving reasons that acknowledge the respondent's action and explain the practices of standardization (e.g., "I know you have given an answer, but I have to read all the categories so that you hear all your choices") or explain a silence (e.g., "Just one moment while I write that down").
- How to acknowledge or respond to apologetic utterances or laughter by the respondent.
- Which micro-adjustments by the interviewer are common in conversation and responsive but do not deviate from standardization in ways that might affect the data. These may include acknowledgments (or "feedback") such as "okay," requests for confirmation, repeating codable answers, making linkages across questions (e.g., "and," "next,"), some forms of mitigating talk (e.g., "just"), and apologetic utterances.
- Whether, when, and how to use verifications. This issue is particularly vexing for standardization, because it is useful but also conflicts with strict standardization. As yet there is no research about the boundaries within which this practice must stay in order to maintain data quality.

Future experimental research using criteria of data quality and future study participation is needed before deciding whether and how particular practices should be incorporated into interviewer training.

One difficulty in any study of interaction is the heterogeneity in actions performed by particular behaviors, or by their absence. Conversational practices are learned through everyday interaction and may vary across people. Although "mhmm" may indicate engagement when it shows listening, it may also simulate attention, and an absence of such utterances does not necessarily indicate a lack of engagement. Similarly, unreciprocated laughter may mean a variety of things, including poor conversational skills or a desire to keep the interview moving along. Although we examine interactional behaviors turn by turn, it is plausible that, in order to say something conclusive about the consequences of these behaviors, the entire interview must be analyzed and baseline standards of talk within interviewers and respondents developed. This is particularly important given that the mutual influence of the interviewer's responsiveness and the respondent's engagement complicates their study. For example, the interviewer's responsiveness may increase the motivation of respondents to work hard and display their uncertainties or misunderstandings—leading to more nonparadigmatic interactions and thus more sites for interviewers to display further their responsiveness.

This study describes behaviors that might constitute rapport and resulting challenges for the interviewer in maintaining the practices of standardization. Despite the potential limitations noted above, the ability to identify responsive behaviors by the interviewer and engaged behaviors by the respondent has potential applications in future studies—for example, focusing field efforts in longitudinal studies, so that recruitment specialists could focus on respondents who displayed little engagement in the prior interview. Markers of respondent engagement could also be used to analyze concurrent data, possibly providing an adjustment for data quality, if such a use is supported by future studies with validation data that examine the association between respondents' engagement and measurement error. Finally, understanding how interviewers import conversational practices into standardized interviews is critical for designing interviewing methods and training interviewers in these methods. More studies are needed to examine which small accommodations and adjustments in interviewers' behaviors might be both compatible with practices of standardization and interactionally responsive, thus meeting measurement goals while maintaining the respondent's engagement and motivation.

Acknowledgments

Earlier versions of this research were presented at the annual meeting of the American Association for Public Opinion Research in 2013 and 2014, the Interviewer-Respondent Interaction workshop in 2013, the Social Psychology and Microsociology seminar in the Department of Sociology at the University of Wisconsin–Madison in 2013, and the IRC seminar at the University of Wisconsin Survey Center in 2014. We thank Ellen Dinsmore and Bo Hee Min for their research assistance, the anonymous reviewers for their comments, and participants in each of the conferences and seminars noted above for their feedback. We also thank the participants in the Wisconsin Longitudinal Study (WLS) for their generous contributions of time and information over many years. The opinions expressed herein are those of the authors.

This research uses data from the WLS of the University of Wisconsin–Madison. Since 1991, the WLS has been supported principally by the National Institute on Aging (grants R01AG009775, P01AG021079 and R01AG033285), with additional support from the Vilas Estate Trust, the National Science Foundation, the Spencer Foundation, and the Graduate School of the University of Wisconsin–Madison. Since 1992, data have been collected by the University of Wisconsin Survey Center. A public-use file of data from the WLS is available from the WLS, University of Wisconsin–Madison, 1180 Observatory Drive, Madison, WI 53706 and at http://www.ssc.wisc.edu/wlsresearch/data/.

Funding

This research was supported by an award to Dana Garbarski from the Charles Cannell Fund in Survey Methodology in the Institute for Social Research at the University of Michigan. In addition, this research was supported by a pilot grant from the Center for Demography of Health and Aging at the University of Wisconsin–Madison to Nora Cate Schaeffer (grant P30AG017266) and a grant to Schaeffer by the UW-Madison Graduate School Research Committee (grant 140581). Additional support was provided by the Center for Women's Health and Health Disparities Research at the University of Wisconsin–Madison (grant T32HD049302), the Center for Demography and Ecology at the University of Wisconsin–Madison (grants R24HD047873, T32HD07014, and P2CHD047873), the Wisconsin Longitudinal Study: Tracking the Life Course (grant P01AG021079), the Sewell Bascom Professorship, and by the University of Wisconsin Survey Center.

Notes

1. The module includes questions that pose potential difficulties for respondents in formulating their answers because they cover abstract topics, use varying reference periods and question structures, have long response options, and occur at the end of a lengthy interview (Schaeffer and Dykema 2011a).
2. Questions about end-of-life planning may be sensitive or threatening if respondents have not done any planning and perceive that the questions imply that they should have done so. Questions may also be perceived as intrusive or raise painful or stressful feelings for the respondent (Schaeffer 2000; Tourangeau and Yan 2007). Furthermore, the end-of-life planning module comes at the end of a lengthy interview in which the respondent has revealed considerable personal information to an interviewer who is usually much younger than the respondent. These asymmetries in disclosure and age may exacerbate the sensitivity of the topic.

3. Of 240 cases that were selected initially, we dropped 15 cases, and their pair members, because of missing audio. In the 210 cases selected, there are 97 interviewers.
4. To construct the propensity score, we estimated a logistic regression model of providing a saliva sample (vs. not) using a variety of predictors of participation, including sex, participation in past waves, high school test scores and class rank, educational attainment, self-rated health, religious attendance, household income and net worth, and number of contact attempts by survey staff members in the 2004 wave of data collection (Hauser 2005; McFadden's pseudo-R^2 = .08).
5. Although verification conflicts with some established versions of standardization, many survey organizations train their interviewers in this practice, having noted its usefulness (e.g., Kovar and Royston 1990, who used the term *confirm*). The WLS investigators authorized verification so that respondents would know they had been heard and to reduce burden on respondents. When verifying, interviewers were to read the entire question, state the respondent's candidate answer, and then request confirmation (WLS, *Phone Booth Manual*, undated).
6. In the 24 instances of verification we observed in these data, the respondent repaired the interviewer's proffered answer only once but ended up agreeing with the interviewer's verification after further follow-up. This may indicate that interviewers are efficiently obtaining codable answers by attending to what the respondent said previously, that the verifications are worded to prefer agreement regardless of substance (Pomerantz and Heritage 2013), or both. The validation data that would be required to sort this out do not exist.

References

Belli, Robert F., Ipek Bilgen, and Tarek Al Baghal. 2013. "Memory, Communication, and Data Quality in Calendar Interviews." *Public Opinion Quarterly* 77(S1): 194–219.

Belli, Robert F., James M. Lepkowski, and Mohammed U. Kabeto. 2001. "The Respective Roles of Cognitive Processing Difficulty and Conversational Rapport on the Accuracy of Retrospective Reports of Doctors' Office Visits." Pp. 197–203 in *Seventh Health Survey Research Methods Conference Proceedings*. Chicago: University of Illinois–Chicago.

Brennan, Susan E., and Herbert H. Clark. 1996. "Conceptual Pacts and Lexical Choice in Conversation." *Journal of Experimental Psychology: Learning, Memory, and Cognition* 22(6):1482–93.

Broome, Jessica. 2015. "How Telephone Interviewers' Responsiveness Impacts Their Success." *Field Methods* 27(1):66–81.

Cannell, Charles F., and Morris Axelrod. 1956. "The Respondent Reports on the Interview." *American Journal of Sociology* 62(2):177–81.

Cannell, Charles F., Floyd J. Fowler, Jr., and Kent H. Marquis. 1968. "The Influence of Interviewer and Respondent Psychological and Behavioral Variables on the Reporting in Household Interviews." *Vital and Health Statistics: Data Evaluation and Methods Research*, Series 2, No. 26. Washington, DC: U.S. Department of Health and Human Services.

Cannell, Charles F., Peter V. Miller, and Lois Oksenberg. 1981. "Research on Interviewing Techniques." Pp. 389–437 in *Sociological Methodology*, Vol. 12, edited by S. Leinhardt. San Francisco, CA: Jossey-Bass.

Cappella, Joseph N. 1990. "On Defining Conversational Coordination and Rapport." *Psychological Inquiry* 1(4):303–305.

Cassell, Justine, and Peter Miller. 2008. "Is It Self-administration if the Computer Gives You Encouraging Looks?" Pp. 161–78 in *Envisioning the Survey Interview of the Future*, edited by F. G. Conrad, and M. F. Schober. Hoboken, NJ: John Wiley.

Clark, Colin, Paul Drew, and Trevor Pinch. 2003. "Managing Prospect Affiliation and Rapport in Real-life Sales Encounters." *Discourse Studies* 5(1):5–31.

Clark, Herbert H. 1996. *Using Language*. Cambridge, United Kingdom: Cambridge University Press.

Clark, Herbert H., and Susan E. Brennan. 1991. "Grounding in Communication." Pp. 127–49 in *Perspectives on Socially Shared Cognition*, edited by L. B. Resnick, J. M. Levine, and S. D. Teasley. Washington, DC: American Psychological Association.

Dijkstra, Wil. 1987. "Interviewing Style and Respondent Behavior: An Experimental Study of the Survey Interview." *Sociological Methods and Research* 16(2):309–34.

Drew, Paul. 1984. "Speakers' Reportings in Invitation Sequences." Pp. 129–51 in *Structures of Social Action: Studies in Conversation Analysis*, edited by J. M. Atkinson, and J. Heritage. Cambridge, United Kingdom: Cambridge University Press.

Drew, Paul, and John Heritage, eds. 1992. *Analyzing Talk at Work: An Introduction*. Cambridge, United Kingdom: Cambridge University Press.

Dundon, Tony, and Paul Ryan. 2010. "Interviewing Reluctant Respondents: Strikes, Henchmen, and Gaelic Games." *Organizational Research Methods* 13(3):562–81.

Dykema, Jennifer, Nora Cate Schaeffer, and Dana Garbarski. 2012. "Effects of Agree-disagree Versus Construct-specific Items on Reliability, Validity, and Interviewer-respondent Interaction." Presented at the annual meeting of the American Association for Public Opinion Research, Orlando, FL.

Dykema, Jennifer, Nora Cate Schaeffer, and Dana Garbarski. 2013. "Associations between Interactional Indicators of Problematic Questions and Systems for Coding Question Characteristics." Presented at the annual meeting of the American Association for Public Opinion Research, Boston, MA.

Dykema, Jennifer, Nora Cate Schaeffer, Dana Garbarski, Erik V. Nordheim, Mark Banghart, and Kristen Cyffka. Forthcoming. "The Impact of Parenthetical Phrases on Interviewers' and Respondents' Processing of Survey Questions." *Survey Practice*.

Dykema, Jennifer, Nora Cate Schaeffer, Dana Garbarski, Rick Nordheim, and Kristen Cyffka. 2013. "Effects of Question, Respondent, and Interviewer Characteristics on Interactional Indicators of Respondent and Interviewer Processing of Health-related Questions." Presented at the Interviewer-Respondent Interaction Workshop, Boston, MA.

Foucault Welles, Brooke. 2013. "Behavioral Correlates of Rapport in Survey Interviews." Presented at the Interviewer-Respondent Interaction Workshop, Boston, MA.

Fowler, Floyd J., Jr., and Thomas W. Mangione. 1990. *Standardized Survey Interviewing: Minimizing Interviewer-related Error*. Newbury Park, CA: Sage.

Garbarski, Dana, Nora Cate Schaeffer, and Jennifer Dykema. 2011. "Are Interactional Behaviors Exhibited When the Self-reported Health Question Is Asked Associated with Health Status?" *Social Science Research* 40(4):1025–36.

Garfinkel, Harold. 1967. *Studies in Ethnomethodology.* Englewood Cliffs, NJ: Prentice Hall.

Goudy, Willis J., and Harry R. Potter. 1975. "Interview Rapport: Demise of a Concept." *Public Opinion Quarterly* 39(4):529–43.

Groves, Robert M., and Lars Lyberg. 2010. "Total Survey Error: Past, Present, and Future." *Public Opinion Quarterly* 74(5):849–79.

Haan, Marieke, Yfke Ongena, and Mike Huiskes. 2013. "Interviewers' Questions: Rewording Not Always a Bad Thing." Pp. 173–94 in *Interviewers' Deviations in Surveys: Impact, Reasons, Detection and Prevention,* edited by P. Winker, N. Menold, and R. Porst. Bern, Switzerland: Peter Lang.

Hak, Tony. 2002. "How Interviewers Make Coding Decisions." Pp. 449–70 in *Standardization and Tacit Knowledge: Interaction and Practice in the Survey Interview,* edited by D. W. Maynard, H. Houtkoop-Steenstra, N. C. Schaeffer, and J. van der Zouwen. New York: John Wiley.

Hauser, Robert M. 2005. "Survey Response in the Long Run: The Wisconsin Longitudinal Study." *Field Methods* 17(1):3–29.

Heath, Christian, and Paul Luff. 1996. "Documents and Professional Practice: 'Bad' Organisational Reasons for 'Good' Clinical Records." Pp. 354–63 in *Proceedings of the 1996 ACM Conference on Computer Supported Cooperative Work.* New York: Association for Computing Machinery.

Heritage, John. 2011. "Territories of Knowledge, Territories of Experience: Empathic Moments in Interaction." Pp. 159–83 in *The Morality of Knowledge in Conversation,* Vol. 29, edited by T. Stivers, L. Mondada, and J. Steensig. Cambridge, United Kingdom: Cambridge University Press.

Heritage, John, and Marja Leena Sorjonen. 1994. "Constituting and Maintaining Activities across Sequences: And Prefacing as a Feature of Question Design." *Language in Society* 23(1):1–29.

Horgan, Mike. 2005. "What Would You Say? Some Practices for Answering Questions in Telephone Survey Interviews." Master's thesis, Department of Sociology, University of Wisconsin–Madison.

Houtkoop-Steenstra, Hanneke. 2000. "Establishing Rapport." Pp. 128–53 in *Interaction and the Standardized Survey Interview: The Living Questionnaire,* by H. Houtkoop-Steenstra. Cambridge, United Kingdom: Cambridge University Press.

Hyman, Herbert H. [1954] 1975. *Interviewing in Social Research.* Chicago: University of Chicago.

Jefferson, Gail. 1984. "On the Organization of Laughter in Talk about Troubles." Pp. 346–69 in *Structures of Social Action,* edited by J. M. Atkinson, and J. Heritage. Cambridge, United Kingdom: Cambridge University Press.

Jefferson, Gail. 1987. "On Exposed and Embedded Correction in Conversation." Pp. 86–100 in *Talk and Social Organisation,* edited by G. Button, and J.R.E. Lee. Clevedon, United Kingdom: Multilingual Matters.

Kitzinger, Celia. 2013. "Repair." Pp. 229–56 in *The Handbook of Conversation Analysis*, edited by J. Sidnell, and T. Stivers. Chichester, United Kingdom: Blackwell.

Kovar, Mary Grace, and Patricia Royston. 1990. "Comment on 'Interactional Troubles in Face-to-face Survey Interviews'." *Journal of the American Statistical Association* 85(409):246–47.

Lavin, Danielle, and Douglas W. Maynard. 2001. "Standardization vs. Rapport: Respondent Laughter and Interviewer Reaction During Telephone Surveys." *American Sociological Review* 66(3):453–79.

Levinson, Stephen C. 2013. "Action Formation and Ascription." Pp. 103–30 in *The Handbook of Conversation Analysis*, edited by J. Sidnell, and T. Stivers. Chichester, United Kingdom: Blackwell.

Lindström, Anna, and Marja-Leena Sorjonen. 2013. "Affiliation in Conversation." Pp. 350–60 in *The Handbook of Conversation Analysis*, edited by J. Sidnell, and T. Stivers. Chichester, United Kingdom: Blackwell.

Maynard, Douglas W., Jeremy Freese, and Nora Cate Schaeffer. 2010. "Calling for Participation: Requests, Blocking Moves, and Rational (Inter)Action in Survey Introductions." *American Sociological Review* 75(5):791–814.

Maynard, Douglas W., and Courtney L. Marlaire. 1992. "Good Reasons for Bad Testing Performance: The Interactional Substrate of Educational Exams." *Qualitative Sociology* 15(2):177–202.

Miller, Bill, and Steve Rollnick. 2002. *Motivational Interviewing: Preparing People for Change*. New York: Guilford.

Moore, Robert J., and Douglas W. Maynard. 2002. "Achieving Understanding in the Standardized Survey Interview: Repair Sequences." Pp. 281–312 in *Standardization and Tacit Knowledge: Interaction and Practice in the Survey Interview*, edited by D. W. Maynard, H. Houtkoop-Steenstra, N. C. Schaeffer, and J. van der Zouwen. New York: John Wiley.

O'Muircheartaigh, Colm, and Pamela Campanelli. 1998. "The Relative Impact of Interviewer Effects and Sample Design Effects on Survey Precision." *Journal of the Royal Statistical Society, Series A*, 161(1):63–77.

Pomerantz, Anita, and John Heritage. 2013. "Preference." Pp. 210–28 in *The Handbook of Conversation Analysis*, edited by J. Sidnell, and T. Stivers. Chichester, United Kingdom: Blackwell.

Ruusuvuori, Johanna. 2013. "Emotion, Affect and Conversation." Pp. 330–49 in *The Handbook of Conversation Analysis*, edited by J. Sidnell, and T. Stivers. Chichester, United Kingdom: Blackwell.

Schaeffer, Nora Cate. 1991. "Conversation with a Purpose—or Conversation? Interaction in the Standardized Interview." Pp. 367–92 in *Measurement Errors in Surveys*, edited by P. P. Biemer, R. M. Groves, L. E. Lyberg, N. A. Mathiowetz, and S. Sudman. New York: Wiley.

Schaeffer, Nora Cate. 2000. "Asking Questions about Threatening Topics: A Selective Overview." Pp. 105–22 in *The Science of Self-Report: Implications for Research and Practice*, edited by Christine A. Bachrach, Arthur A. Stone, Jared B. Jobe, Howard S. Kurtzman, and Virginia S. Cain. Mahwah, NJ: Lawrence Erlbaum Associates.

Schaeffer, Nora Cate, and Hong-Wen Charng. 1991. "Two Experiments in Simplifying Response Categories: Intensity Ratings and Behavioral Frequencies." *Sociological Perspectives* 34(2):165–82.

Schaeffer, Nora Cate, and Jennifer Dykema. 2011a. "Questions for Surveys: Current Trends and Future Directions." *Public Opinion Quarterly* 75(5):909–61.

Schaeffer, Nora Cate, and Jennifer Dykema. 2011b. "Response 1 to Fowler's Chapter: Coding the Behavior of Interviewers and Respondents to Evaluate Survey Questions." Pp. 23–39 in *Question Evaluation Methods: Contributing to the Science of Data Quality*, edited by J. Madans, K. Miller, A. Maitland, and G. Willis. Hoboken, NJ: John Wiley.

Schaeffer, Nora Cate, Jennifer Dykema, and Douglas W. Maynard. 2010. "Interviewers and Interviewing." Pp. 437–70 in *Handbook of Survey Research*, 2nd ed., edited by P. V. Marsden, and J. D. Wright. Bingley, United Kingdom: Emerald Group.

Schaeffer, Nora Cate, Dana Garbarski, Jeremy Freese, and Douglas W. Maynard. 2013. "An Interactional Model of the Call for Participation in the Survey Interview: Actions and Reactions in the Survey Recruitment Call." *Public Opinion Quarterly* 77(1):323–51.

Schaeffer, Nora Cate, and Douglas W. Maynard. 1996. "From Paradigm to Prototype and Back Again: Interactive Aspects of Cognitive Processing in Survey Interviews." Pp. 65–88 in *Answering Questions: Methodology for Determining Cognitive and Communicative Processes in Survey Research*, edited by N. E. Schwarz, and S. Sudman. San Francisco, CA: Jossey-Bass.

Schaeffer, Nora Cate, and Douglas W. Maynard. 2008. "The Contemporary Standardized Survey Interview for Social Research." Pp. 31–57 in *Envisioning the Survey Interview of the Future*, edited by F. G. Conrad, and M. F. Schober. Hoboken, NJ: John Wiley.

Schnell, Rainer, and Frauke Kreuter. 2005. "Separating Interviewer and Sampling-point Effects." *Journal of Official Statistics* 21(3):389–410.

Schober, Michael F., and Frederick G. Conrad. 1997. "Does Conversational Interviewing Reduce Survey Measurement Error?" *Public Opinion Quarterly* 61(4): 576–602.

Smit, Johannes H., Wil Dijkstra, and Johannes van der Zouwen. 1997. "Suggestive Interviewer Behaviour in Surveys: An Experimental Study." *Journal of Official Statistics* 13(1):19–28.

Stivers, Tanya. 2013. "Sequence Organization." Pp. 191–209 in *The Handbook of Conversation Analysis*, edited by J. Sidnell, and T. Stivers. Chichester, United Kingdom: Blackwell.

Suchman, Lucy, and Brigitte Jordan. 1990. "Interactional Troubles in Face-to-face Survey Interviews." *Journal of the American Statistical Association* 85(409):232–53.

Tartter, Vivien C., and David Braun. 1994. "Hearing Smiles and Frowns in Normal and Whisper Registers." *Journal of the Acoustical Society of America* 96(4):2101–7.

Thornton, Arland, Deborah S. Freedman, and Donald Camburn. 1982. "Obtaining Respondent Cooperation in Family Panel Studies." *Sociological Methods and Research* 11(1):33–51.

Tickle-Degnen, Linda, and Robert Rosenthal. 1990. "The Nature of Rapport and Its Nonverbal Correlates." *Psychological Inquiry* 1(4):285–93.

Tourangeau, Roger, Lance J. Rips, and Kenneth A. Rasinski. 2000. *The Psychology of Survey Response*. Cambridge, United Kingdom: Cambridge University Press.

Tourangeau, Roger, and Ting Yan. 2007. "Sensitive Questions in Surveys." *Psychological Bulletin* 133(5):859–83.

van der Zouwen, Johannes, and Wil Dijkstra. 2002. "Testing Questionnaires Using Interaction Coding." Pp. 427–48 in *Standardization and Tacit Knowledge: Interaction and Practice in the Survey Interview*, edited by D. W. Maynard, H. Houtkoop-Steenstra, N. C. Schaeffer, and J. van der Zouwen. New York: John Wiley.

Viterna, Jocelyn S., and Douglas W. Maynard. 2002. "How Uniform Is Standardization? Variation within and across Survey Research Centers Regarding Protocols for Interviewing." Pp. 365–401 in *Standardization and Tacit Knowledge: Interaction and Practice in the Survey Interview*, edited by D. W. Maynard, H. Houtkoop-Steenstra, N. C. Schaeffer, and J. van der Zouwen. New York: John Wiley.

Weiss, Carol H. 1968. "Validity of Welfare Mothers' Interview Responses." *Public Opinion Quarterly* 32:622–33.

Weiss, Carol H. 1970. "Interaction in the Research Interview: The Effects of Rapport on Response." Pp. 17–20 in *Proceedings of the Social Statistics Section of the American Statistical Association*. Denver, CO: American Statistical Association.

Author Biographies

Dana Garbarski is an assistant professor in the Department of Sociology at Loyola University Chicago. Her research interests comprise a mix of substantive and methodological issues related to social inequalities, health, and the life course, including complex longitudinal relationships between social factors and health, the measurement of self-rated health, and interviewer-respondent interaction. Her work has been published in sociological, methodological, and public opinion research journals.

Nora Cate Schaeffer is the Sewell Bascom Professor of Sociology and the faculty director of the University of Wisconsin Survey Center at the University of Wisconsin–Madison. Her current research focuses on instrument design issues as well as interaction when the sample member is recruited and during the interview.

Jennifer Dykema is a senior scientist and survey methodologist at the University of Wisconsin Survey Center at the University of Wisconsin–Madison. Her current research interests focus on questionnaire design, interviewer-respondent interaction, and methods to increase response rates. She has published work on these topics in methodological, sociological, and public opinion journals as well as in books.

Sociological Methodology
2016, Vol. 46(1) 39–52
© American Sociological Association 2016
http://sm.sagepub.com

COMMENT: FINDING EVIDENCE FOR RAPPORT OVER SIX DECADES

Peter V. Miller*

*U.S. Census Bureau, Washington, DC, USA
Corresponding Author: Peter V. Miller, peter.miller@census.gov
DOI: 10.1177/0081175016644900

Consider this fragment from an interview transcript and the comment that follows:

```
I:   Did you have some favorite subjects [in school]?
R:   Yes, typing.
I:   Good, and how is your speed in typing?
R:   Well, when I finished school I had between 57 and 60
     words per minute.
I:   I see.
R:   Of course, I haven't been doing too much typing.
     With my job it wasn't required, and so I'm a little
     rusty right now.
I:   Sure. That's something you have to keep up on and do
     every day.
R:   Yes, once you get your speed up there, I find, if you
     stop, it goes down to about 50. I went all summer,
     and when I started school it was down to about 35,
     but then it came right up again.
```

The information obtained here . . . reflects the permissive relationship which has been established. Even in the collection of such factual information as the number of words per minute that the respondent can type, the attitude of the respondent toward the interview and the interviewer is relevant. In this case the respondent is sufficiently secure to be very frank in her evaluation of her own typing speed. (Kahn and Cannell 1957:306)

Like Garbarski, Schaeffer, and Dykema (this volume, pp. 1–38), Kahn and Cannell (1957), in *The Dynamics of Interviewing*, sought to characterize interviewer-respondent relationships through an examination of transcribed interview exchanges. The similarity between the two efforts ends there. The interviews Kahn and Cannell analyzed were in no way "standardized." Interviewers were free to ask questions as they saw fit, to determine the order and wording, to interject viewpoints of their own. The success of their approaches was judged by how forthcoming respondents were in answering questions. Forthcomingness included the amount and relevance of detail in responses and the avoidance of "socially desirable" answers.

In the above excerpt (from an employment interview), the respondent volunteered details about her typing skills that could be viewed negatively. Kahn and Cannell (1957) attributed the level of disclosure and the self-deprecation to the "permissive" relationship the interviewer and respondent had established. They noted that the respondent felt "secure" enough to reveal things that could hurt her chances of getting a job. The interviewer had created a nonjudgmental environment that fostered such responses. She had accomplished this, Kahn and Cannell argued, through asking open questions and following with "nondirective, permissive probe questions" (p. 305).

To my mind, this relationship is one of *rapport*. The respondent feels at ease, without fear of negative judgment, willing to supply the details requested, trusting in the interviewer's bona fides and accepting demeanor. Where Garbarski et al. see rapport in the interviewer's attempts to respect conversational norms that sometimes conflict with standardization, Kahn and Cannell (1957) saw it in the interviewer's ability to establish an interpersonal environment in which the respondent would provide candid and thorough answers. They did not use the word *rapport*, however; they used *permissive relationship* instead. Cannell, a student of the clinical psychologist Carl Rogers, transported this term from Rogers's (1942) writings on counseling and psychotherapy.

Rapport, by the late 1950s, had acquired a mixed reputation, following Hyman et al.'s (1954) argument that it can connote a superficial "friendliness" that discourages candor. (Weiss [1968] later called this a *kaffeeklatsch* atmosphere.) Kahn and Cannell (1957), nevertheless, said that even in the collection of factual information, "the attitude of the respondent toward the interview and the interviewer is relevant" (p. 306). They found evidence for this attitude in a respondent's full and candid answers.

Kahn and Cannell's (1957) analysis of the interview exchange must be regarded as impressionistic when compared with the detailed coding framework presented by Garbarski et al. These authors explore the complex interplay of conversational norms and standardized interview requirements. Taking standardized interview rules as given, they observe how interviewers depart from the questionnaire script to attend to conversational expectations. A common issue in the standardized interview is how to ask questions that appear to have been answered already. Garbarski et al. note, for example, how interviewers sometimes modify the scripted question in order to indicate to respondents that they are paying attention to what has been said. The authors regard these modifications as evidence of rapport. They argue that some modifications for the sake of conversational maintenance violate standardization rules, while others do not. Looking at the other side of the interaction, Gabarski et al. code instances in which respondents show that they have learned how the standardized interview interaction is supposed to go, by adhering to the demands of the protocol, as evidence for rapport. One might say that rapport in these observations amounts to each party's recognizing the unusual interaction demands posed by the standardized interview and showing deference for the other's position.

In the decades following the publication of *The Dynamics of Interviewing*, Charles Cannell and his colleagues (particularly Floyd J. [Jack] Fowler) developed the standardized interview technique. They argued against rapport-building as a method for acquiring valid information in the survey interview (see Cannell, Miller, and Oksenberg 1981). Adopting the *kaffeeklatsch* definition of rapport, they asserted that seeking it is apt to lead to interviewer variance and inaccurate responses. In the course of developing the standardized approach, however, the conflict with conversational norms became apparent. In 1980, the standardized interview was examined in a national health study (Thornberry 1987). The interviewer training course for that study rehearsed circumstances in which the interview script presented conversational awkwardness and gave interviewers some rules for dealing with them. The rules resembled, in some cases, the methods devised by interviewers in the study analyzed by Garbarski et al. For example, interviewers were permitted to recognize the respondent's previous answers when asking new but overlapping questions.

Does this sort of behavior constitute evidence of rapport? It seems that it may, to the extent that it positively influences the respondent's attitude toward the interviewer and the interview and increases the validity of responses. If respondents are more at ease and willing to provide complete, unvarnished information when interviewers pay attention to conversational norms, then, following Kahn and Cannell (1957), the "good" sort of rapport seems to have been established.

Garbarski et al. argue, instead, that the effect of the interviewer's blending conversational norms and standardized interviewing can be seen in the respondent's future survey participation. The evidence they present in a first effort to test this idea is mixed. As future tests are developed, it seems useful to examine also the effect of rapport, as they define it, on response quality in the interview in which it is established.

Author's Note

The author prepared this work within the scope of his employment with the U.S. Census Bureau. The views expressed on statistical, methodological, technical, or operational issues are those of the author and not necessarily those of the U.S. Census Bureau.

References

Cannell, Charles F., Peter V. Miller, and Lois Oksenberg. 1981. "Research on Interviewing Techniques." Pp. 389–437 in *Sociological Methodology*, Vol. 12, edited by S. Leinhardt. San Francisco, CA: Jossey-Bass.

Hyman, Herbert, with William J. Cobb, Jacob J. Feldman, Clyde W. Hart, and Charles Herbert Spencer. 1954. *Interviewing in Social Research*. Chicago: University of Chicago Press.

Kahn, Robert L., and Charles F. Cannell. 1957. *The Dynamics of Interviewing*. New York: John Wiley.

Rogers, Carl R. 1942. *Counseling and Psychotherapy*. Boston: Houghton Mifflin.

Thornberry, Owen T. 1987. "Experimental Comparison of Telephone and Personal Health Interview Surveys." *Vital and Health Statistics*. No. 106. Washington, DC: United States Public Health Service.

Weiss, Carol H. 1968. "Validity of Welfare Mothers' Interview Responses." *Public Opinion Quarterly* 32(4):622–38.

Author Biography

Peter V. Miller is a senior researcher for survey measurement at the U.S. Census Bureau. He joined the staff of the Census Bureau in 2011. He is a member of the Federal Committee on Statistical Methodology and cochairs the adaptive design interest group. Before arriving at the Census Bureau, Miller spent 29 years at Northwestern University, where he holds an appointment as professor emeritus. Miller was editor-in-chief of *Public Opinion Quarterly* from 2001 to 2008. He has held several elective offices in the American Association for Public Opinion Research (AAPOR), most recently serving as president in 2009 and 2010. He received the Harry W. O'Neill Award for Outstanding Achievement from the New York Chapter of AAPOR in 2012. He was also named a fellow of the Midwest Chapter of AAPOR in 2012. In 2015, he was named a fellow of the American Statistical Association. Miller was born in Pontiac, Michigan, and earned AB and PhD degrees at the University of Michigan.

COMMENT: RAPPORT IN SURVEY INTERACTIONS

*Michael F. Schober**

*Department of Psychology, The New School for Social Research, New York, NY, USA
Corresponding Author: Michael F. Schober, schober@newschool.edu
DOI: 10.1177/0081175016644897

Garbarski, Schaeffer, and Dykema (this volume, pp. 1–38) are quite right to point out the slipperiness of the concept of rapport between interviewers and respondents in social research. Their overview is admirable in laying out the complexities and what has led the concept into disrepute, and their focus on transcripts of question-answer sequences in standardized telephone interviews to illustrate their account of rapport makes the issues concrete. The interdependencies of interviewer and respondent behavior Garbarski et al. highlight, and their exposition of the resulting ambiguities and possible alternative interpretations, add nuance to our notions of rapport and of how we conceive of standardized interviews.

Another useful contribution is Garbarski et al.'s coding scheme for analyzing transcripts of these interviews. The scheme adds some new distinctions that have not been made in similar coding schemes (Dijkstra and Ongena 2006; Schober et al. 2012, among others) and that are amenable to quantitative assessment of coding reliability. Some of the new distinctions made in this coding scheme also seem to have predictive value: because Garbarski et al. know about future participation or nonparticipation by these respondents, they are able to make some intriguing observations (at least for this sample)—for example, that an interviewer's likelihood of prefacing question-reading with "and" and a respondent's use of "sorry" can correlate with that respondent's future nonparticipation. As such, they add to a growing literature on how these and other nonverbal behaviors and microbehaviors in research interactions can collectively affect participation, data quality, observable features of respondent engagement, and respondents' feelings about participation (e.g., Conrad et al. 2013, 2015; Schober and Conrad 2015).

As I see it, the conversation analyses and coding scheme outlined here raise as many questions about the nature of rapport in survey interviews as they answer. A first set of questions is about the generalizability of the coding scheme and findings to other survey topics and populations and to other modes of interviewing (and, more broadly, to other kinds of nonresearch interaction). The interviews coded here are standardized surveys about a particularly fraught topic, end-of-life planning, in a very particular population of respondents: people from one area of the United States who have been participating in a longitudinal study, over multiple years, that (eventually) includes high-imposition physiological measures like saliva samples. Would the hesitations, laughs, and uses of "sorry" and of "and" before a question-reading have the same implications in a younger or differently distributed geographical sample, or in a sample of respondents with different norms for interpreting pauses or laughter or discourse markers? In what ways would these behaviors have effects in survey interactions when the questions are about other sensitive topics that are regularly part of social surveys—say, lifetime drug use, sexual behaviors, or income—or about less sensitive behaviors or opinions? To what extent do the effects of the audio behaviors evident in the transcripts analyzed here translate to face-to-face interviews, in which respondent audio behaviors such as speech disfluencies can be complemented by (or trade off with) visual behaviors such as gaze aversion (Schober et al. 2012)?

Understanding the role of rapport-related behaviors in other kinds of interviews is particularly pressing given the evidence that human connection in survey interviews about some sensitive topics (embarrassing and illegal behaviors) can actually be detrimental to disclosure and data quality. The evidence is that, in large national samples, people sometimes disclose more—report more sex partners, more drug use—when they respond to questions in "self-administered" modes (in online textual or audio surveys or in paper-and-pencil questionnaires; for example, see Tourangeau and Smith 1996), and their answers are more likely to be accurate compared with presumably accurate administrative records (Kreuter, Presser, and Tourangeau 2008). Disclosure is reduced by live human interviewers relative to textual responding in a Web survey; it can even be reduced by the video presence of an animated virtual interviewer who is clearly nonhuman (Lind et al. 2013). And disclosure even to a human interviewer can increase when interviews are texted rather than spoken (Schober et al. 2015). Although it is, of course, unclear what counts as rapport with a nonhuman interviewer (Cassell and Miller 2007), the possibility that rapport with a human interviewer is not always best for data quality, or what respondents most want, is important to consider. Perhaps in some kinds of interviews, or for some respondents, privacy and social distance are more desirable.

A second set of questions raised by this study is practical: Just how trainable are the interactional skills advocated here? One possibility is that anyone—or at least anyone who manages to hold on to an interviewing job—can improve in his or her skills through careful behavioral training and monitoring and that responsivity and close listening are learnable. Another possibility is that these skills are part of interviewers' basic conversational repertoire from other parts of their lives and that if interviewers do not already have sensitivity and interactional competence about what it feels like to build rapport, and how differently it must be built with different respondents, it will not be trainable. To put it another way, is rapport an almost magical element that only sometimes happens to "hit" or emerge in a pairing or interaction, or can it be brokered by research managers through judicious pairings of interviewers and respondents (without creating new interviewer effects) or specialized training of interviewers (without harming the goals of standardization)? How survey centers should proceed depends on answers to these questions.

A third set of questions raised by this study is more fundamental. Although Garbarski et al.'s methodological move of boiling rapport down to observable behaviors is sensible, it runs the risk of stretching the definition of rapport far enough that it could end up including just about any behavior by interviewer and respondent. Is a respondent's being "interactionally cooperative" the same thing as having rapport? Respondents might not all agree that they had good rapport with an interviewer simply because they were fully compliant with the rules of the game, nor might they believe that the interviewer's script-following felt like rapport; I expect that a question-answer sequence that looks smooth ("paradigmatic") could range in tone from warmth to iciness. Garbarski et al. are careful not to make psychological claims about rapport (even if, I would argue, their coding distinctions implicitly embody a few), but I would propose that the notion requires some affective component if it is to capture the ordinary intuition that we know when we "click" with a conversational partner and when we do not. Of course, Garbarski et al. are trying to go beyond intemperate psychologizing and ill-defined everyday notions, and the implicit account here—that the particular combination of an interviewer's taken and missed opportunities for responsivity with a particular respondent build up to the level of rapport that the respondent experiences—is plausible. But I think an important future research step will be to connect the behavioral elements captured here with respondents' and interviewers' and observing researchers' judgments—all of which could differ—about the interview's affective tone and participants' degree of connection with each other.

References

Cassell, Justine, and Peter Miller. 2007. "Is It Self-administration if the Computer Gives You Encouraging Looks?" Pp. 161–78 in *Envisioning the Survey Interview of the Future*, edited by Frederick G. Conrad and Michael F. Schober. Hoboken, NJ: Wiley.

Conrad, Frederick G., Jessica S. Broome, José R. Benkí, Frauke Kreuter, Robert M. Groves, David Vannette, and Colleen McClain. 2013. "Interviewer Speech and the Success of Survey Invitations." *Journal of the Royal Statistical Society, Series A (Statistics in Society)* 176(1):191–210.

Conrad, Frederick G., Michael F. Schober, Matt Jans, Rachel A. Orlowski, Daniel Nielsen, and Rachel Levenstein. 2015. "Comprehension and Engagement in Survey Interviews with Virtual Agents." *Frontiers in Psychology* 6(October):1578.

Dijkstra, Wil, and Yfke Ongena. 2006. "Question-answer Sequences in Survey-interviews." *Quality and Quantity* 40(6):983–1011.

Kreuter, Frauke, Stanley Presser, and Roger Tourangeau. 2008. "Social Desirability Bias in CATI, IVR, and Web Surveys: The Effects of Mode and Question Sensitivity." *Public Opinion Quarterly* 72(5):847–65.

Lind, Laura H., Michael F. Schober, Frederick G. Conrad, and Heidi Reichert. 2013. "Why Do Survey Respondents Disclose More When Computers Ask the Questions?" *Public Opinion Quarterly* 77(4):888–935.

Schober, Michael F., and Frederick G. Conrad. 2015. "Improving Social Measurement by Understanding Interaction in Survey Interviews." *Policy Insights from the Behavioral and Brain Sciences* 2(1):211–19.

Schober, Michael F., Frederick G. Conrad, Christopher Antoun, Patrick Ehlen, Stefanie Fail, Andrew L. Hupp, Michael Johnston, Lucas Vickers, H. Yanna Yan, and Chan Zhang. 2015. "Precision and Disclosure in Text and Voice Interviews on Smartphones." *PLoS ONE* 10(6):e0128337.

Schober, Michael F., Frederick G. Conrad, Wil Dijkstra, and Yfke P. Ongena. 2012. "Disfluencies and Gaze Aversion in Unreliable Responses to Survey Questions." *Journal of Official Statistics* 28(4):555–82.

Tourangeau, Roger, and Tom W. Smith. 1996. "Asking Sensitive Questions: The Impact of Data Collection Mode, Question Format, and Question Context." *Public Opinion Quarterly* 60(2):275–304.

Author Biography

Michael F. Schober is a professor in the Department of Psychology at The New School for Social Research and associate provost for research at The New School. His research examines shared understanding (and misunderstanding), perspective-taking, and coordinated action in studies of casual conversations, standardized interviews, and musical performances and improvisations. He also studies how modes of communication (e.g., remote video chat, asynchronous texting, social media broadcasting) and interactions with human and automated partners affect interaction and understanding. His work has been published in journals and books across a range of fields, from psychology and cognitive science to public opinion and human-computer interaction; from 2005 to 2015, he was editor of the multidisciplinary journal *Discourse Processes*, and he coedited (with Fred Conrad) the 2008 John Wiley volume *Envisioning the Survey Interview of the Future*.

COMMENT: THE CONSTITUENTS OF RAPPORT IN THE STANDARDIZED SURVEY INTERVIEW

*Johanna Ruusuvuori**

*University of Tampere, Tampere, Finland
Corresponding Author: Johanna Ruusuvuori, johanna.ruusuvuori@staff.uta.fi
DOI: 10.1177/0081175016644898

The starting point of Garbarski, Schaeffer, and Dykema (this volume, pp. 1–38) is the important discussion of how "rapport" between an interviewer and a respondent in a standardized survey interview may benefit or harm (1) the quality of responses and (2) future survey participation. The authors adapt the concept of rapport to the context of standardized interviewing and to actors' institutional roles by discerning two actor-specific concepts: the interviewer's responsiveness and the respondent's engagement. The stated main purpose of the analysis is qualitative: to describe the complexity of the chosen "constituents" of rapport with regard to the rules of standardized interviews, but they also develop a coding scheme drawing on the analysis. Their additional aim is to explore the analytic potential of their research design with regard to future study participation.

The qualitative analysis is conducted using a combination of content and conversation analysis, though not in a strict sense, as the authors state themselves. The concepts of *the interviewer's responsiveness* and *the interviewee's engagement* seem to be developed top down, with regard to the participants' institutional roles in a standardized interview. The starting point is how well the examined behaviors support the institutional task at hand—keeping to the rules of standardization rather than the participants' own orientations to potential problems in reaching intersubjective understanding on the topic discussed (cf. Wilkinson 2014). So the concept of responsiveness is operationalized (1) as fitting the interviewer's response to the respondent's talk but also (2) as fitting it in a way that attends to four dimensions: the survey task, rules of standardization, content, and conversational practices.

The solution to operationalize responsiveness as the fit of the responses by the interviewer to the respondent's preceding turns (usually answers to the interviewer's question) is insightful. It makes it possible to describe the ways in which the interviewers orient to maintaining intersubjectivity between the respondent and themselves. This could be a well-functioning approach to analyze rapport in a more general sense, considering different ways in which the interviewers' turns of talk are attentive to the respondents' preceding utterances. There is thus some consideration of the position of the utterances within sequences. With the addition of the aspect of standardization and survey task, however, the issue gets more complicated. Examining the ways in which interviewers' turns of talk fit to the rules of standardization and the institutional task seems an effort that is based on somewhat different analytic premises than analyzing the fit to the respondent's previous utterances. The point of view is no longer the achievement of intersubjectivity in interaction but an adaptation to institutional norms.

In a closer reading of the described dimensions of responsiveness, we can suggest that the analytic logic on which the classification is based can be further developed. In the scheme described, there seem to be acts of responsiveness orienting to the relational and affective aspects of interaction and those that orient to maintaining mutual understanding on what was said. These dimensions would be transparently understandable as "responsiveness" and "rapport." In addition to this, however, there are dimensions that are deduced drawing on the norms of standardization, not with regard to the actual fit between interviewers' and respondents' utterances. Consider dimension 5, for instance. It includes behaviors that facilitate efficient progress through the interview, for example, reading questions exactly as scripted. The presumed connection of this act with the phenomenon of responsiveness or rapport is not straightforward. Rather than rising from observing the fit of the interviewers' turns of talk to the respondents' previous turns, reading questions exactly as scripted is an interview technique deduced from the rules of standardization. In this sense, it seems not to connect straightforwardly with "responsiveness" to respondents, although it may alleviate achieving the institutional task. As stated by the authors, the status of "rapport" may be ambivalent with regard to the institutional task at hand. The fit of utterances to rules of standardization seems to derive from somewhat different analytic premises than observable interaction.

I agree with the authors about the possibilities nested in the quantification of qualitative results. And I agree with them when they state that this effort is not without problems. One such question is the choice of the aforementioned aspect of reading questions exactly as scripted to illustrate the analytic potential of the coding scheme. It is plausible that this practice is key to standardization and that it does facilitate progress through the interview. But the idea that properties such as efficient and professional question reading would serve to motivate the respondent to continue taking part in follow-up interviews because these constitute a pleasant scientific experience is not immediately transparent. The observation that the exact reading of questions is associated with increased odds of future participation is interesting, but its analytic potential in connection with rapport will need clarification.

To my mind, the main contribution of the article is the qualitative analysis that is structured by describing a series of situations at which the respondent's actions occasion possible problems with the rules of standardization. The analysis illuminates the sort of complications that may ensue and gives insights for further study on maintaining a balance between practices of standardization and responsiveness.

Reference

Wilkinson, Ray. 2014. "Intervening with Conversation Analysis in Speech and Language Therapy: Improving Aphasic Conversation, Research on Language and Social Interaction." *Research on Language and Social Interaction* 47(3):219–38.

Author Biography

Johanna Ruusuvuori is a professor of social psychology in the School of Social Sciences and Humanities, University of Tampere, Finland. Her research interests include qualitative methodology in social sciences, ethnomethodology and conversation analysis, the interview as a data-gathering method, and the dynamics of social interaction at work.

REJOINDER: RESPONSE TO COMMENTS ON "INTERVIEWING PRACTICES, CONVERSATIONAL PRACTICES, AND RAPPORT: RESPONSIVENESS AND ENGAGEMENT IN THE STANDARDIZED SURVEY INTERVIEW"

Dana Garbarski*
Nora Cate Schaeffer[†]
Jennifer Dykema[†]

*Loyola University Chicago, Illinois, USA
[†]University of Wisconsin–Madison, USA
Corresponding Author: Dana Garbarski, dgarbarski@luc.edu
DOI: 10.1177/0081175016651074

We would like to extend our thanks to former editor Tim Futing Liao and current editor Duane F. Alwin for allowing our work to be part of a scholarly dialogue. We would also like to thank our colleagues for their insightful commentaries, which discuss the implications of our study in terms of the conceptual and operational definitions of rapport—what is included and excluded—and criteria to examine the validity of rapport and its dimensions. We discuss each of these in turn in the following.

First, each of the three commentaries discuss—in different ways—what is or is not rapport, revealing the tension between our definition of rapport as something situated in the interaction and other understandings of rapport as something intuitional, affective, or psychological. Further, the commentaries by both Ruusuvuori (this volume, pp. 47–49) and Schober (this volume, pp. 43–46) note that our definition of rapport may stretch the concept too thin across multiple analytic goals. Finally, the varying conceptual and operational definitions discussed in our paper and in the three commentaries have implications for interviewer training (as noted in our paper and by Schober) as well as the generalizability of this particular study to other respondents, topics, and modes (as noted by Schober).

Our review of the literature finds that conceptual and operational definitions of rapport are varied, but previous studies have not examined how rapport might actually be constituted within what Ruusuvuori notes is the specific institutional context of standardization. Each of the commentators raises important issues that are implied by or excluded by our conceptual and operational definitions of the dimensions of rapport, and we appreciate this contribution to the conversation. We still contend, however, that researchers should consider a definition of rapport that moves beyond unobservable affective components and considers responsiveness and engagement not just among the interactional partners but also with respect to the other goals (the "institutional norms") that underpin the nature and shape of that interaction: the survey task, rules of standardization, content of the talk, and conversational practices. As Miller's (this volume, pp. 39–42) reminder about Kahn and Cannell's (1957) analysis of a different style of interview suggests, before we decide what rapport is not, we need to consider more broadly what it might be, particularly within the constraints of a standardized survey interview. In our study, we chose a particular approach to consider what rapport might be—something that is mutually constituted by both parties in the interaction and grounded in what is actually observable in the survey interview.

It is an empirical and unresolved question whether all the dimensions of the interviewer's responsiveness and the respondent's engagement that we identified in our study pass muster with respect to criterion validity. Our criterion of future participation was one that we were able to examine with our particular data, but we agree that it is incomplete with respect to examining the impact of the dimensions of rapport. As we note in our paper and as is further echoed by Miller, future studies must consider data quality as a criterion of interest—in terms of the study of rapport, it is often the central justification for either promoting or inhibiting rapport (depending on the conceptual definition of rapport being used).

In addition, both Miller and Schober discuss linking the behaviors from the survey interview (the interactional dimensions of rapport) to more psychological-based dimensions, such as the degree of connection between the interviewer and respondent or the attitude of the respondent toward the interview and interviewer (e.g., feelings of ease and trust). We agree that this would be an important next step for future research, as long as there is conceptual clarity as to whether these psychological dimensions are a component or result of rapport in the survey interview. One analytic issue that arises is the measurement of these psychological dimensions in terms of when and by whom they are measured. Indeed, the degree of connection between interactional partners judged by an outside observer (through audio or visual channels) may be a component of rapport, while asking the respondent their feelings after their interview may be a result of rapport in the survey interview.

We close with a brief discussion of other methodological issues involved in studies of rapport in the survey interview and in studies that examine interviewer-respondent interaction more generally. We describe in the discussion (Section 6) and in Appendix B of our paper (available in the online journal) some of the issues involved in moving from a more qualitative description of rapport in particular interactions to a quantitative analysis. Inspired by Schober's comment that behaviors may "build up to the level of rapport," we point out that additional analytic issues include choosing the level of analysis, recognizing the development of rapport across time, and aggregating measures across turns of talk and questions. These issues lead us to pose the following questions: Should researchers count the number of types of behaviors and aggregate across questions? Should the patterns of behaviors be described and coded for the entire interview, select question-answer sequences, or some other subset of the interview? How should the many individual and patterned behaviors exhibited by interviewers and respondents be incorporated in a multivariate analysis? In translating to a more quantitative analysis of rapport, it is unclear whether rapport is simply the sum of its component parts, and if so, what the most valid and parsimonious analytic strategy to employ is.

It is an honor to have our work discussed by our colleagues. We look forward to continued dialogue about rapport within the survey interview, moving the concept forward in terms of both conceptual clarity and analytic refinement.

Reference

Kahn, Robert L., and Charles F. Cannell. 1957. *The Dynamics of Interviewing.* New York: John Wiley.

Author Biographies

The author biographies can be found on p. 38 of this volume.

Sociological Methodology
2016, Vol. 46(1) 53–83
© American Sociological Association 2016
DOI: 10.1177/0081175016632804
http://sm.sagepub.com
$SAGE

$\mathfrak{S}$ 2 $\mathfrak{E}$

ELICITING FRONTSTAGE AND BACKSTAGE TALK WITH THE ITERATED QUESTIONING APPROACH

Laura Robinson*
Jeremy Schulz[†]

Abstract

This article advances interviewing methods by introducing the authors' original contribution: the iterated questioning approach (IQA). This interviewing technique augments the interviewer's methodological arsenal by exploiting insights from symbolic interactionism, particularly Goffman's concepts of frontstage and backstage. IQA consists of sequenced iterations of a baseline question designed to elicit multiple forms of talk. The approach consists of four distinct steps: (1) establishing the baseline iterated question, (2) eliciting frontstage talk, (3) going backstage, and (4) eliciting backstage talk. To illuminate IQA's versatility, transcript excerpts are reproduced from interviews with two very different populations: disadvantaged high school students and business professionals. IQA promises to invigorate future interview-based inquiry by offering significant advantages compared with conventional interviewing procedures. IQA's theoretically informed question design offers a more formalized and structured approach to gather interview data on identity-relevant themes. Capitalizing on Goffman's dramaturgical framework,

*Santa Clara University, Santa Clara, CA, USA
[†]Institute for the Study of Societal Issues, University of California, Berkeley, CA, USA

Corresponding Author:
Laura Robinson, Santa Clara University, Department of Sociology, 500 El Camino Real, Santa Clara, CA 95053, USA
Email: laura@laurarobinson.org

IQA produces readily classifiable forms of talk that correspond to frontstage and backstage self-presentations. As a result, IQA ensures replicability and allows interviewers to systematically analyze comparable talk within the same interview as well as across multiple respondents. For these reasons, IQA promises to be an innovative interviewing technique that pushes forward the methodological frontier.

Keywords

iterated questioning approach, IQA, interviewing, question design, qualitative methods, symbolic interaction, Goffman, backstage, frontstage, dramaturgy

1. INTRODUCING IQA

The time is ripe for new contributions to the craft of interviewing as an important sociological method. Recent debates have pushed this venerable staple of sociological research to the center of an ongoing methodological conversation (Pugh 2013). The emergent consensus holds that interviewing is an indispensable technique to gather data on identity broadly defined. In their programmatic statement, Lamont and Swidler (2014) affirmed the value of in-depth interviewing for analyzing identities and lived experiences. According to a National Science Foundation report on qualitative research, the strength of in-depth interviewing resides in its capacity to capture the contours of identity work, self-conceptions, emotions, and meanings (Ragin, Nagel, and White 2003).

Indeed, as Pugh's (2012, 2015) theoretical and methodological contributions underscore, in-depth interviewing remains an essential instrument of unearthing narratives providing insight into identity-relevant themes and revealing multiple layers of meaning and emotion. Skillful in-depth interviewing makes it possible to elicit multidimensional interview talk and to probe how interviewees grapple with their social and cultural environments. However, neither interview-based studies nor the methodological literature on interviewing practice offers a roadmap to meeting this goal. To meet this need, we propose just such a roadmap consisting of concrete questioning procedures that enable the interviewer to systematically gather multiple forms of interview talk.

We call this interviewing technique the iterated questioning approach (IQA). IQA offers significant advantages compared with conventional interviewing procedures. First, IQA provides theoretically informed techniques rooted in symbolic interactionism. Capitalizing on Goffman's (1959) dramaturgical framework, IQA takes advantage of the distinction between self-presentations unfolding in the interactional *frontstage* and self-presentations taking place in the *backstage*. IQA's specially formulated questions allow the interviewer to systematically elicit both frontstage talk calibrated for public consumption and backstage talk produced for private audiences (Rubin and Rubin 1995; Tovares 2007). Second, IQA's sequencing of questions produces readily classifiable forms of talk that correspond to frontstage and backstage self-presentations. IQA's four distinct steps allow the interviewer to easily distinguish between identity performances regarding the same themes and to transparently differentiate the talk produced for different audiences. Third, IQA allows interviewers to gather data in a predictable fashion within the same interview as well as across material from multiple respondents. IQA is thereby designed to ensure replicability by systematically evoking interview talk corresponding to multiple identity positions. IQA offers a formalized and structured approach that exploits the strengths of in-depth interviewing as an interpretive practice carried out by the interviewer in collaboration with an active interviewee (Holstein and Gubrium 1995). Finally, these innovations strengthen the "interviewing partnership" (Weiss 1994:99) by enhancing rapport, boosting interactivity, and leveling power differentials between interviewers and respondents.

To present the advances afforded by IQA, this article is organized as follows. In Section 1, conventional interviewing practices are put in dialogue with the symbolic interactionist framework undergirding IQA. Section 2 outlines IQA's four-step iterated question design. Section 3 reproduces transcript materials from two very different respondent groups (disadvantaged high school students and business professionals) to show how IQA techniques have been fruitfully used with an array of respondents. Section 4 discusses the conditions under which IQA may be a vital tool available to the interviewer, as well as the methodological challenges that arose in the field trials and strategies employed to successfully address them. Finally, Section 5 concludes the article by illustrating the advantages of IQA vis-à-vis conventional interviewing, as well as on its own terms.

1.1. *Rapport-based Approaches to Interviewing*

Rapport plays a central role in conventional in-depth interviewing, particularly when researchers wish to trigger self-disclosure and activate multiple identity positions (Holstein and Gubrium 1995; Mishler 1986). Strong rapport can evoke spontaneous visceral narratives in addition to more scripted honorable narratives (Pugh 2015). Thus, many guides to interviewing underscore the key role of rapport in encouraging respondents to open up about experiences and feelings (Weiss 1994), as well as in eliciting identity-relevant narratives (Alvesson 2011).

Rapport-building strategies include multiple interviews and verbal affirmations that acknowledge respect for the interviewee (Weiss 1994). Rapport also results from practices such as active listening in which the interviewer remains attentive, sustains eye contact, and conveys nonjudgmental empathy by mirroring the respondent's emotions (Wengraf 2001). Physical signaling such as leaning forward to convey emotional understanding and respect may also contribute to rapport (Rubin and Rubin 1995). Others call for the interview to resemble what has been called a therapeutic event (Lee 1993) in which the interviewer takes on the role of trusted confidant. In sum, rapport facilitates the movement between different levels of self-disclosure in which the interviewee shifts self-presentation from a public self to an emotional self (Weiss 1994).

Where existing interview studies are concerned, strong rapport has served as the primary catalyst for facilitating shifts in self-presentations and accompanying multidimensional narratives. Pugh's (2012) work on varying cultures of commitment demonstrates the capacity of high-rapport, in-depth interviewing to garner multiple forms of talk regarding the same identity-relevant themes. The strong rapport evident in Pugh's interviews functions to elicit honorable narratives highlighting dominant cultural norms, as well as visceral narratives conveying culturally subversive content. For example, in discussing their work and private lives, respondents' honorable narratives rehearse dominant American cultural scripts celebrating self-sufficiency and emotional invulnerability. By contrast, in their emotionally loaded visceral narratives, the very same interviewees disclose their sense of vulnerability to betrayal by self-interested others (Pugh 2012). Here we see how visceral narratives reveal respondents' private self-presentations sometimes at odds with their public social faces. In this way, Pugh's interviewing demonstrates the capacity of high-rapport in-depth interviewing to garner multiple forms of talk regarding the same identity-relevant themes.

Building strong and sustained rapport will always be an important means of facilitating the movement between different levels of self-disclosure. However, rapport is a fragile and fickle condition that relies wholly on the skills of the interviewer rather than on question design. In contrast, IQA's replicable techniques make both data collection and analysis more predictable. For these reasons, we propose IQA to exploit the potential for specially formulated and sequenced questions as a means of systematically eliciting multiple forms of interview talk thematizing the same topic.

1.2. *A Goffmanian Approach to Interviewing*

Although valuable, methodological works highlighting rapport fail to provide guidelines by which the interviewer may activate multiple identity positions in a replicable or systematic way. Existing work on interviewing question design overlooks Goffman's insights regarding the presentation of self frontstage and backstage. To fill these gaps, IQA integrates Pugh's insights into honorable and visceral narratives with Goffman's dramaturgical concepts of frontstage and backstage. This synthesis enables interviewers to formulate replicable questions that predictably elicit frontstage and backstage talk. Drawing from Goffman, the terms *frontstage talk* and *backstage talk* are used to show how particular interviewing techniques can produce complementary narratives regarding the same identity-relevant theme. For Goffman (1959), in the frontstage, when "the individual presents himself before others, his performance will tend to incorporate and exemplify the official accredited values" (p. 35). Referencing Robert Ezra Park, Goffman (1967) also maintained that the social actor's frontstage represents "the conception we have formed of ourselves—the role we are striving to live up to" (p. 19). When answering questions frontstage, interviewees construct face-building talk (Dominici and Littlejohn 2006) often expressed through Pugh's honorable narratives conforming to cultural expectations. In contrast, backstage interactions allow interviewees to express face-threatening narratives (Dominici and Littlejohn 2006). Safely backstage, interviewees can "drop their front" and "step out of character" (Goffman 1959:112). According to Prasad (2005), "The backstage is also a place where we hide a self that, if revealed, might prove to be awkward, embarrassing, or even suspect" (p. 46–47). Backstage talk is often expressed through visceral narratives that challenge cultural expectations and fall out of alignment with honorable narratives expressed frontstage.

In applying Goffmanian concepts to the craft of interview question design, IQA blazes a new trail. To date, Goffman's concepts have primarily been used to interpret bodies of data subsequent to their collection (Alvesson and Sköldberg 2009). However, interviewers have not incorporated Goffmanian insights into question design. Myers and Newman (2007) applied a Goffmanian lens to understand the dynamics of the interview encounter. Schwalbe and Wolkomir (2003) called attention to facework as it occurs in interviews with male respondents. Demant and Järvinen (2006) analyzed the interview encounter in terms of their respondents' impression management. Tannen and Wallat (1987) made use of Goffman's concepts of footing and frame in interaction. Although important, none of these studies reveals how the interviewer can use a Goffmanian framework to build interview questions designed to elicit frontstage and backstage talk.

In contrast, IQA's systematic approach realizes the full potential of Goffman's (1959, 1967) extended dramaturgy metaphor and Pugh's (2015) multidimensional narratives with purposefully designed interview questions. Approaching interviewing from a symbolic interactionist perspective, interviewers can gather rich data by drawing out different facets of interviewees' self-presentations. With IQA, the interviewer can design questions that explicitly elicit data revealing interviewees' conceptions of the "I" and the "me" (Mead 1934). In Goffmanian terms, by using IQA, the interviewer can strategically uncover individuals' frontstage and backstage presentations of self.

2. IQA: NUTS AND BOLTS

In this section we outline the nuts and bolts of iterated questioning. IQA simulates exchanges with interlocutors drawn from interviewees' personal communities (Spencer and Pahl 2006). Posing permutations of the same question—formulated in very specific ways—at different points during the interview invites interviewees to perform a credible frontstage self-presentation that articulates honorable narratives while allowing them to produce potentially divergent visceral narratives in the safety of backstage. By answering the iterated questions in sequence, the interviewee progresses through frontstage performances before taking the interviewer backstage. Once backstage, the respondent is liberated from the need to maintain a consistent self-presentation and is shielded from identity threats due to shamefacing (Goffman 1967).

IQA was developed through interviews with a total of 77 individuals drawn from studies carried out with two different groups of respondents:

disadvantaged high school students and business professionals. The initial field trials involved interviews with 21 respondents and were used to refine IQA's four steps. Refined techniques used with an additional 56 respondents confirmed that no further changes were needed to IQA. This section presents the four finalized steps of the IQA sequence as it emerged from this process of field testing and validation.

2.1. *Step 1: Establishing the Baseline Iterated Question*

After background questions have been asked, the interviewer initiates the IQA sequence. In step 1, the interviewer introduces the baseline form of the iterated question on one of the interview's central themes. To create the baseline question, the interviewer uses a straightforward approach recommended in traditional interviewing practice (Weiss 1994). Once the interviewee has answered the baseline question, the interviewer indicates that the interaction has been successfully concluded. Often a brief "okay" or "yes" will do. The interviewer should then pursue other related questions so that an interval elapses before moving to the next step.

2.2. *Step 2: Eliciting Frontstage Talk*

In step 2, the interviewer asks a variation of the iterated question to elicit frontstage talk intended for a third-party interlocutor, posing a question as it may have been discussed by the respondent with a third party not present at the interview. Frontstage questions are especially effective when they feature institutional representatives, acquaintances, or any other untrusted others who are not a confidant or intimate of the respondent—audiences for which the respondent would produce honorable narratives. For example, if the interviewer is interested in institutional identities, the third party should represent the institution. Subsequently, the interviewer proceeds with other related questions so that an interval elapses before moving to the next step.

2.3. *Step 3: Going Backstage*

In step 3, the interview moves backstage, where interviewees may safely share visceral narratives. As with previous steps, the interviewer continues to craft this iteration of the baseline question to maintain thematic alignment. The interviewer asks two questions in sequence: "Tell me the name of someone you trust—a confidant with whom you could talk freely about [theme of iterated question]" and "Tell me about [interviewee's trusted third party]." Through these questions, the interviewer invites the interviewee to craft a self-presentation geared toward trusted audiences. Once the interviewee has named at least one trusted third party, the interviewer then asks, "Tell me about [trusted third party]." In asking these questions, the interviewer is explicitly asking permission to be taken backstage, where it will be safe for the respondent to invoke visceral narratives. In step 3, as with the rest of the IQA sequence, the interviewer continues to signal respect for the sincerity of the interviewee's self-presentations. However, unlike previous steps that require an interval, after the two questions in step 3 have been asked in sequence and answered, the interviewer should immediately proceed to step 4 without an interval.

2.4. *Step 4: Eliciting Backstage Talk*

Step 4 continues the backstage steps in the sequence. Here, the interviewer asks a two-part variation of the iterated question. The interviewer reformulates the iterated question using the following script: "Imagine you are talking to [trusted third party] alone where no one can hear you. [Trusted third party] will keep what you say private. If [trusted third party] asked you [iterated question], how would you answer?" and "How would [trusted third party] respond?" If the interviewee had mentioned more than one trusted person in step 3, the interviewer should follow up with each named interlocutor in step 4, as all of the trusted interlocutors are potential audiences for visceral narratives. Respondents will answer in one of two ways. They will either verbalize an imagined conversation with their trusted interlocutor, or they will replay an actual conversation that has already occurred with that trusted person. In either case, the interviewee articulates valuable backstage talk. Finally, in step 4, some interviewees may produce talk that bridges their frontstage and backstage self-presentations. If the interviewee does not initiate bridging talk, the interviewer may wish to extend step 4 by gently inviting the respondent to bring frontstage and backstage talk into dialogue with one another. Here, hypothetical formulations may prove particularly useful. When inviting the interviewee to articulate frontstage and backstage talk alongside one another, the interviewer continues to signal that the interaction has been successfully concluded.

3. IQA EXEMPLARS: REPRESENTATIVE CASES

This section reviews the application of IQA with four exemplars, each illustrating an analytically revealing permutation of IQA. Transcript excerpts show how IQA succeeds in eliciting comparable frontstage and backstage talk from two very different populations: high school students and business professionals. The first two exemplars are Hunter and Marta, economically disadvantaged high school seniors in a blue-collar town, who are asked about their college plans, a highly self-relevant theme. Will and Kim, successful business professionals in a large metropolitan city, serve as the second pair of exemplars. Their interviews are focused on the self-relevant theme of work-family balance. Please note, all names and places are pseudonyms to protect anonymity.

3.1. *High School Student Hunter*

Hunter, the first exemplar, plans on attending a local community college following high school graduation. Whereas Hunter's strategic frontstage self-presentation draws on institutionally validated scripts about financial prudence, his backstage talk presents an institutionally nonconformist self averse to the idea of attending a four-year college on his own. Without his loyal "posse," Hunter fears the loss of his academic lifeline, as well as his social support network. Hunter's transcript demonstrates how IQA creates a safe interactional space conducive to the disclosure of nonconformist self-presentations.

3.1.1. Step 1: Establishing the Baseline Iterated Question. In step 1 of IQA, Hunter is invited to articulate honorable narratives in the form of institutionally sanctioned scripts about college planning. Hunter answers the baseline question by appealing to financial rationales as the driving force behind his plan to attend community college.

Interviewer:	What are your plans after graduation?
Hunter:	I'm going to Jefferson [local community college].
Interviewer:	Why is this a good choice for you?
Hunter:	Money. Jefferson's cheap.
Interviewer:	Did you apply to any four-year schools?

Hunter:	Yeah. Cal State Long Beach.
Interviewer:	Were you accepted?
Hunter:	Uh huh.
Interviewer:	So, you're going to Jefferson instead of CSU Long Beach?
Hunter:	Yup. Goin' to Jefferson. . . . Like I said, it's cheap.

3.1.2. *Step 2: Eliciting Frontstage Talk.* In step 2, the interviewer asks Hunter a variation of the iterated question designed to elicit frontstage talk. Hunter is asked to envision discussing his college plans with a teacher acting in the capacity of an institutional representative. In response, Hunter imagines a hypothetical conversation with a teacher, as well as the teacher's validation of this frontstage talk.

Interviewer:	Imagine that you are talking to a teacher. If the teacher asked you "Why are you going to Jefferson?" how would you answer?
Hunter:	Like I'd bring up the money. You know those four-years are really expensive.
Interviewer:	Anything else?
Hunter:	Hmmm, probably about Coach Rivera coming into our class and it makin' me think about how goin' to Jefferson is gonna save me a lot of money.
Interviewer:	And what do you think the teacher would say to you?
Hunter:	Dunno. "Good for you" or something.

3.1.3. *Step 3: Going Backstage.* In step 3, the interviewer directs Hunter's attention to someone who would be considered a confidant, a trusted interlocutor rather than an institutional representative. Introducing Baxter, Hunter takes the interviewer backstage.

Interviewer:	Tell me the name of someone you trust— a confidant with whom you could talk freely about your future plans.
Hunter:	That'd be Baxter.
Interviewer:	Tell me about Baxter.
Hunter:	He's like my bud. I've known him forever.

3.1.4. *Step 4: Eliciting Backstage Talk and Bridging Talk.* In step 4, the interviewer reiterates the baseline question as if asked by Baxter, Hunter's trusted friend. In response, Hunter imagines a hypothetical conversation in which he is free to express visceral narratives that run counter to institutional scripts. Hunter's backstage talk reveals a nonconformist self with very different priorities than the frontstage self he presented to teachers as institutional representatives. Backstage, Hunter is primarily concerned with avoiding the risks posed by separation from his "posse": missing out on good times, loss of his academic safety net, and potential stigmatization as a college dropout should he go "off solo" and fail. The interviewer closes the IQA sequence with a gentle query inviting Hunter to self-reflectively compare his frontstage and backstage narratives. To do so, the interviewer poses a hypothetical question that allows Hunter to maintain face while voicing both kinds of talk.

Interviewer:	Imagine you are talking to Baxter alone where no one can hear you. Baxter will keep what you say private. If Baxter asked you "Why are you going to Jefferson?" how would you answer?
Hunter:	If I'm talking to Baxter, yeah, I'd probably say, "Like why go to Long Beach when we can take it easy here?" [laughs]
Interviewer:	How would Baxter respond?
Hunter:	Yeah, man. Yeah, he'd get it. Baxter'd say that like it sucks to go off somewhere alone when everybody's stayin' here and havin' a good time together.
Interviewer:	Tell me more about how Baxter would "get it."
Hunter:	Yeah, like he got into [CSU] Fresno but he's not going. So, yeah, we're gonna be chillin' at Jefferson. No stressin', dude [laughs].
Interviewer:	[Laughs] What about your teachers? Would they "get it"?
Hunter:	They'd get it for sure about the money but maybe not the other part.
Interviewer:	The "other part"?
Hunter:	You know, the chillin' part.
Interviewer:	So imagine that you and Baxter both got full-ride scholarships to the same college and could "chill" together at say Cal State Long Beach or Fresno. What would happen then?
Hunter:	Dude, like we'd be on our way!
Interviewer:	Yeah?
Hunter:	Yeah, not like Rick. You know, the dude who like crashed and burned 'cause he went off solo. Like dude, I'm not gonna do a Rickster an' bomb out of a four-year just 'cause I don't have my posse. I'll be with my posse at Jefferson.

3.2. *High School Student Marta*

The second exemplar, Marta, differs in important ways from Hunter as a respondent. An academically ambitious young woman aspiring to be the first person in her family to attend college, Marta has earned a scholarship that will pay for her education at a four-year college. In contrast to Hunter, who is held fast by ties to his personal community, Marta's friends are dispersing upon graduation. Further unlike Hunter, Marta feels ready to tackle the academics at a four-year college. However, she doubts her capacity to thrive in an intimidating social environment with "rich kids." Here, the power of IQA unlocks a vulnerable self that is comfortable expressing self-doubt and inner conflict in the safety of the backstage.

3.2.1. *Step 1: Establishing the Baseline Iterated Question.* In step 1, Marta answers the baseline question with honorable narratives about her intended program of study, as well as fiscal responsibility.

Interviewer:	What are your plans after graduation?
Marta:	Going to a four-year! [CSU] Northridge here I come!
Interviewer:	Why is this a good choice for you?
Marta:	They have my major, and the price is right. I think it's good.
Interviewer:	Did you apply to other any four-year schools?
Marta:	Mostly Cal States.
Interviewer:	Which ones?
Marta:	Uhh . . . Northridge, Chico, and San Diego.
Interviewer:	Were you accepted?
Marta:	Well, I got in almost everywhere I applied.
Interviewer:	And you're going to Northridge?
Marta:	That's right.
Interviewer:	How did you make your decision?
Marta:	They've got my program, and I got good financial aid.

3.2.2. *Step 2: Eliciting Frontstage Talk.* In step 2, the interviewer invites Marta to articulate frontstage talk in which she imagines conversations with multiple teachers. Frontstage, Marta dutifully recites institutionally validated scripts.

Interviewer:	Imagine that you are talking to a teacher. If the teacher asked you "Why are you going to Northridge?" how would you answer?
Marta:	I guess I would tell them that I got good financial aid.

Interviewer:	Anything else?
Marta:	That I like Northridge 'cause they have my major.
Interviewer:	And what do you think the teacher would say to you?
Marta:	Mostly they would want to make sure about financial aid and all of that.

3.2.3. *Step 3: Going Backstage.*

In step 3, Marta selects her mentor, Mrs. Lafitte, as her trusted interlocutor.

Interviewer:	Tell me the name of someone you trust—a confidant with whom you could talk freely about your future plans.
Marta:	OK, [pause] Mrs. Lafitte.
Interviewer:	Tell me about Mrs. Lafitte.
Marta:	She comes into where I work. She's *really* cool.
Interviewer:	Oh yeah? Tell me more about why she's cool.
Marta:	She always wants to know how I'm doing and makes time to talk to me.

3.2.4. *Step 4: Eliciting Backstage Talk.*

In step 4, Marta reveals her vulnerabilities when she recounts an actual conversation she had with her mentor, Mrs. Lafitte. Marta airs her innermost fears about her college plans in visceral narratives. Her backstage talk unveils the emotional burdens weighing on a young woman from a poverty-level school—burdens she must surmount if she is to earn a college degree and put herself on a path to upward mobility. Mrs. Lafitte functions as an indispensable interlocutor as she invites Marta into her own backstage by sharing her own struggles as a first-generation college student. Encouraged by Mrs. Lafitte's own backstage revelation, Marta conquers her fears and feels emboldened to "go ahead" and do her best. Step 4 closes with Marta juxtaposing her frontstage and backstage answers with a hybrid response joining both honorable and visceral narratives.

Interviewer:	Imagine that you are talking to Mrs. Lafitte alone where no one can hear you. Mrs. Lafitte will keep what you say private. If Mrs. Lafitte asked you "Why are you going to Northridge?" how would you answer?
Marta:	Well, she did talk to me about it when she came into my work. I was kinda dealing with this whole college thing, having a hard time with it.
Interviewer:	How did Mrs. Lafitte respond?

Marta:	She really listened to me. You know I was totally nervous. And talking to her made me know that it's OK to be scared. And that's what was holding me back.
Interviewer:	Tell me more.
Marta:	I mean I thought maybe I won't fit in. Or maybe everyone will be different. Like they'll have lots of money and stuff. And won't want to be friends with people like me.
Interviewer:	Did you talk about this to Mrs. Lafitte?
Marta:	Yeah, totally. She told me to go look at the college website and get a feel for the students. She told me that it wouldn't be just a bunch of rich kids and how she was the first person in her family to go to college and how it was hard, but I just had to go ahead. And I had to do my best.
Interviewer:	So what happened next?
Marta:	I looked at the website. And I found these stories about students. And there are like a bunch of students like me at Northridge. And I was like surprised 'cause like not all of their families had went [*sic*] to college and some of them had hard stuff going on. But they made it, and they were like graduating. So I think it'll be good for me to go there too.
Interviewer:	It sounds like you have really thought things out. So please recap it for me. What makes going to Northridge good for you? The kinds of students or financial aid or your major?
Marta:	Well, all of them. I mean you gotta have the money or like you can't go. But if you got a choice, you need to feel good about where you're going or like you might not stay—even if they've got your major.

3.3. *Business Professional Will*

The third exemplar is Will, a corporate attorney who often works 11-hour days during the week and a few hours on the weekends. Will has a wife and children who compete for time and attention with his all-consuming professional obligations. Will's interview offers an illuminating foil to the other exemplars. Only Will overtly identifies with his frontstage audience composed of likeminded peers. Thus, Will's frontstage talk is an authentic expression of his own views, even as it replicates the scripts sanctioned by his workplace. However, Will chooses a critical friend as his backstage interlocutor, leading to an imagined conversation in which he confesses his unease. Will's interview shows what

IQA can accomplish when respondents present their idealized selves in their frontstage talk, while presenting a more vulnerable self backstage.

3.3.1. Step 1: Establishing the Baseline Iterated Question. In answering the baseline question, Will enlists the honorable narrative regarding breadwinning that conforms to cultural expectations. He attempts to reconcile two idealized yet competing roles: devoted father and dedicated professional.

| Interviewer: | Currently, at your present job, how are you handling the demands of work and family? |
| Will: | It's tough. I want to be there and spend time with my daughter. So, we really make it a priority for us to eat dinner as a family on the weekends. As long as I'm not under the gun, I work from home on the weekends and spend some time with Hailie. Then I can still put in a few hours on my laptop to be ready to go for Monday. |

3.3.2. Step 2: Eliciting Frontstage Talk. In step 2, in his frontstage talk with colleagues as institutional representatives, Will rationalizes prioritizing investment in his career rather than in his parenting by defending his choice in terms that his audience is likely to understand.

| Interviewer: | Imagine talking to a colleague. If your colleague asked you "How are you handling the demands of work and family right now?" how would you answer? |
| Will: | It's a demanding line of work, but it's a choice that every single one of us made willingly. Seriously, we're playing in the big leagues, and that's what it takes to be the best. You've got to work out a long-term trajectory. It may mean spending less time with your kids when they're little. But that's a small price to pay for making partner. And then you can be more of a regular father. For now, I'm lucky to have my wife. She's gotten used to controlling the home environment. Once I make partner, I'll have a more manageable schedule and spend more time with my kids when they're older . . . and I'll have a bigger paycheck to do it right. |

3.3.3. Step 3: Going Backstage. In step 3, Will identifies Lenny as his trusted interlocutor and goes backstage. Unlike other exemplars, Will's backstage self-disclosures are catalyzed by an imagined conversation with a critical friend who challenges his decisions.

Interviewer:	Tell me the name of someone you trust—a confidant with whom you could talk freely about managing the demands of work and family.
Will:	I have a friend, Lenny. We go way back.
Interviewer:	Tell me about Lenny.
Will:	Lenny and I were at school together—been friends ever since. He's also Hailie's godfather and never misses an opportunity to let me know everything I'm doing wrong [chuckle].

3.3.4. *Step 4: Eliciting Backstage Talk.* In step 4, Will's backstage interlocutor, Lenny, takes Will to task for not living up to his parenting ideals. In this imagined conversation, Will's visceral narrative reveals his "guilt" over allowing work commitments to monopolize his attentions and divert time and energy from his daughter. The sequence concludes when Will bridges his frontstage and backstage talk by worrying that one day it "will just be too late" to realize his parental aspirations. Here, he owns up to the pain resulting from the parental bonding opportunities he has sacrificed on the altar of his career.

Interviewer:	Imagine you are talking to Lenny alone where no one can hear you. Lenny will keep what you say private. If Lenny asked you "How are you handling the demands of work and family right now?" how would you answer?
Will:	I'd say that I probably could be at home a little more than I am right now.
Interviewer:	And how would Lenny respond?
Will:	Hah! Lenny is not exactly my role model. You see, he made a career switch himself and says it was the best thing that he ever did for his family. So I'm sure he would say that I should do the same thing.
Interviewer:	And what would you say to Lenny?
Will:	I'd ask him if he's offering to pay my mortgage [chuckle]. No, seriously, on a certain level, he may be right [pause].
Interviewer:	How so?
Will:	On a good day, I'll make it home in time for dinner, and things seem to be working. But other days I'll come home late, and Hailie will already be in bed or she'll say something that guilts me out. Those days, I wonder if I can really make up for the time later, or if it will just be too late [sigh].

3.4. *Business Professional Kim*

Kim, the fourth exemplar, is a high-flying female executive at a Silicon Valley company who travels extensively for work. Married with children, Kim's family competes for time and attention with her all-consuming professional obligations. At the beginning of the sequence, Kim uses humor to evince skepticism toward the idea of reconciling work and family obligations. During the frontstage phase, she delivers a strategic self-presentation crafted for her frontstage audience of male colleagues. Alone among the exemplars, Kim's case illustrates the potential of conversations with real or hypothetical strangers. With a sympathetic stranger, Kim sheds her tough exterior and speaks freely about the toll her work has exacted on her family life. In Kim's case, the IQA sequence evokes frontstage and backstage talk from a self-reflective respondent who needs a safe and anonymous interactional space in which to unveil a vulnerable self.

3.4.1. *Step 1: Establishing the Baseline Iterated Question.* In step
1, Kim offers a humorous response to the baseline question, intimating her discomfort with practices prevailing in her male-dominated professional world. Whereas her male counterpart, Will, began the sequence by touting his ability to reconcile the demands of professional work and parenting, Kim's tongue-in-cheek tone serves to mark her skepticism about the ideal of balancing work and parenting obligations.

Interviewer:	Currently, at your present job, how are you handling the demands of work and family?
Kim:	Family, what family? Oh yeah, I have sons! I'm hoping to get some quality time with my boys before they go off to college [laughs]. No, really, things are as good as it gets.

3.4.2. *Step 2: Eliciting Frontstage Talk.* In step 2, Kim's frontstage
talk is calibrated for her workplace audience. Her talk conforms to her male colleagues' expectations, namely that she can and will operate as they do, adopting a single-minded focus on work without allowing family matters to interfere. Unlike Will, who expresses his affinity for his workplace ideology, Kim signals her distance from institutional scripts by sarcastically parroting the "party line."

Interviewer:	Imagine that you are talking to a colleague. If your colleague asked you "How are you handling the demands of work and family right now?" how would you answer?
Kim:	I'd say everything is fine, thank you very much. Husband is fine, check. Kids are fine, check. Family is fine, check. It's all good.
Interviewer:	How would this colleague respond?
Kim:	Respond? I wouldn't expect a response. Where I work, "How are you?" is strictly a rhetorical question. . . . Look, I mostly work with high-octane guys. There's no place for uber mom. So it's pretty much don't ask, don't tell. Sick kids? Got a nanny for that, right? The guys just don't want to hear about it . . . at work, it's all work. The party line is that everything is running smoothly at home in nannyland, and so let's get on the deadline, gentlemen.

3.4.3. *Step 3: Going Backstage.* Kim's backstage scenario marks a contrast with the backstage scenarios of the other three exemplars. Rather than someone from her personal community, Kim prefers to go backstage with a nonjudgmental stranger who offers a sympathetic ear as well as the assurance of privacy and confidentiality.

Interviewer:	Tell me the name of someone you trust—a confidant with whom you could talk freely about managing the demands of work and family.
Kim:	Oddly enough it's easier to talk to strangers. One thing I love about traveling is that you meet people and can say anything you want because you'll never see them again. You know, ships passing in the night.
Interviewer:	Please elaborate.
Kim:	Strangers—people you will never see again, so you can tell them how it really is—the unvarnished truth. Like you, in fact—I'll probably never see you again after this interview.

3.4.4. *Step 4: Eliciting Backstage Talk.* In her backstage talk, Kim confronts the hard realities standing in the way of reconciling her work and family lives. Articulating a heartfelt and visceral narrative, Kim admits to struggling with the impossible challenge of meshing devoted motherhood with a successful professional career in the family-unfriendly tech world.

Interviewer:	OK! Imagine you are talking to one of these strangers—one of these ships passing in the night as you say—alone where no one can hear you. This stranger will keep what you say private. If the stranger asked you "How are you handling the demands of work and family right now?" how would you answer?
Kim:	So like I was saying . . . strangers. . . . One time, I was in Vegas for this show . . . what goes on in Vegas stays in Vegas . . . [laughs] and I met Linda in the sauna. So, we're just sitting there in the sauna in our towels, just the two of us, and we get to talking and I pretty much unload. . . . When I'm on deadline it is such a rush, but then I come down, and this little voice asks if I'm failing them as a mom [sighs].
Interviewer:	How did Linda respond?
Kim:	She knew what I meant. We both grew up thinking we could do it all . . . bring home the bacon, fry it up in a pan [pause], but you have to choose. I mean I can't very well say, "Hey, I'll take half the prestige for half of the hours." In my world, it's all or nothing. So, do I want to keep the career I worked so hard for? Or do I give it up for momdom? Is that even a choice?

3.5. *The Four Exemplars: Comparable Talk and Multidimensional Narratives*

Each exemplar renders apparent the strengths of IQA as a means of eliciting rich yet readily classifiable interview talk. IQA's theoretically informed sequential questions make it possible for the interviewer to interrogate the same theme manifested in narratives corresponding to different self-presentations. Starting with the baseline question in step 1, progressing to the frontstage question in step 2, and concluding with the backstage questions in steps 3 and 4, the IQA sequence facilitates the comparison of responses both within a single interview as well as across multiple interviewees. Because IQA works in the same way for diverse respondents, it affords opportunities for systematic and meaningful comparisons of parallel interview talk across different individuals. As cross-respondent comparisons are a staple of interview-based research, this strength offers important payoffs in the analysis of interview material.

These four exemplars, all typical of multiple cases within the data set, demonstrate the versatility of IQA as a technique applicable across multiple themes and respondents. Despite the idiosyncrasies of the respondents' life circumstances, the sequence moves predictably from frontstage talk to backstage talk in all four representative cases. Moreover, IQA works well with prospective themes such as imagined educational futures (Frye 2012), seen with Hunter and Marta, as well as retrospective themes such as work-family choices, evident in interviews with Will and Kim. In the backstage part of the sequence, interviewees enter a safe interactional space where they can give expression to their nonconformist selves, vulnerable selves, and flawed selves, all of which are hidden in the frontstage. The sequence as a whole enables respondents to give voice to sometimes discordant narratives without feeling the discomfort that often accompanies the expression of inconsistent views in traditional interviewing.

The payoffs of IQA become readily apparent through examination of Hunter and Marta as representative cases. For this pair of interviewees, IQA makes it an easy matter to contrast respondents in terms of their frontstage self-presentations containing honorable narratives and their backstage self-presentations drawing on visceral narratives. Without IQA, the analyst would perforce compare interview talk from Hunter and Marta without necessarily knowing how to classify their talk in terms of identity positions, self-presentations, or types of narrative.

In a non-IQA interview, it would appear that Hunter and Marta mirror one another in their focus on financial prudence as the deciding factor in college plans. However, once the IQA sequence moves into the backstage phase, it becomes clear that Hunter and Marta harbor radically different orientations toward their postsecondary trajectories. Both have become fluent in the honorable narratives demanded by the school, even though they have drifted far apart in terms of how they view the college planning process in their more private moments. In Hunter's case, the presence of a likeminded and affirming backstage interlocutor brings to the surface an aversion to risks absent from his frontstage talk. Backstage, Hunter confesses an unwillingness to strike out on his own to attend a four-year school if it means separation from the academic and social support provided by his personal community. By contrast, Marta's backstage talk—triggered by an encounter with an interventionist mentor—conveys a genuine desire to follow the institutionally sanctioned postsecondary trajectory that vies with her fear of social ostracism at the hands of economically advantaged peers. In these ways, IQA reveals how Marta stands apart from Hunter.

The cross-respondent comparison between Will and Kim, two business professionals, is also aided by the parallel structures of IQA sequences. Once backstage, Will and Kim, like Hunter and Marta, switch self-presentations predictably in response to imagined or remembered conversational scenarios built into IQA's question formulation. Unlike Marta and Hunter, Will and Kim diverge in terms of their affinity for the frontstage talk prompted at the beginning of the IQA sequence. Will illustrates the success of IQA with an interviewee who (1) buys into the honorable narrative encapsulated in his frontstage talk and (2) identifies closely with the frontstage interlocutor. Will typifies respondents in other IQA interviews with male business professionals. By contrast, Kim's frontstage talk is representative of other respondents who feel compelled to at least pay lip service to institutional scripts. Kim's flippant humor is aimed at male colleagues with whom she does not identify. Kim illustrates the analytically important type of respondent who puts forth a strategic frontstage narrative.

Contrary to Marta and Hunter, Kim and Will converge in their backstage talk. Despite choosing significantly different backstage interlocutors, friend versus total stranger, their talk converges at the final stage of the sequence. Here, both Will and Kim unveil the problematic aspects of their work-family decisions and express visceral narratives full of self-doubt and guilt. The IQA sequences reveal that the backstage experience of sacrifice on behalf of one's career carries very similar emotional resonance for these two different interviewees. This commonality between the two business professionals with different life circumstances might well have eluded conventional interviewing techniques. But this convergence is easily captured with IQA, which renders visible the full spectrum of identity positions and narratives for each respondent, allowing targeted comparisons of parallel frontstage and backstage talk across respondents.

4. LESSONS FROM THE FIELD

The presentation of IQA's strengths in the previous section draws on interviews using the finalized version of the technique. In contrast, this section reveals lessons from the field trials indicating the conditions under which IQA strengthens the interviewer's arsenal. First, two assumptions inform IQA: (1) the interview's themes concern self-identity, and (2) standards of good interviewing practice are observed. Second, the field trials indicate the importance of setting strict parameters for the formulation of backstage questions. In honing IQA techniques in the course of these field trials, it became clear that when these conditions are satisfied, IQA improves the craft of interviewing. Third, in addition to these conditions, some IQA interviews require two related backstage troubleshooting strategies: acknowledging potentially disruptive emotions and creating a safe space by restating the questions using the respondent's own words.

4.1. *Assumptions Informing IQA: Themes and Practices*

The field trials made clear that although identity-relevant themes constitute fertile terrain for the application of IQA, the technique will fall on fallow soil where such themes are absent. In the absence of self-relevant themes, self-presentations and identity performances do not differ across the frontstage-backstage divide, and respondents are likely to articulate identical frontstage and backstage talk. For example, IQA should not be used in interviews emphasizing factual details of particular events or phenomena that have no bearing on respondents' self-presentation (Rubin and Rubin 1995; Weiss 1994).

In tandem, for IQA to succeed, the interviewer must observe standard interviewing practices. Rapport is just as critical a prerequisite for the effectiveness of IQA as it is for less structured forms of in-depth interviewing. The field trials confirmed that it is critical for the IQA interviewer to continually maintain rapport by signaling successful interaction. In the initial field trials, when the interviewer failed to indicate that the interaction had been successfully concluded, interviewees waited for the interviewer and said things like "Okay?" or "Are we good?" In contrast, when the interviewer affirmed the respondent's answer to the questions with a simple "yes" or "okay," this signaled successful interaction and greased the wheels of the interviewing partnership (Weiss 1994). Such rapport building is particularly important for interviews dealing with potentially delicate topics that IQA is designed to handle because respondents often reveal vulnerabilities and discuss aspects of their lives which can elicit feelings of shame or guilt. To use Goffmanian terminology, the interviewer's affirmation allows the interviewee to "maintain face" (Goffman 1967:6). All of these standard interviewing practices aimed at building rapport support the interviewing partnership.

4.2. *Going Backstage: Question Formulation*

As the initial field trials indicate, IQA works best with the strictly formulated backstage questions in steps 3 and 4. In the initial field trials, more loosely posed questions had to be reformulated with much greater specificity to facilitate the transition to the backstage. More specifically, when the theme of the interview was not clearly specified as part of the question, the interviewees talked through their answers to weigh prospective confidants' trustworthiness. For example, the following exchange appeared in an early field trial with a business professional in what would become step 3.

Interviewer:	Tell me the name of someone you trust.
Cynthia:	Well, someone I trust. Hmm . . . that might be my husband if it had to do with the kids. On the other hand, if it had to do with work, then it would be my colleague Rose. But you know, if it is about my parents, then it would be my sister. So what kind of trust?
Interviewer:	Since we're talking about managing work and family, someone you trust about that.
Cynthia:	That's a hard one. I'd talk to Rose, but someone at work might hear us. I'd talk to my sister, but she might blab.

In response to these issues arising from the field trials, the questions in step 3 were reformulated in the following way: "Tell me the name of someone you trust—a confidant with whom you could talk freely about [theme of iterated question]" and "Tell me about [interviewee's trusted third party]."

In parallel, a few respondents voiced concerns about speaking to a trusted third party and asked the interviewer for more clarity. Interviewees' questions included "Can anyone hear us talking?" and "Will they [trusted confidant] share what I say?" As a result, we refined step 4 questions to add the needed specificity. These reformulations proved successful in subsequent field trials. The final version of the questions reads, "Imagine you are talking to [trusted third party] alone where no one can hear you. [Trusted third party] will keep what you say private. If [trusted third party] asked you [iterated question] how would you answer?" and "How would [trusted third party] respond?"

4.3. *Troubleshooting Strategies*

The field trials indicated that backstage troubleshooting strategies were necessary when visceral narratives evoked deeply felt emotions. Two related strategies were successful: (1) acknowledging strong emotions and (2) restating the question using the respondent's own terminology. These strategies were effective in the case of Tomás, a business professional, who teared up when imagining a conversation with his wife about his inability to harmonize work and family demands.

Interviewer:	Imagine you are talking to your wife Jodi alone where no one can hear you. She will keep what you say private. If Jodi asked you "How are you handling the demands of work and family right now?" how would you answer?
Tomás:	[Pause.] It's not so good right now. She knows that. We've talked about it. She knows . . . [long pause, tears up].
Interviewer:	Take all the time that you need.
Tomás:	[Strangled voice.] I may need a break here.
Interviewer:	I understand. We can stop if you like [pause]. You know, talking about work and family—especially people we care about—can evoke strong emotions. These things matter. That's why we are doing these interviews because we think these things are really important [long pause].
Tomás:	OK, I'm good. Let's go.
Interviewer:	So you were saying that you've shared with Jodi that things aren't so good right now. What did you say? And how did Jodi respond?

Acknowledging emotions and restating the question using the respondent's own terminology also worked well with respondents like high school student Samantha:

Interviewer:	Tell me the name of someone you trust.
Samantha:	Dunno. I don't like to talk about my business. It's like private.
Interviewer:	I understand. OK. So in general what qualities would make someone trustworthy?
Samantha:	It's not easy. They can't be tryin' to trip you out.

Interviewer:	OK. What else?
Samantha:	They've gotta keep it tight. No loose lips.
Interviewer:	I get it. So imagine that you know someone who cares about you and is not there to trip you out. And this person is going to keep it tight, as you say. OK?
Samantha:	Yeah.
Interviewer:	OK, keep someone like this in mind.
Samantha:	Yeah. OK.
Interviewer:	Now, imagine you are talking to this person alone where no one can hear you. This person will keep what you say private—they're going to keep it tight. If this person asked you "Why are you going to Jefferson?" how would you answer provided there were no loose lips?
Samantha:	Yeah. I see. Well, if they would keep it tight, I would say . . . .

In both cases, acknowledging strong emotion and restating the question are effective troubleshooting strategies that create a safe space backstage.

5. CONCLUSIONS: THE DIVIDENDS OF IQA

In-depth interviewing continues to prove its merits as an indispensable research method. Nonetheless, innovative techniques are needed to enrich this venerable method of social science inquiry (Lamont and Swidler 2014). IQA constitutes one such innovative technique that offers significant advantages compared with conventional interviewing procedures. By augmenting rapport and interactivity, while at the same time minimizing power differentials, IQA strengthens the interviewing partnership (Weiss 1994). These advances are accomplished through IQA's Goffmanian approach that delivers theoretically informed questioning techniques, readily classifiable forms of talk, and greater replicability.

5.1. *Strengthening the Interviewing Partnership with IQA*

In conventional interviewing, interviewers who seek to delve into respondents' identity-relevant experiences and feelings often face a challenge in generating both shallow and deep narratives (Rubin and Rubin 1995). Meeting this challenge typically necessitates the cultivation of strong and sustained rapport with interviewees in order to encourage full self-disclosure. Building rapport has clearly paid dividends, especially when the interviews raise painful topics to the surface (Blair-Loy 2003; Pugh 2015; Silva 2013). Strong and sustained rapport can eventuate in the evocation of spontaneous visceral narratives in addition to more scripted honorable narratives, exemplified in Pugh's (2012, 2015) work on cultures of commitment. But rapport often lies beyond the control of the interviewer and can fluctuate dramatically and unpredictably within the course of an interview. Because rapport is an elusive condition kept alive by the artfulness of the interviewer (Wengraf 2001), it can easily break down. Furthermore, even the most ambitious interviewer will acknowledge that rapport building can consume temporal and emotional resources that the interviewee may not be able to afford (Alvesson 2011). For these reasons interviewers can benefit from IQA to peel back interviewees' surface self-presentations and unearth self-disclosure and visceral narratives.

IQA also advances conventional approaches by augmenting the scope of interactivity within the interview (Alvesson 2011). IQA simulates interactions between interviewees and varied members of their personal communities—ranging from institutional representatives to confidants—in an imagined or remembered way. These simulated conversations make it possible for the interviewee to express frontstage and backstage forms of talk about the same themes. When interviewees are invited to imagine themselves in conversation with various interlocutors with whom they have widely divergent relationships, the range of interactivity is enlarged. In amplifying interactivity in this way, IQA improves conventional interviewing by deepening the interviewing partnership and expanding the inclusiveness of the interview as a dialogical and relational event (Holstein and Gubrium 1995). Moreover, by inviting respondents to enter into simulated conversations, IQA creates additional opportunities to elicit talk that more closely approximates natural conversation (Luker 2009).

At the same time, IQA lessens power differentials between the interviewer and the interviewee and increases trust between them. During the backstage portion of the sequence in particular, the interviewee engages trusted others in conversation rather than a stranger cloaked in the power and authority of the interviewer. This has the effect of mitigating some of the potential estrangement resulting from the frequent asymmetry between a more powerful interviewer and a less powerful interviewee (Mishler 1986). IQA also catalyzes trust, which is indispensable to facilitating self-disclosure regarding "personally threatening and potentially painful" issues (Lee 1993:98). In these ways, IQA deepens the interviewing partnership by transforming the interview into a relational event (Holstein and Gubrium 1995) involving members of the interviewee's personal community.

5.2. *The Advantages of IQA's Goffmanian Approach*

Incorporating Goffmanian theoretical insights into the formulation of questions and drawing on the Goffmanian distinctions between frontstage and backstage self-presentations (Hester and Francis 1994; Weiss 1994), IQA advances interviewing practice. Rooted in the dramaturgical insights of symbolic interactionism, IQA's replicable sequenced questions give interviewers a new way of predictably and reliably delivering multidimensional narratives corresponding to both frontstage and backstage self-presentations. IQA does so by triggering the foregrounding of both a self-reflexive me-for-others (Goffman 1959) in the frontstage, as well as a more spontaneous "I" inhabiting the safety of the backstage.

By enabling the interviewer to awaken identity performances in a parallel manner, IQA's question design also produces readily classifiable forms of talk that correspond to particular self-presentations. Sequencing the iterated questions evokes parallel responses corresponding to frontstage and backstage forms of talk along with the honorable and visceral narratives they contain. In the analysis phase, IQA's careful sequencing of questions makes the movement between identity positions and accompanying narratives predictable and thereby alleviates the interviewer's interpretive burden involved in teasing apart public and private identity positions.

Moreover, IQA supplies the interviewer with new tools to tackle the issues of replicability that have been raised with regard to interviewing in particular, as well as qualitative research in general (Luker 2009; Mishler 1986). Because IQA can be applied systematically across a group of respondents, it facilitates comparisons of like with like. For example, equipped with IQA, the analyst is in a position to juxtapose comparable answers to institutional representative questions, as well as comparable answers to trusted interlocutor questions. Thus, interview talk obtained with IQA may be analyzed in a comparative way within the same interview, in multiple interviews with the same respondent, or across material from multiple respondents.

5.3. *Summary of Contributions and Future Interviewing with IQA*

Finally, to summarize, IQA offers significant advantages compared with conventional in-depth interviewing. IQA strengthens the interviewing partnership on three fronts: (1) enhancing rapport, (2) augmenting inter-activity, and (3) leveling power differentials. IQA does so thanks to its Goffmanian approach, which provides theoretically informed, replicable questioning techniques designed to gather interview data on identity-relevant themes. These techniques produce readily classifiable forms of talk that correspond to frontstage and backstage self-presentations. As a result, IQA resolves issues of replicability by providing a tool to sys-tematically gather and analyze comparable talk. For all of these reasons, IQA moves the field forward.

We close this article with thoughts on the potential of IQA to spur methodological innovation beyond the qualitative realm. For example, IQA could enhance the vignettes approach in survey research. In sur-veys employing vignette techniques, respondents are presented with stories and invited to consider how they would respond to hypothetical situations centered on the actions of imaginary characters or "dramatis personae" (Finch 1987). Although vignettes vary in their complexity and specificity from one-sentence descriptions of situations to extremely elaborate and highly specified scenarios, the vignettes approach is designed to uncover norms, ideals, and understandings that elude con-ventionally formulated attitudinal questions (Ganong and Coleman 2006). Both IQA and the vignettes approach encourage the respondent to step outside of the immediate interactional context imposed by the interview or survey and to envision what they would do or say outside of this context. However, IQA affords respondents the opportunity to formulate their answers in the company of concrete interlocutors popu-lating their own lives. In contrast, survey vignettes invite respondents to imagine themselves making decisions and judgments about hypothetical individuals and scenarios. Should survey researchers choose to incorpo-rate IQA techniques into survey-based questions, they could appropriate some of the more dynamic elements of IQA by crafting a set of vign-ettes corresponding to both frontstage and backstage situations.

Looking forward, we consider the applicability of IQA to enrich future sociological inquiry. Our exemplars have demonstrated that IQA may be fruitfully used with diverse respondent groups as well as with identity-relevant themes relating to traditional sociological topics such as work and family or educational and career aspirations. Therefore, IQA holds much potential for a variety of additional sociological subfields. IQA promises to open new doors for the sociology of culture, where interviewing has anchored pioneering work on cultural assumptions and boundary-drawing (Lamont 1992), emotion work (Hochschild 1983), and devotional schemas (Blair-Loy 2003). Life-course research could also benefit from IQA, as it yields both frontstage and backstage accounts of biographical experiences. Digital sociology researchers tackling identity-relevant dimensions of social media or mediated communication will also find IQA a valuable addition to their methodological toolkits. Using IQA techniques can enhance interviews dealing with sensitive topics such as gender, ethnic identity, health, sexuality, or criminality, as well as highly polarizing subjects with identity-relevant dimensions such as gun control, immigration, and abortion. Given the flexibility and versatility of IQA as a means of exploring a range of topics and themes, IQA promises to be an innovative interviewing technique that advances the methodological frontier.

Acknowledgments

We thank Tim Liao for his editorial guidance and the four anonymous reviewers selected by *Sociological Methodology*, as well as Arlie Hochschild, Michèle Lamont, Robert Weiss, and Allison Pugh, for offering commentary and insights.

References

Alvesson, Mats. 2011. *Interpreting Interviews*. Thousand Oaks, CA: Sage.
Alvesson, Mats, and Kaj Sköldberg. 2009. *Reflexive Methodology: New Vistas for Qualitative Research*. Thousand Oaks, CA: Sage.
Blair-Loy, Mary. 2003. *Competing Devotions*. Cambridge, MA: Harvard University Press.
Demant, Jakob, and Margaretha Järvinen. 2006. "Constructing Maturity through Alcohol Experience—Focus Group Interviews with Teenagers." *Addiction Research and Theory* 14(6):589–602.
Dominici, Kathy, and Stephen Littlejohn. 2006. *Facework: Bridging Theory and Practice*. Thousand Oaks, CA: Sage.
Finch, Janet. 1987. "Research Note: The Vignette Technique in Survey Research." *Sociology* 21(1):105–14.
Frye, Margaret. 2012. "Bright Futures in Malawi's New Dawn." *American Journal of Sociology* 117(6):1565–1624.
Ganong, Lawrence, and Marilyn Coleman. 2006. "Multiple Segment Factorial Vignette Designs." *Journal of Marriage and Family* 68(2):455–68.
Goffman, Erving. 1959. *The Presentation of Self in Everyday Life*. New York: Anchor.

Goffman, Erving. 1967. *Interaction Ritual*. New York: Anchor.

Hester, Stephen, and David Francis. 1994. "Doing Data: The Local Organization of the Sociological Interview." *British Journal of Sociology* 45(4):675–95.

Hochschild, Arlie. 1983. *The Managed Heart*. Berkeley: University of California Press.

Holstein, James, and Jaber Gubrium. 1995. *The Active Interview*. Thousand Oaks, CA: Sage.

Lamont, Michèle. 1992. *Money, Morals, and Manners*. Chicago: University of Chicago Press.

Lamont, Michèle, and Ann Swidler. 2014. "In Praise of Methodological Pluralism." *Qualitative Sociology* 37(2):153–71.

Lee, Raymond. 1993. *Doing Research on Sensitive Topics*. Thousand Oaks, CA: Sage.

Luker, Kristin. 2009. *Salsa Dancing into the Social Sciences: Research in an Age of Info-glut*. Cambridge, MA: Harvard University Press.

Mead, George Herbert. 1934. *Mind, Self, and Society*. Chicago: University of Chicago Press.

Mishler, Elliot. 1986. *Research Interviewing: Context and Narrative*. Cambridge, MA: Harvard University Press.

Myers, Michael D., and Michael Newman. 2007. "The Qualitative Interview in IS Research: Examining the Craft." *Information and Organization* 17(1):2–26.

Prasad, Pushkala. 2005. *Crafting Qualitative Research: Working in the Postpositivist Traditions*. London: M. E. Sharpe.

Pugh, Allison. 2012. "What Good Are Interviews for Thinking about Culture? Demystifying Interpretive Analysis." *American Journal of Cultural Sociology* 1(1): 1–27.

Pugh, Allison. 2013. "The Divining Rod of Talk: Emotions, Contradictions and the Limits of Research." *American Journal of Cultural Sociology* 2(1):159–63.

Pugh, Allison. 2015. *The Tumbleweed Society*. Oxford, UK: Oxford University Press.

Ragin, Charles, Joane Nagel, and Patricia White. 2003. "Workshop on Scientific Foundations of Qualitative Research." Washington, DC: National Science Foundation. Retrieved March 1, 2015 (http://www.nsf.gov/pubs/2004/nsf04219/nsf04219.pdf).

Rubin, Herbert J., and Irene S. Rubin. 1995. *Qualitative Interviewing: The Art of Hearing Data*. Thousand Oaks, CA: Sage.

Schwalbe, Michael L., and Michelle Wolkomir. 2003. "Interviewing Men." Pp. 55–72 in *Inside Interviewing: New Lenses, New Concerns*, edited by James Holstein, and Jaber Gubrium. Thousand Oaks, CA: Sage.

Silva, Jennifer. 2013. *Coming Up Short*. Oxford, UK: Oxford University Press.

Spencer, Liz, and Ray Pahl. 2006. *Rethinking Friendship*. Princeton, NJ: Princeton University Press.

Tannen, Deborah, and Cynthia Wallat. 1987. "Interactive Frames and Knowledge Schemas in Interaction: Examples from a Medical Examination Interview." *Social Psychology Quarterly* 50(2):205–16.

Tovares, Alla. 2007. "Family Members Interacting while Watching TV." Pp. 283–309 in *Family Talk: Discourse and Identity in Four American Families*, edited by Deborah Tannen, Shari Kendall, and Cynthia Gordon. Oxford, UK: Oxford University Press.

Weiss, Robert. 1994. *Learning from Strangers*. New York: Free Press.

Wengraf, Tom. 2001. *Qualitative Research Interviewing: Biographic Narrative and Semi-structured Methods.* Thousand Oaks, CA: Sage.

Author Biographies

Laura Robinson is an assistant professor at Santa Clara University. Robinson has served as chair of the Communication and Information Technologies section of the American Sociological Association, visiting assistant professor at Cornell University, visiting scholar at Trinity College Dublin, and postdoctoral researcher at the University of Southern California Annenberg. She earned her PhD from the University of California, Los Angeles, where she held a Mellon Fellowship in Latin American Studies and received a Bourse d'Accueil at the École Normale Supérieure. Her work has appeared in journals including *Sociology, Qualitative Sociology*, and *Information, Communication & Society*. Robinson's award-winning research examines digital and informational inequalities, interaction and identity work, and digital media in Brazil, France, and the United States.

Jeremy Schulz is a visiting scholar at the Institute for the Study of Societal Issues, University of California, Berkeley. He earned his PhD from the University of California, Berkeley, and held a National Science Foundation postdoctoral position at Cornell University. He has published on a broad range of topics, including consumption, work, family, culture, and inequalities. His recent publications include "Talk of Work," published in *Theory and Society*, and "Shifting Grounds and Evolving Battlegrounds," published in the *American Journal of Cultural Sociology*. His article "Winding Down the Workday," published in *Qualitative Sociology*, received the Shils-Coleman Award from the American Sociological Association Theory Section. His current research examines peer-to-peer consumption, wealth trajectories, indebtedness, and innovative set-theoretic methods.

Sociological Methodology
2016, Vol. 46(1) 84–120
© American Sociological Association 2015
DOI: 10.1177/0081175015599808
http://sm.sagepub.com

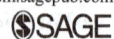

ASSESSING THE EFFECTIVENESS OF ANCHORING VIGNETTES IN BIAS REDUCTION FOR SOCIOECONOMIC DISPARITIES IN SELF-RATED HEALTH AMONG CHINESE ADULTS

Hongwei Xu*
Yu Xie[†]

Abstract

The authors investigate how reporting heterogeneity may bias socioeconomic and demographic disparities in self-rated general health, a widely used health indicator, and how such bias can be adjusted by using new anchoring vignettes designed in the 2012 wave of the China Family Panel Studies (CFPS). The authors find systematic variation by sociodemographic characteristics in thresholds used by respondents in rating their general health status. Such threshold shifts are often nonparallel in that the effect of a certain group characteristic on the shift is stronger at one level than another. The authors find that the resulting bias of measuring group differentials in self-rated health can be too substantial to be ignored. They demonstrate that the CFPS anchoring vignettes prove to be an effective survey instrument in obtaining bias-adjusted estimates of health disparities not only for the CFPS sample but also for an independent sample from the China Health and

*University of Michigan, Ann Arbor, MI, USA
[†]Princeton University, Princeton, NJ, USA, and Peking University, Beijing, China

Corresponding Author:
Hongwei Xu, University of Michigan, 426 Thompson Street, ISR 2459, Ann Arbor, MI 48104, USA
Email: xuhongw@umich.edu

Retirement Longitudinal Study. Effective adjustment for reporting heterogeneity may require vignette administration only to a small subsample (20 percent to 30 percent of the full sample). Using a single vignette can be as effective as using more in terms of anchoring, but the results are sensitive to the choice of vignette design.

Keywords

anchoring vignettes, reporting heterogeneity, self-rated health, socioeconomic status

1. INTRODUCTION

Because of the robust predictive power of self-rated health (SRH) for mortality (Benjamins et al. 2004; House et al. 2000; Idler and Benyamini 1997), its strong association with morbidity and physical functioning (Goldberg et al. 2001; Singh-Manoux et al. 2006), and the simplicity and low cost associated with its collection, it has been used as a general health indicator in numerous social surveys (Chen, Yang, and Liu 2010; Tandon, Zhuang, and Chatterji 2006; Wen et al. 2010; Zimmer and Kwong 2004). A large literature consistently documents a positive association between socioeconomic status (SES) and SRH for the United States and Europe (Huisman, Kunst, and Mackenbach 2003; Kakwani, Wagstaff, and van Doorslaer 1997; Knesebeck et al. 2003; Mirowsky and Ross 2008; Ross and Wu 1995; Willson, Shuey, and Elder 2007). However, several studies in many developing countries, including China, Thailand, and the Philippines, have found either no significant positive association between SES and SRH or an inverse relationship (Luo and Wen 2002; Pei and Rodriguez 2006; Whyte and Sun 2010; Zimmer and Amornsirisomboon 2001; Zimmer et al. 2000).

Rather than providing evidence of cross-country differences in SES-based health inequalities, these findings may instead reflect reporting heterogeneity; that is, respondents of varying SES backgrounds may adopt systematically different frames of reference in rating their overall health. For example, the peer comparison theory predicts that high-SES respondents are likely to compare themselves with their peers and hence adopt a higher standard for "excellent" health, whereas low-SES respondents may apply a lower standard, resulting in an inflated level of SRH relative to that of high-SES respondents, despite the latter group's advantage in true health status (Dowd and Todd 2011; Schnittker 2005). Alternatively, the health optimism/pessimism theory predicts that high-SES respondents, believing that their affluence confers well-being, will systematically boost their self-ratings of health (Ferraro 1980), whereas low-SES respondents are more pessimistic about their health in the face of limited resources (Ferraro 1993), resulting in an overestimated SES gradient.

The methodology of anchoring vignettes—brief descriptions of hypothetical people or situations that survey respondents are asked to evaluate on the same scale as they use to assess their own situations—has been proposed to address the problem of cross-group reporting heterogeneity. This approach allows a comparison of the respondents' self-assessments with the assessments they assign to the hypothetical others on the same questions. Vignettes fix the categorical levels of interest so that variation in responses is adjusted by heterogeneity in thresholds, or cut points, in respondents' evaluation of health.

Several studies have reevaluated intergroup health inequalities using the vignettes methodology for SRH. However, most of these studies focused on cross-country comparisons (Jürges 2007; Murray et al. 2003; Salomon et al. 2004), and the few that looked at response bias in health inequalities by SES, focused mainly on American and European elderly populations, largely because of the availability of vignettes data from the Health and Retirement Study (HRS) and its sibling surveys in Europe (Bago d'Uva, O'Donnell, and van Doorslaer 2008a; Dowd and Todd 2011; Grol-Prokopczyk, Freese, and Hauser 2011). This focus limits the generalizability of results given that the elderly may tend to self-assess their health differently than do younger populations (Schnittker 2005) and that Westerners may respond differently to the vignettes methodology than would respondents in non-Western countries.

Using anchoring vignettes to assess health may be less effective in China or other non-Western societies for two reasons. First, survey responses in East Asian regions are characterized by a strong tendency to agree (high acquiescence) and a weak tendency to disagree (low disacquiescence) with any item, regardless of content, and by a strong preference for middle over polar response categories on ratings scales (Harzing 2006). These reporting behaviors reduce the amount of information available for differentiating true health status. Second, vignettes require that respondents evaluate the health of a hypothetical person on the basis of a text description, a cognitive burden that may prove taxing to respondents in developing societies in which the average educational attainment is relatively low (Bago d'Uva et al. 2008b).

Many health surveys have followed the World Health Organization (WHO) World Health Survey (WHO-WHS) in collecting multiple domain-specific health ratings (e.g., mobility, pain, cognition, vision, sleep, self-care, and affect) instead of a single general health rating, with each domain using nearly identical vignettes. This is not a feasible approach in general-purpose household surveys, in which, because of time constraints, only a single question about self-rated general health is typically asked. However, to the best of our knowledge, no other surveys have designed vignettes to anchor self-rated general health.

Given these limitations in the previous research, we seek in this study to address the following questions:

1. Does reporting heterogeneity bias the measurement of health disparities by SES among Chinese adults?
2. Are anchoring vignettes effective in correcting such bias?
3. Can vignette adjustment estimated from a population-based sample be generalized to other surveys?

This study reports the application of vignettes we designed to anchor self-rated general health in the China Family Panel Studies (CFPS). We evaluate the effectiveness of our anchoring vignettes in obtaining more accurate estimates of health disparities by SES in a national sample of Chinese adults and thereby help reconcile previous findings of an inverse association between SES and health. In addition, we evaluate the cost-effectiveness of the vignettes methodology and assess the validity of extrapolating vignette adjustments estimated from our sample to another national sample of middle-aged and older Chinese adults in an effort to demonstrate the broad utility of the vignettes methodology.

2. VIGNETTES METHODOLOGY

2.1. *Intergroup Reporting Heterogeneity*

In considering the vignettes methodology, it is important to distinguish between adjusting for individual-level and for group-level reporting heterogeneity. The ubiquitous population heterogeneity in social science research dictates that individual reporting heterogeneity cannot be naively discarded as a mere nuisance or measurement error by assuming reporting behaviors are essentially the same within a subpopulation (Xie 2013). Unfortunately, it is impossible to estimate individual reporting

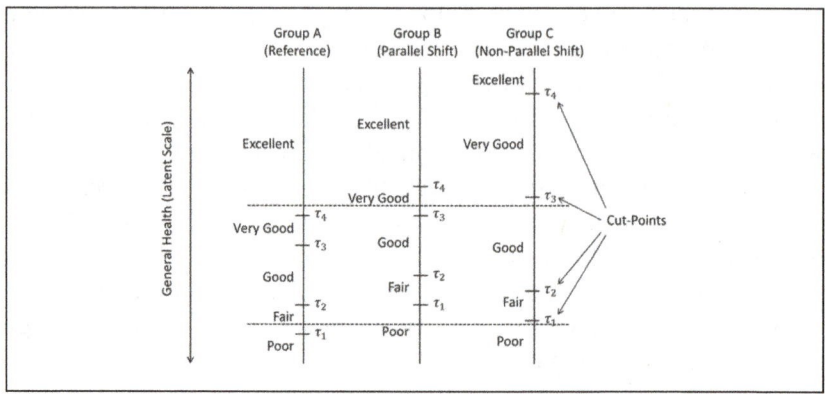

Figure 1. Illustration of cut-point shifts on the latent response scales of self-rated health.

heterogeneity without administering multiple vignettes in the full range of the latent construct (latent true health in this study) to each respondent, whose corresponding vignette assessments also provide enough support for estimating the full-scale individual cut points from low to high. Not only would such a practice constitute an expensive data collection option in a multipurpose survey, it would also be extremely challenging to design multiple vignettes that cover the full range of the latent construct and to ensure each respondent's assessments in accordance with the intended vignette ranking. To our knowledge, only one study has estimated individual reporting heterogeneity by pooling 15 vignettes across three different domains and assuming a common response scale across these domains (Kapteyn, Smith, and van Soest 2007). Similar efforts with fewer vignettes have not been successful (Bago d'Uva et al. 2011a, 2011b). Being unable to estimate individual-level reporting heterogeneity, we essentially follow in this article a conventional practice in empirical social research by focusing on group-level differences (Xie 2013).

Intergroup reporting heterogeneity may assume two patterns on latent response scales: (1) parallel cut-point shift or (2) nonparallel cut-point shift (see Figure 1). For the former, cut points shift up or down in parallel for each of the comparison groups, providing evidence that the covariates affect all cut points equally and supporting the hypothesis that different groups may simply assume higher or lower thresholds in self-evaluating their health. In the case of nonparallel shift, intergroup differences are seen in unaligned upward or downward cut-point shifts varying with covariates.

2.2. Parametric Model

Identifying cut-point shifts among groups cannot be done using a conventional ordered probit (or logit) model of SRH, because it requires data such as objective health measures for the latent scale. Anchoring vignettes provide such auxiliary data without the high cost of taking objective health measures. For our analyses, we estimate hierarchical ordered probit (HOPIT) models that draw on anchoring vignettes to purge reporting heterogeneity and attain intergroup comparable SRH (King et al. 2004; Tandon et al. 2003).

Because a vignette is a description of a hypothetical person's health status presented to all respondents in the same way, we should expect no systematic variation (apart from random error) in the ratings of the vignette by different respondents, except that they may apply different cut points, if they perceive the vignette in the same way and on the same unidimensional scale, a feature known as the *vignette equivalence assumption* (King et al. 2004). In other words, respondents' characteristics influence their assessments of the health condition of a vignette only through affecting cut points.

Formally, let $y_{i,j}^{v*}$ denote the continuous latent true health of vignette j as perceived by respondent i. It can be modeled as a linear combination of an intercept α_j and random measurement error $\epsilon_{i,j}^{v}$:

$$y_{i,j}^{v*} = \alpha_j + \epsilon_{i,j}^{v}, \epsilon_{i,j}^{v} \sim N(0, 1), \tag{1}$$

with the normalization $\alpha_1 = 0$ for identification. Respondent i translates the continuous latent health of vignette j into one of K ordered response categories, in this case poor (1), fair (2), good (3), very good (4), or excellent (5), through a mapping mechanism:

$$y_{i,j}^{v} = k, \text{ if } \tau_i^{v,k-1} \leq y_{i,j}^{v*} < \tau_i^{v,k}, k = 1, \ldots, 5, \tag{2}$$

where $\tau_i^{v,k}$ denotes the cut point for respondent i to rate the latent true health of the vignettes as in one of the K categories, and $\tau_i^{v,0} < \tau_i^{v,1} < \tau_i^{v,2} < \ldots < \tau_i^{v,5}$, $\tau_i^{v,0} = -\infty$, and $\tau_i^{v,5} = \infty$. Assuming reporting heterogeneity, we allow the cut points to vary as a linear function of covariates X_i, plus individual heterogeneity $u_i^{v,k}$:

$$\tau_i^{v,k} = \gamma_0^{v,k} + X_i\gamma^{v,k} + u_i^{v,k}, k = 1, \ldots, 4, \tag{3}$$

where $\gamma_0^{v,k}$ are the intercepts in the respective cut points for the vignettes. As mentioned earlier, identification of $u_i^{v,k}$ requires rich data from multiple vignettes that capture the full range of latent health, which are not available to us. We therefore follow the prevailing practice in the literature by restricting our attention to identifying group-specific cut points. Reporting homogeneity results from imposing $\gamma^{v,k} = 0$. Parallel cut-point shift arises when $\gamma^{v,k} = \gamma^v$ for $k = 1, \ldots, 4$; that is, the impact of a covariate on shifting the cut-point location is the same for all the cut points. Conversely, $\gamma^{v,k} \neq \gamma^v$ represents nonparallel shift.

The SRH component of a HOPIT model takes a form similar to that of the vignette component. Let y_i^{s*} denote the continuous latent true health for respondent i; it can then be modeled as a linear combination of the SES variables and other control variables, denoted together by X_i, and an independent normal error term ϵ_i:

$$y_i^{s*} = \beta_0 + X_i\beta + \epsilon_i, \epsilon_i \sim N(0, \sigma^2), \tag{4}$$

where β_0 is the intercept. The measurement model divides y_i^{s*} into K ordinal response categories of SRH y_i^s through a mapping mechanism similar to that found in equation (2):

$$y_i^s = k, \text{ if } \tau_i^{s,k-1} \leq y_i^{s*} < \tau_i^{s,k}, k = 1, \ldots, 5, \tag{5}$$

where $\tau_i^{s,k}$ denotes the cut point for respondent i to report his or her health status as in one of the K categories, and $\tau_i^{s,0} < \tau_i^{s,1} < \tau_i^{s,2} < \ldots < \tau_i^{s,5}$, $\tau_i^{s,0} = -\infty$, and $\tau_i^{s,5} = \infty$. Again, we allow the cut points for SRH to vary as a linear function of observed covariates Z_i, plus individual heterogeneity $u_i^{s,k}$:

$$\tau_i^{s,k} = \gamma_0^{s,k} + Z_i\gamma^{s,k} + u_i^{s,k}, k = 1, \ldots, 4, \tag{6}$$

where $\gamma_0^{s,k}$ are the intercepts in the respective cut points, and Z_i can include the same covariates as X_i. We again choose not to identify individual reporting heterogeneity here for practical data limitation. However, without the auxiliary information provided by the vignettes, the above model is underidentified in that we cannot simultaneously estimate β (the effects of SES and other covariates on SRH), $\gamma^{s,k}$ (the effects of SES and other covariates on cut points in response styles), and σ^2. Model identification is achieved by assuming *response consistency* (King et al. 2004), meaning that respondents rate their own health in the same way that they assess all the vignettes. Formally, the response consistency assumption amounts to setting

$$\tau_i^{s,k} = \tau_i^{v,k}. \tag{7}$$

In other words, the vignette component and the SRH component are linked through shared cut points in survey reporting. The group-specific cut points are estimated from the vignettes data, provided that the response consistency assumption holds, thereby purging out reporting heterogeneity in estimating group differences in SRH.

Let $P_{i,k}^s$ denote the probability of respondent i reporting his or her own health as in category k, and let $P_{i,j,k}^v$ denote the probability of the same respondent rating vignette j as in category k. The log-likelihood of the HOPIT model in this case (two vignettes and SRH, all on five response categories of health status) is defined as the sum of the respondent's SRH and ratings of the vignettes:

$$\ln L = \sum_{i=1}^{N} \sum_{k=1}^{5} I(y_i^s = k) \ln P_{i,k}^s + \sum_{i=1}^{N} \sum_{j=1}^{2} \sum_{k=1}^{5} I(y_{i,j}^v = k) \ln P_{i,j,k}^v, \tag{8}$$

where $I(y_i^s = k)$ and $I(y_{i,j}^v = k)$ are two indicator functions. Parameter estimates can be attained by maximizing this joint log-likelihood.

2.3. *Cost-effectiveness of the Vignettes Methodology*

The degree of cost-effectiveness of the vignettes methodology hinges on two implementation choices. First, cost-effectiveness can be enhanced by administering multiple vignettes to only a subsample of respondents, from which the anchored group-level response patterns can be generalized to the entire study sample (King et al. 2004). For example, vignette data were collected in only 10 percent to 16 percent of the full samples in the first wave of the Survey of Health, Ageing and Retirement in Europe (SHARE), and about 25 percent to 50 percent of the full samples in the WHO Multi-country Survey Study on Health and Responsiveness 2000–2001 (WHO-MCSS) (Bago d'Uva et al. 2008b). Because most analyses have used data from the small subsamples with the multiple vignettes (often five or more), it is unclear to what extent vignette adjustment seen in these subsamples can be applied to the rest of the sample, although in principle administering vignettes to the full sample only improves statistical efficiency.

Second, it may be possible to enhance the method's cost-effectiveness while maintaining its capacity to correct for group-level reporting heterogeneity by using one rather than multiple vignettes, as long as there is enough within-group variation for all response categories. Using more than one vignette may add only to statistical efficiency while increasing survey design and implementation costs, as well as respondent burden. The literature does not address this issue—and little empirical evidence supports the effectiveness of bias reduction with only one vignette—but we note reductions in the number of vignettes used per self-assessment in some recent surveys. In SHARE, for example, the number of vignette questions for each health domain was reduced from three in the first wave to one in the second wave (Peracchi and Rossetti 2013; Voňková and Hullegie 2011).

We investigate the cost-effectiveness for these two variations on the vignettes method and report the findings after the main results. Assuming comparable group-level reporting behaviors across population-based samples, we further use vignette results from one survey to correct for reporting heterogeneity in another contemporaneous survey. Such cross-sample adjustment would mean a substantial reduction in data collection cost because borrowing existing vignette results from an external source would cost much less than collecting new data.

3. EMPIRICAL DATA

3.1. *CFPS*

The primary data source for this study comes from the CFPS, a nationally representative longitudinal survey of Chinese communities, families, and individuals. The studies focus on the well-being of the Chinese population, collecting a wealth of information on economic activities, education outcomes, family dynamics and relationships, and health. The CFPS tracks all members of the sampled families in the 2010 baseline through biennial follow-up surveys.

The 2010 nationwide CFPS baseline survey successfully interviewed 14,960 households from 635 communities, including 33,600 adults and 8,990 children, located in 25 designated provinces. The approximate response rate was 81.3 percent at the household level and 84.1 percent at the individual level, most of the nonresponse being due to noncontact. CFPS's stratified multistage sampling strategy ensured that the sample represented 95 percent of the total population in China in 2010 (Xie 2012; Xie and Hu 2014). The first full-scale follow-up survey was conducted in 2012, with more than 80.6 percent of the baseline respondents reinterviewed. This study relies on the data from the 2012 follow-up survey, for which we designed our own anchoring vignettes for SRH that were administered to all the adult respondents during in-person interviews.

3.1.1. SRH and Vignettes. The dependent variable, SRH, was collected by asking respondents to rate their overall health status at the time of interview by selecting one of five categories: poor, fair, good, very good, or excellent. Every respondent who rated his or her own health was then administered the following two vignettes in random order, on the same response scale, about the health status of a hypothetical person with a common Chinese male or female name matched to the respondent's sex. The health vignettes were designed to reflect two substantially different health statuses, thereby providing greater power to differentiate the varying cut points applied by respondents to assessing their own health status.

> *Vignette 1:* Sun Jun [male]/Li Mei [female] has no problem with walking, running, or moving his/her limbs. He/she jogs 5 km twice a week. He/she does not remember the last time when he/she felt sore, which was not within the past year. He/she never feels sore after physical labor or exercise. How would you rate his/her health status?
> *Vignette 2:* Zhao Gang [male]/Wang Li [female] has no problem walking 200 meters. He/she feels tired, however, after walking 1 km or climbing several flights of stairs. He/she has no problem with daily activities such as bringing home vegetables from market. He/she has a headache once every month, but gets better after taking medicine. Even while feeling the headache, he/she can still do daily work. How would you rate his/her health status?

3.1.2. SES Indicators. Education is measured in years of schooling. Economic resources are measured by employment status and annual family income per capita. Political capital is measured by one's own as well as other family members' cadre and/or party membership.

We also include a measure of cognitive functioning, known as episodic memory, as part of general fluid intelligence. Cognitive ability not only affects respondents' mental comprehension and assessment of the hypothetical vignette scenario but also correlates with a variety of educational and labor market outcomes (Herrnstein and Murray 1996; Marks 2013). Like respondents in other Chinese household surveys (e.g., the China Health and Retirement Longitudinal Study [CHARLS]), the CFPS-2012 respondents were read a randomly selected list of 10 simple nouns, then immediately asked to recall as many of those words as possible in any order. After 31 questions concerning subjective well-being, the respondents were again asked to recall as many of the original words as possible. Following the literature (Hu et al. 2012; McArdle, Smith, and Willis 2011), we calculated the final score of episodic memory by averaging the number of successes between the immediate and delayed word recalls.

Other control variables include age and its square, gender, marital status, rural or urban residence, *hukou* (household registration) status, and region of origin. Age is centered at mean and divided by 10 to facilitate the interpretation of the parameter. All the other control variables are discrete in nature and entered into regression models as dummies.

We focus on adults aged 16 to 70 years in 2012 ($N = 30,774$), excluding about 4 percent of this sample who had missing data on SRH or at least one of the two vignettes and about 15 percent of the remaining sample who gave ratings inconsistent with the designed rank ordering of the two vignettes, thereby violating the vignette equivalence assumption (King et al. 2004). As a group, this 15 percent of respondents had significantly lower SES (e.g., lower educational attainment, worse memory, and lower income) and reported poorer health compared with those whose ratings of the vignettes were consistent with the survey design (results not shown). Therefore, our results may underestimate the true SES disparities in health. After excluding these respondents, the sample size was 25,141 and was further reduced to 23,207 after listwise deletion of cases with missing data on covariates.

3.2. CHARLS

To demonstrate the utility and external validity of our newly designed health vignettes for the CFPS-2012, we apply the estimated cut-point shifts to estimate health disparities in the CHARLS, a nationally representative longitudinal survey of adults aged 45 years and older and their spouses if available. Launched in 2011, the CHARLS national baseline survey interviewed 17,708 respondents, with a response rate of 80.5 percent (Zhao et al. 2014). Unlike the CFPS, the CHARLS did not interview all the members in a sampled household. For comparison, we constructed most of the same SES indicators and control variables in the CHARLS as those in the CFPS, except for cadre status and/or party membership among other family members. Only a random subsample of 8,712 CHARLS respondents were asked to rate their general health using the same response categories as those in the CFPS-2012, and 7,129 of them were 45 to 70 years of age, within the age range of our CFPS analytical sample. After excluding respondents with missing data on covariates, we had 5,928 cases from the CHARLS sample.

4. MAIN RESULTS

4.1. *Descriptive Statistics*

Table 1 presents frequency distributions of SRH and vignette ratings in the CFPS-2012 and CHARLS-2011 samples. In the CFPS sample, the responses to self-assessment were more or less evenly distributed, with about one third of the respondents considering themselves in fair or poor health, another third in good health, and the rest in very good or excellent health. As expected given the vignette design, the majority of respondents rated the person in the first vignette as in very good or excellent health and the person in the second vignette as in poor health. In the CHARLS sample, the entire distribution of SRH was shifted downward, as the respondents were on average much older than those in the CFPS. Fewer than 10 percent of the sample considered themselves in very good or excellent health, whereas nearly three quarters considered themselves in fair or poor health.

Table 2 summarizes the descriptive statistics of the independent variables in the CFPS and CHARLS samples. Our CFPS analytical sample is evenly split between men and women, with an average age of about 43 years and more than 80 percent of the respondents being married. The average for years of schooling was 7.6, and respondents on average recalled about 4 of the 10 words in the episodic memory test. Nearly two thirds of the sample was employed, with an average per capita annual family income of 14,490 RMB (about $2,415), more than six times above the new poverty line in rural China (2,300 RMB; see Zhang et al. 2012). About 7.7 percent of the respondents were members of the Communist Party of China (CPC) and/or cadres of various government

Table 1. Frequency Distributions of SRH and Vignette Ratings

Category	CFPS-2012			CHARLS-2011
	SRH (%)	Vignette 1 (%)	Vignette 2 (%)	SRH (%)
Poor	16.4	.0	60.6	26.1
Fair	18.4	4.5	23.6	48.1
Good	34.8	27.3	15.0	16.5
Very good	20.5	40.0	.8	8.7
Excellent	10.0	28.1	.0	.7
n	23,207	23,207	23,207	5,928

Note: Vignette 1 describes a person in better health status compared with vignette 2 by design. CFPS = China Family Panel Studies; CHARLS = China Health and Retirement Longitudinal Study; SRH = self-rated health.

agencies and public institutions, and 13.8 percent had at least one family member who was a CPC member or cadre. In terms of residential and migration status, just over half of the sample consisted of rural nonmigrants or rural-to-rural migrants (hereafter referred to collectively as rural residents); 18.7 percent migrated from rural to urban areas; fewer than 5 percent were urban-to-rural migrants; and about 25 percent were urban nonmigrants or urban-to-urban migrants (hereafter referred to as urban residents).

Our CHARLS analytical sample is, unsurprisingly, older than the CFPS sample but still roughly gender balanced. Being older, the CHARLS sample had on average higher rates of marriage and widowhood, lower educational attainment, worse memory, a slightly higher employment rate but lower income, and lower percentages of cadres or CPC members relative to the CFPS sample. The CHARLS sample had a similar distribution of rural-urban residence and *hukou* status as that in the CFPS sample but a less even distribution across regions.

4.2. *Reporting Heterogeneity*

We assess parallel versus nonparallel cut-point shift by estimating two nested models and performing Wald tests against parallel shift. The results are reported in Table 3. Bear in mind that, generally speaking, lower (downward shift) and higher (upward shift) cut points would deflate and inflate group differentials in health, respectively, without vignette adjustment. Assuming parallel shift, cut points would decline

Table 2. Descriptive Statistics of Independent Variables

Variable	CFPS-2012		CHARLS-2011	
	Mean	*SD*	Mean	*SD*
Age (years)	42.7	14.7	56.5	7.1
Male (%)	49.4	—	47.3	—
Marital status (%)				
Single	14.5	—	.9	—
Married/cohabitation	81.2	—	91.2	—
Divorced/widowed	4.4	—	8.0	—
Years of education	7.6	4.7	5.9	4.4
Episodic memory	4.3	1.9	3.7	1.7
Employed (%)	72.8	—	74.0	—
Family income (RMB)	14,490	24,795	6,048	9,478
Cadre/party member (%)	7.7	—	4.4	—
Had a family member as cadre/party (%)	13.8	—	—	—
Residence and *hukou* status (%)				
Rural resident with rural *hukou*	51.9	—	57.3	—
Rural-to-urban migrant	18.7	—	20.2	—
Rural resident with urban *hukou*	4.6	—	2.1	—
Urban resident with urban *hukou*	24.9	—	20.4	—
Region (%)				
Northeast	14.7	—	7.8	—
Northern coast	12.3	—	14.8	—
Eastern coast	10.9	—	8.8	—
Southern coast	10.1	—	8.2	—
Yellow River middle reach	18.3	—	20.2	—
Yangtze River middle reach	8.8	—	16.8	—
Southwest	13.5	—	19.6	—
Northwest	11.4	—	3.8	—
n	23,207		5,928	

Note: CFPS = China Family Panel Studies; CHARLS = China Health and Retirement Longitudinal Study.

with increases in respondent age ($\beta = -.019$), and the rate of decline would increase with age. Men applied significantly higher cut points ($\beta = .103$) and hence tended to underrate the same level of true health compared with women. Compared with being married or cohabiting, being single was associated with lower cut points. Better educated respondents had higher cut points ($\beta = .009$), whereas those with better episodic memory had lower cut points ($\beta = -.022$). The relationship between family income and cut-point shift was nonlinear in that those in the third quartile tended to have significantly higher cut points compared

Table 3. Coefficient Estimates of Cut-point Shift (Parallel versus Nonparallel) from the Vignette Data: CFPS-2012

| Variable | Parallel Shift | Nonparallel Shift | | | | Wald Test of Parallel Shift ($df = 3$) |
		Poor–Fair	Fair–Good	Good–Very Good	Very Good–Excellent	
Age (mean centered and divided by 10)	−.019***	−.022**	−.054***	.005	−.013	68.960***
Age^2	−.019***	−.035***	−.029***	−.011*	−.002	33.670***
Male (reference: female)	.103***	.135***	.121***	.092***	.056***	15.400**
Marital status (reference: married/cohabitation)						
Single	−.080***	−.036	−.149***	−.103***	−.059	16.660***
Divorced/widowed	.014	.081*	.029	−.014	−.043	7.390
Years of education	.009***	−.010***	−.003	.028***	.023***	260.620***
Episodic memory	−.022***	−.032***	−.039***	−.009*	−.010*	46.020***
Employed (reference: no)	−.006	.012	−.015	−.006	−.021	2.970
Family income quartiles (reference: poorest)						
Second	−.003	−.011	.007	.014	−.022	3.570
Third	.045**	.040*	.072***	.056**	.018	6.090
Fourth (richest)	.031	.030	.027	.052*	.011	3.530
Cadre/party member (reference: no)	−.083***	−.098***	−.101***	−.113***	−.018	9.820*
Family cadre/party member (reference: no)	−.020	−.043*	−.036	−.017	.017	4.620
Residence and *hukou* status (reference: rural)						
Rural-to-urban migrant	.032*	−.048*	.046*	.044*	.103***	42.130***
Rural resident with urban *hukou*	.019	−.041	.029	.034	.074	6.920
Urban resident with urban *hukou*	−.038*	−.158***	−.018	.035	.018	73.600***

Note: Estimates of ancillary parameters are not shown. CFPS = China Family Panel Studies.

*$p < .05$. **$p < .01$. ***$p < .001$.

with the poorest, although the richest also had a significantly higher cut point between good and very good health. Being a cadre or CPC member was related to downward shifted cut points ($\beta = -.083$), although other family members' political status did not matter. Compared with rural residents, rural-to-urban migrants had higher cut points and urban residents had lower ones.

The Wald tests provide statistical evidence in favor of a nonparallel cut-point shift for most of the aforementioned covariates except family income. For example, a higher level of education was associated with an upward cut-point shift at the higher end of health but a downward shift at the lower end when the assumption of parallel shift was relaxed. To gain a better understanding of this complex pattern, Figure 2 plots predicted cut points for five different levels of education and varying migration status, respectively, holding everything else constant. It is clear that better educated respondents tended to apply lower cut points when considering what constitutes poor health. The cut point between poor and fair was roughly -2.43 for college graduates as opposed to -2.27 for those without any schooling. However, the gradient reversed at the high end of health rating: for college graduates and the unschooled, respectively, the cut points between good and very good were approximately $-.2$ and $-.64$, and between very good and excellent, they were $.77$ and $.42$. As a result, for a given level of true health, better educated respondents would be much less likely than respondents with no schooling to report very good or excellent health. With respect to migration status, the pattern is less clear, but two findings stand out. First, urban natives had the lowest cut points at the low end of the health distribution and thus were more likely to report poor or fair health than the other subgroups when they were indeed in poor health. Second, urban natives did not retain a similar high standard at the high end of health rating. Instead, it was the rural-to-urban migrants who held the highest threshold for what constitutes excellent health.

4.3. *Bias Reduction*

To evaluate the performance of vignettes methodology in remedying reporting heterogeneity, we compare group differences in SRH as estimated from three models: (1) a standard ordered probit model, (2) a HOPIT model assuming parallel cut-point shift, and (3) a HOPIT model assuming nonparallel shift. Because of different scaling in these

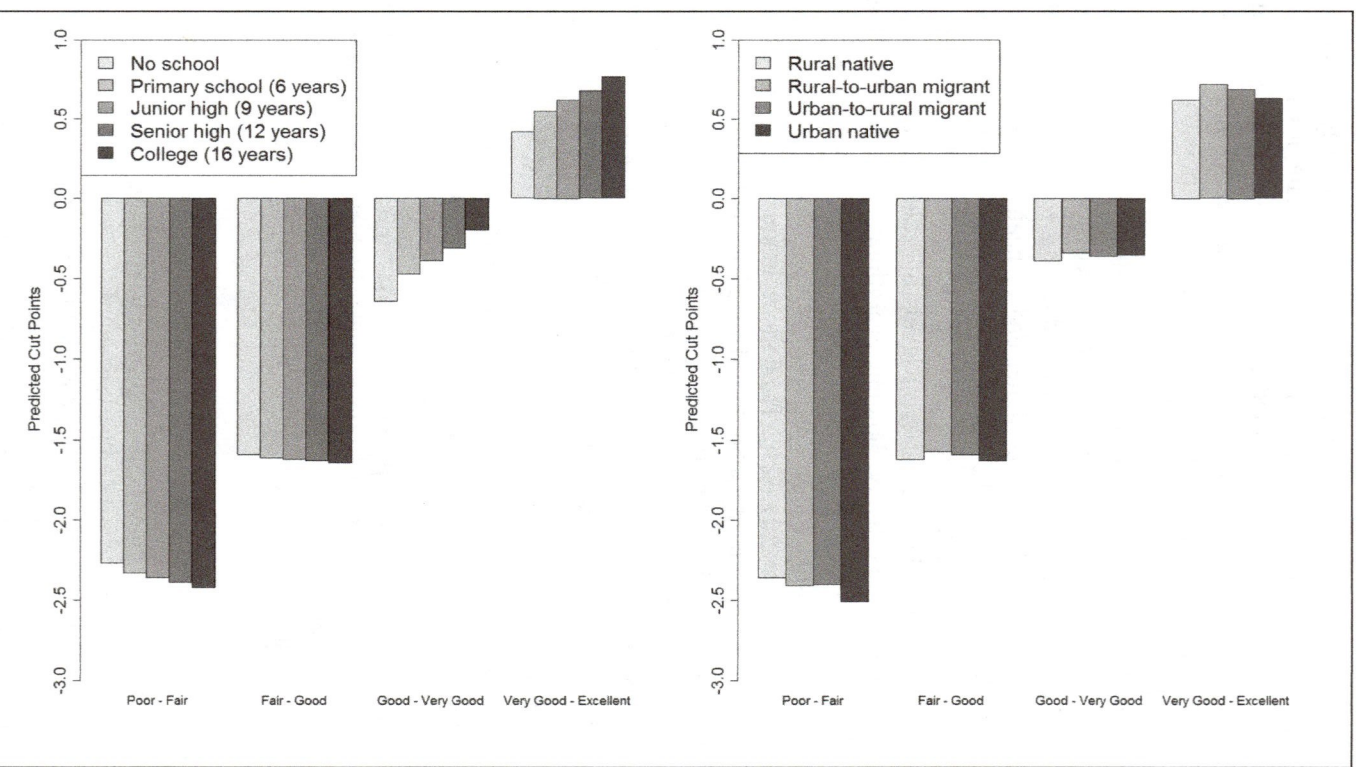

Figure 2. Predicted cut points by levels of education and migration status from the hierarchical ordered probit model assuming nonparallel shift: China Family Panel Studies 2012.

models,[1] we fixed the scale of the HOPIT models by dividing the estimated coefficients by the estimated variance terms, which is equivalent to imposing the same variance as in the ordered probit model (Jones et al. 2007). Table 4 presents the comparable coefficient estimates after rescaling and suggests several related patterns.

First, anchoring vignettes did affect the estimates of health disparities by socioeconomic and demographic groups, as demonstrated by the changes in coefficients between the ordered probit and HOPIT models for every covariate that induced cut-point shift (as shown in Table 3).

Second, the magnitude of some of these changes was substantial. For example, the coefficient for years of education nearly tripled from .004 to .011 after vignette adjustment (assuming nonparallel cut-point shift), whereas the coefficient for the episodic memory dropped by half from .036 to about .017. More strikingly, certain coefficients that were not significant in the ordered probit model became significant in the HOPIT models. For example, none of the coefficients for family income quartiles was significant in the standard ordered probit model. But estimates from both HOPIT models indicated that respondents in the top two quartiles of family income reported significantly better health than those in the bottom quartile. It is also noteworthy that the size of the coefficient associated with family income nearly doubled, from about .03–.04 to .06–.07, after vignette adjustment. For other covariates such as divorce and widowhood, one's own cadre, and CPC membership, significant differences disappeared after vignette adjustment.

Third, the assumption of parallel or nonparallel cut-point shift exerted a limited impact on estimating the SRH component, as evidenced by the very small size and sign changes in the coefficients between the two specifications. Nevertheless, the model specification assuming nonparallel shift revealed a more complex pattern of reporting heterogeneity with respect to many covariates. In practice, we recommend a stepwise and iterative model-building strategy by exploring parallel cut-point shift first and then allowing all the cut points to vary freely. If a statistical test between the two nested models leads us to prefer the parallel shift model, we have a parsimonious model and can easily interpret the results. If the statistical test rejects the parallel shift model, we may still constrain certain cut points to be constant or parallel shift while retaining others to be nonparallel to obtain a more parsimonious model yet with sufficient explanatory power.

Table 4. Coefficient Estimates from Standard Ordered Probit versus HOPIT Models of Self-rated Health: CFPS-2012 and CHARLS-2011

Variable	CFPS-2012 ($N = 23,207$)			CHARLS-2011 ($N = 5,928$)		
		HOPIT			HOPIT	
	Ordered Probit	Parallel Shift	Nonparallel Shift	Ordered Probit	Parallel Shift	Nonparallel Shift
Age (mean centered and divided by 10)	−.279***	−.294***	−.301***	−.252**	−.285**	−.303***
Age^2	.031***	.015**	.015**	.045	.026	.020
Male (reference: female)	.229***	.312***	.308***	.152***	.270***	.298***
Marital status (reference: married/cohabitation)						
Single	−.051*	−.114***	−.095***	−.499**	−.555**	−.557***
Divorced/widowed	−.086*	−.074	−.057	.097	.122*	.162**
Years of education	.004*	.012***	.011***	.010*	.017***	.009*
Episodic memory	.036***	.018***	.017***	.057***	.030**	.021*
Employed (reference: no)	.143***	.138***	.143***	.347***	.313***	.351***
Family income quartiles (poorest)						
Second	.003	.000	.003	.177***	.165***	.172***
Third	.032	.068**	.072**	.254***	.297***	.313***
Fourth (richest)	.037	.061*	.061*	.369***	.402***	.425***
Cadre/party member (reference: no)	.080**	.013	.012	.149*	.061	.065
Family cadre/party member (reference: no)	.066**	.050*	.055*	—	—	—
Residence and *hukou* status (reference: rural)						
Rural-to-urban migrant	.032	.057*	.063**	.151***	.166**	.143***
Rural resident with urban *hukou*	−.018	−.003	.007	.114	.126	.118
Urban resident with urban *hukou*	−.033	−.064**	−.054*	.169***	.108*	.099
σ	—	1.244***	1.231***	—	.914***	.917***

Note: Estimates of ancillary parameters are not shown. All the models control for regional variation. Coefficients in the HOPIT models have been rescaled to be comparable with those in the ordered probit model. CFPS = China Family Panel Studies; CHARLS = China Health and Retirement Longitudinal Study; HOPIT = hierarchical ordered probit.

*$p < .05$. **$p < .01$. ***$p < .001$.

To further gauge the amount of reporting bias reduction achieved by using vignettes, we carried out a simple counterfactual exercise as employed in prior research (Bago d'Uva et al. 2008a). Specifically, we first fixed the latent health status for a reference person,[2] then predicted the probability of reporting very good or excellent health with varying cut points as would be adopted by people with different characteristics while holding everything else constant. We computed the ratio of probabilities (relative probability) with any two different sets of cut points to measure the relative magnitude of the reporting effect. To preserve space, we focused on the effects of education and migration here, as shown in Figure 3. The denominator in the case of education, held constant, is the predicted probability of reporting very good or excellent health when using the cut points of no schooling, while the numerators are calculated in the same way but with cut points shifting from primary schooling to college. The relative probability of reporting very good health dropped from .85 to .62 and for reporting excellent health dropped from .77 to .48 as the associated cut points shifted from those of primary school to those of college education. This means that, given the same latent health for any respondent, the probability of giving an excellent health self-rating with the cut points of college education imposed would be less than half the probability if applying the cut points of no schooling (the denominator of the relative probability). For migration status, the denominator refers to rural natives. The gradient is more evident for the rating of excellent health than for that of very good health. The relative probability of reporting excellent health decreases from .97 to .82 when we replace the corresponding cut points of urban natives with those of rural-to-urban migrants.

In light of China's long-standing rural-urban divide and likely rural-urban difference in health-related reporting patterns, we analyzed the rural and urban CFPS-2012 respondents separately. The main results are reported in Table 5. Three findings are worth highlighting. First, vignette adjustment was more effective in uncovering the education gradient in health in the rural subsample, as the unadjusted coefficient of schooling was not statistically significant. Second, variation in SRH by episodic memory was driven mainly by reporting heterogeneity among urban residents, as the coefficient of episodic memory became insignificant after vignette adjustment. Third, the income gradient in SRH seems to be mainly a rural phenomenon.

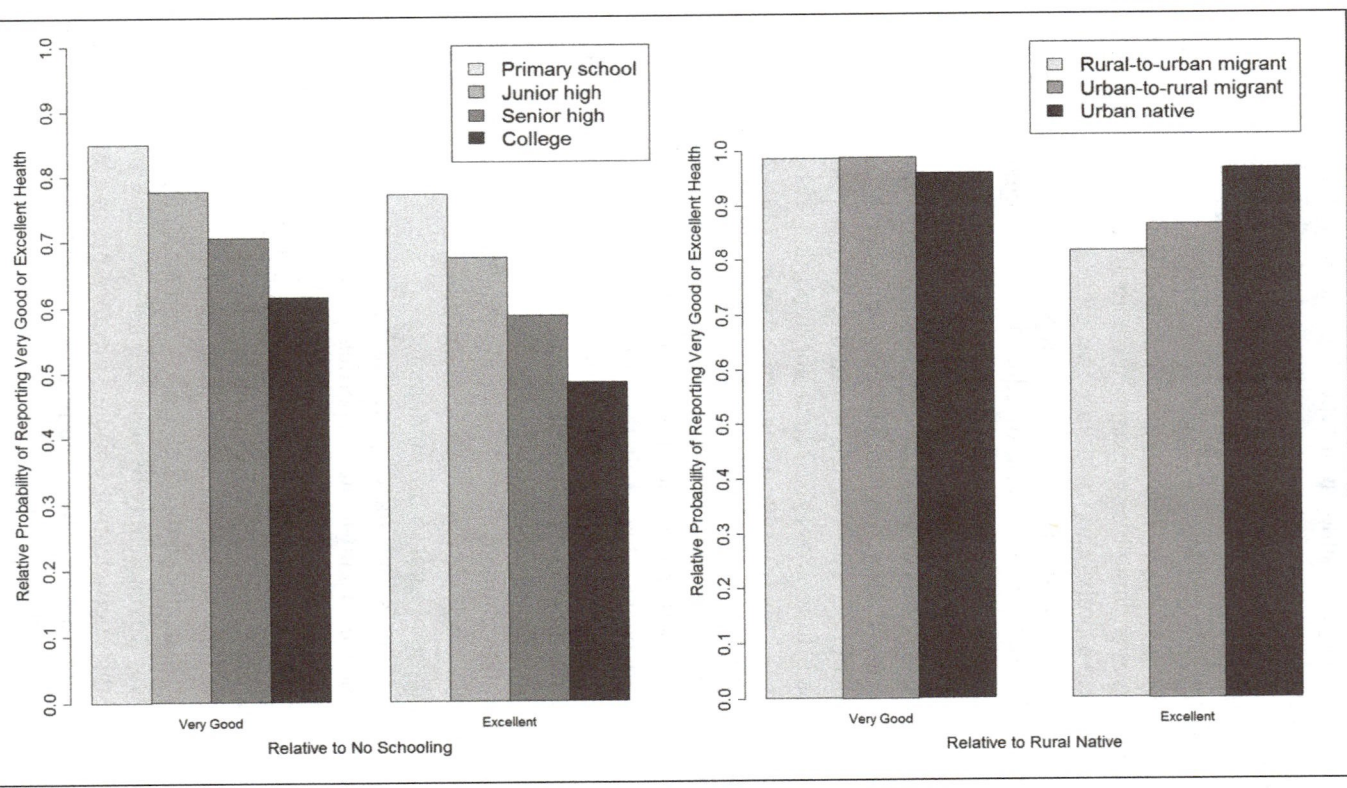

Figure 3. Relative probabilities of reporting very good or excellent health for a reference person's (see note 2) health with varying cut points by levels of education and migration status: China Family Panel Studies 2012.

Table 5. Rural-urban Stratified Coefficient Estimates from Standard Ordered Probit versus HOPIT Models of Self-rated Health: CFPS-2012

| | Rural Sample ($N = 16{,}002$) | | | Urban Sample ($N = 7{,}205$) | | |
| | | HOPIT | | | HOPIT | |
Variable	Ordered Probit	Parallel Shift	Nonparallel Shift	Ordered Probit	Parallel Shift	Nonparallel Shift
Age (mean centered and divided by 10)	−.285***	−.304***	−.312***	−.261***	−.261***	−.265***
Age^2	.022***	.004	.005	.057***	.048***	.042***
Male (reference: female)	.238***	.310***	.306***	.204***	.303***	.296***
Marital status (ref: married/cohabit)						
Single	−.034	−.054	−.034	−.072	−.218***	−.199**
Divorced/widowed	−.065	−.046	−.033	−.102	−.096	−.067
Years of education	.003	.011***	.011***	.009*	.013**	.010*
Episodic memory	.032***	.020**	.019**	.046***	.017	.015
Employed (reference: no)	.164***	.163***	.157***	.165***	.206***	.210***
Family income quartiles (poorest)						
Second	−.005	−.004	.000	.024	−.045	−.041
Third	.017	.068*	.073**	.064	.000	.008
Fourth (richest)	.007	.044	.044	.091*	.035	.038
Cadre/party membership (reference: no)	.098*	.001	.000	.049	.008	.007
Family cadre/party member (reference: no)	.041	.032	.039	.096**	.078*	.076*
Urban hukou (reference: rural *hukou*)	−.020	−.033	−.026	−.083*	−.117**	−.101*
σ	—	1.284***	1.271***	—	1.154***	1.157***

Note: Estimates of ancillary parameters are not shown. All the models control for regional variation. Coefficients in the HOPIT models have been rescaled to be comparable with those in the ordered probit model. CFPS = China Family Panel Studies; HOPIT = hierarchical ordered probit.

*$p < .05$. **$p < .01$. ***$p < .001$.

5. COST-EFFECTIVENESS ANALYSIS

5.1. *Administering Vignettes to Subsample*

Identification of group-level reporting heterogeneity rests on the assumption of significant within-group similarity, or group-specific reporting patterns, apart from additional within-group individual variation. This assumption implies that group-specific cut points estimated from a random subsample (or even an external sample from the same population) can be applied to the full sample. This approach has significant practical implications, as it substantially reduces survey costs and respondent burden. We therefore proceed to perform cross-validation to assess the degree to which vignettes administered to a small subsample can assist in bias reduction for the entire sample. Unlike other large-scale social surveys, the CFPS vignettes data were collected on the same respondents who were administered self-assessments, producing a large sample that ensured enough statistical power for cross-validation. Specifically, we randomly partition the full sample into a relatively small subsample as training data and the remaining larger subsample as validation data. We experiment with a series of partitions, including 10 percent, 20 percent, 30 percent, 40 percent, and 50 percent, for each of which we repeat the random partition 500 times. After each partition, we fit a HOPIT model to the training data and compute out-of-sample predictive cut points and latent health status. We then fit another HOPIT model to the larger subsample validation data and compute in-sample predictive cut points and latent health status. Closeness between the two sets of predictive values, measured by mean squared error, indicates external validity and thereby the cost-effectiveness of extrapolating vignette adjustment obtained from a small subsample to the full sample.

Our cross-validation analyses suggest that a surprisingly small sample would be sufficient to make reasonably good extrapolation of adjustment for reporting heterogeneity. As shown in Figure 4, when using both vignettes available in the CFPS, the mean squared errors between the out-of-sample predictions on the basis of the training subsample and the in-sample predictions on the basis of the validation subsample take the form of exponential decay as the size of training data increases. The decay rate is greater at the lower end, with the largest decline in mean squared errors occurring as the proportion of the full sample used as training data increases from 10 percent to 20 percent. The trend of

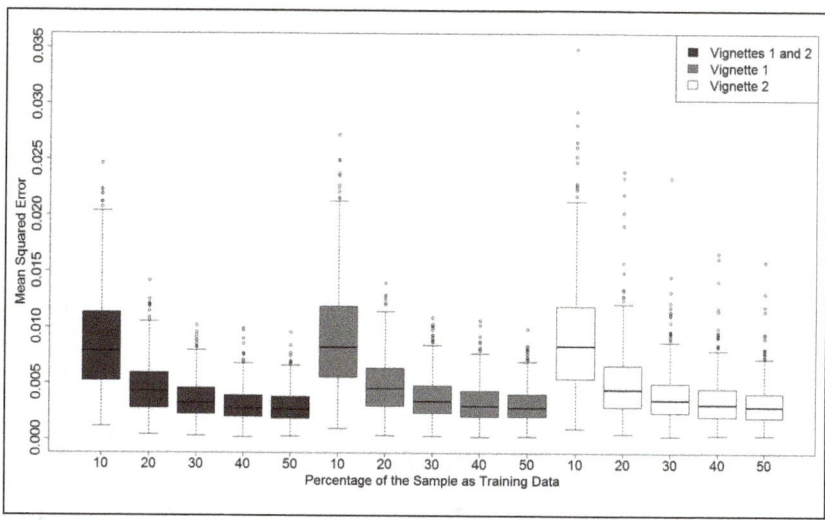

Figure 4. Mean squared error of predicted latent health from cross-validation of hierarchical ordered probit models by randomly selecting a subset of the China Family Panel Studies 2012 sample as training data.
Note: Vignette 1 describes a healthier person than vignette 2.

decline flattens out beyond 30 percent. As shown in Figure 5, the same pattern holds for the mean squared errors between the out-of-sample predictions of cut points on the basis of the training subsample and the in-sample predictions on the basis of the validation subsample.

5.2. Number of Vignettes

In principle, one vignette is sufficient for identifying group-level differences, provided it yields sufficient variation in the vignette ratings, or full support, which enables estimation of the full range of cut points. Adding more vignettes would then improve the estimation efficiency. As shown in Table 1, however, the assumption of full support is not satisfied in the CFPS data, because neither the first nor the second vignette yielded responses in all categories; that is, the rating of poor for the first vignette and excellent for the second vignette received zero responses. Therefore, we expect that using both vignettes complementally is the best solution in this particular scenario.

To demonstrate this, we repeat the HOPIT model estimation by using one vignette at a time to ascertain whether it can attain similar bias

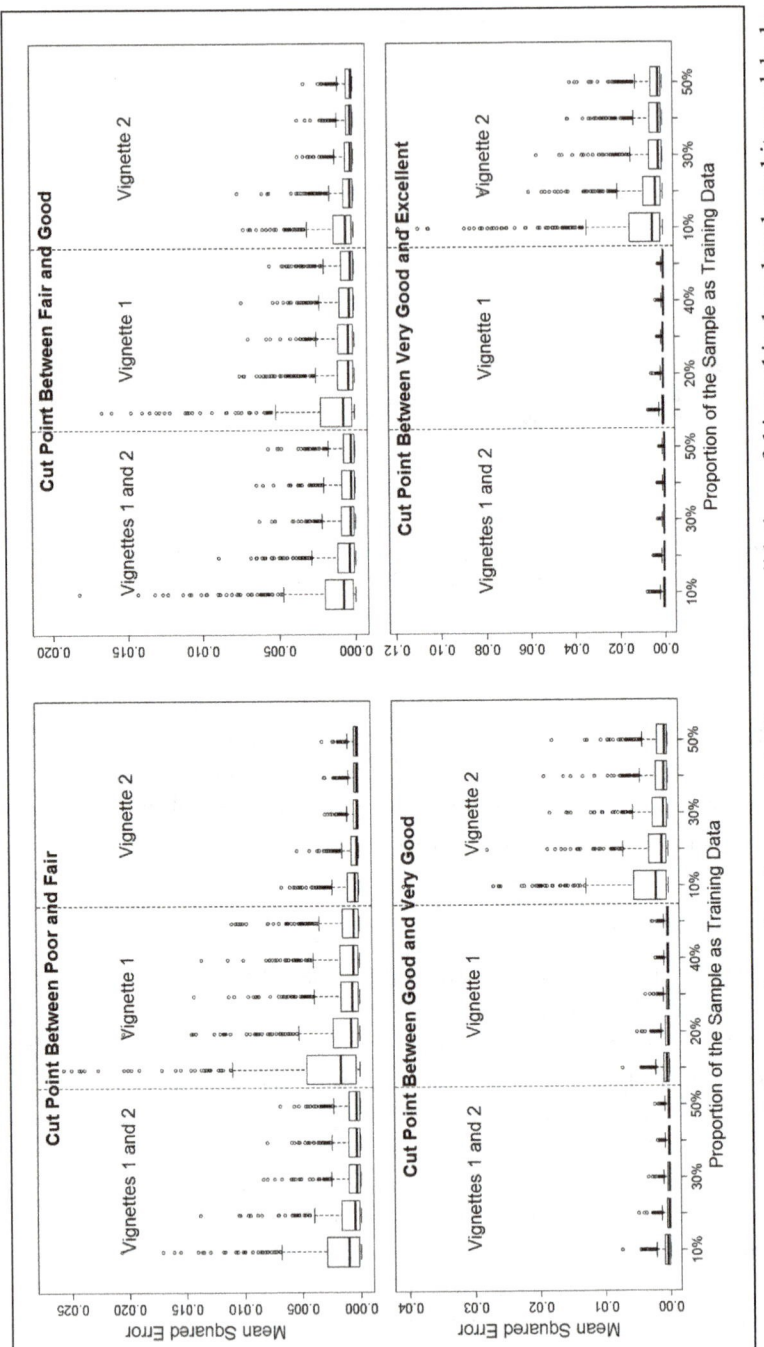

Figure 5. Mean squared error of predicted cut points from cross-validation of hierarchical ordered probit models by randomly selecting a subset of the China Family Panel Studies 2012 sample as training data.

Note: Vignette 1 describes a healthier person than vignette 2.

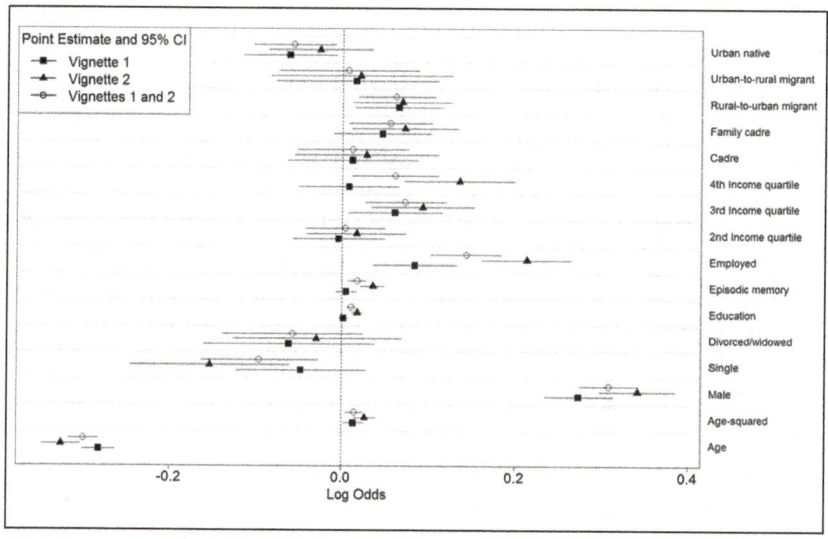

Figure 6. Comparisons of coefficient estimates for the health component of hierarchical ordered probit models when using different vignettes: China Family Panel Studies 2012.

Note: Vignette 1 describes a healthier person than vignette 2. CI = confidence interval.

reduction when using two vignettes. We not only compare coefficient estimates but also examine whether different vignettes lead to similar adjusted SRH (Voňková and Hullegie 2011).

First, we compare coefficient estimates of the associations of covariates with SRH anchored by using different vignettes. Figure 6 plots the point estimates and the associated 95 percent confidence intervals. It is notable that the point estimates when using the second (worse health) vignette are generally bigger in terms of absolute values than the point estimates when using the first (better health) vignette, while the coefficient sizes when using both vignettes fall in between, reflecting a result of smoothing. In most cases, the 95 percent confidence intervals are overlapped for the same covariate, indicating insufficient statistical power to distinguish estimates using different vignettes. However, substantive variation does occur with certain important SES indicators. For example, the 95 percent confidence interval for education covers zero when using the first vignette only but not so when using either the second only or both vignettes. Similar patterns can be observed for episodic memory, top income quartile, and family members' cadre or party membership. To the extent that we expect significant SES disparities in health, it is likely that the second (worse health) vignette is relatively more effective than the first vignette in anchoring reporting behaviors.

Second, we compute pairwise correlation coefficients among the three sets of vignette-anchored SRH data (two sets using one vignette only, and the third set using both vignettes). A correlation coefficient close to 1 indicates a similar adjustment when using different sets. As shown in Table 6, correlation coefficients range between .97 and .99 for predicted latent health and between .87 and .92 for predicted SRH when different vignettes are used. These large positive correlation coefficients lend further support to the robustness of vignette adjustment in our CFPS design, which is not achieved in other surveys (Voňková and Hullegie 2011).

Third, we repeat the above cross-validation procedure using one vignette only to determine whether it is valid to extrapolate a subsample anchoring to the full sample by using a single vignette. As shown in Figure 4, the same pattern of exponential decay in mean squared errors for predicted latent health when using both vignettes holds for using either one of the two vignettes. The mean squared errors experience a substantial decline when the proportion of the full sample used as training data increases from 10 percent to 20 percent, and the decline trend levels off beyond 30 percent. Similar results are retained for mean squared errors related to cut points as plotted in Figure 5. It is worth noting that the mean squared errors for the cut points are greater at the lower end (poor vs. fair and fair vs. good) when using the first vignette, but greater at the higher end (good vs. very good and very good vs. excellent) when using the second vignette. This is not surprising given the first vignette's description of relatively good health, which should provide greater differential power toward the higher end, and the second vignette's description of relatively worse health, which should engender better anchors at the lower end.

5.3. *External Validation in CHARLS*

A working assumption that motivated the CFPS vignette design and the present study is that intergroup reporting heterogeneity is more or less stable. We now generalize the estimated cut-point shifts from the CFPS sample to the CHARLS sample by applying group-specific cut points estimated from the CFPS vignettes to the CHARLS respondents. We

Table 6. Correlation Coefficients among Predicted Self-rated Health Anchored by Different Vignettes: China Family Panel Studies 2012

	Latent Health Scale			Predicted Self-rated Health Based on Modal Item Response Probability		
	Vignette 1	Vignette 2	Vignettes 1 and 2	Vignette 1	Vignette 2	Vignettes 1 and 2
Vignette 1	1	—	—	1	—	—
Vignette 2	.970	1	—	.920	1	—
Vignettes 1 and 2	.990	.990	1	.870	.880	1

Note: Vignette 1 describes a healthier person than vignette 2.

again compare estimated disparities in SRH before and after vignette adjustment, assuming parallel and nonparallel cut-point shifts, respectively (see the right panel in Table 4). The effect of vignette adjustment is evident in several coefficient estimates.

First, the coefficient of education increased by 70 percent from .01 in the ordered probit model to .017 in the parallel shift model, but it dropped back to .009 in the nonparallel shift model, reflecting the complicated bidirectional reporting heterogeneity by education. As was shown in Figure 2, cut-point shift by education was nonparallel not only in magnitude but also in direction. It seems that well-educated people tend to invoke health optimism and consider themselves in fair health—the downward shift of the cut point between poor and fair—when they are indeed in poor health. But when they are indeed in very good or excellent health, they tend to use peer comparison and higher standards—the upward shift of the cut points between good and very good and between very good and excellent—and thus underrate their true health status.

Second, the coefficient of episodic memory was halved, while the coefficient of gender was doubled after adjusting for cut-point shift, either parallel or nonparallel. The income gradient in SRH also increased, albeit to a smaller extent because of the relatively weak impact of income on reporting behaviors. Similarly, the coefficient of employment status remained relatively stable because of no significant cut-point shift by employment (see again Table 3).

Third, being a cadre or CPC member was associated with significantly better SRH when reporting heterogeneity was ignored. After correcting for the negative cut-point shift by cadre and CPC membership, again either parallel or nonparallel, this association was no longer statistically significant. Similarly, the significant difference between rural and urban natives in the ordered probit model was considerably reduced (parallel shift model) or even disappeared (nonparallel shift model) after adjusting for urban natives' lower thresholds in rating their health status. In contrast, largely driven by a higher threshold between poor and fair health, the divorced or widowed respondents turned out to have better SRH compared with those who were married.

6. DISCUSSION

Using vignettes to anchor survey responses is not a new idea. The history of vignette methodology can date back to at least the 1970s, when sociologists employed vignettes to measure social status (Nosanchuk 1972) and racial attitudes (Farley et al. 1978). The statistical methods for analyzing vignette data were developed more than 10 years ago (King et al. 2004; Tandon et al. 2003), and abundant research efforts have been devoted to understanding and refining the assumptions related to model identification (Angelini, Cavapozzi, and Paccagnella 2011; Bago d'Uva et al. 2011b; Kapteyn et al. 2011; Paccagnella 2011; Peracchi and Rossetti 2013). However, the challenge of how to design feasible vignettes and implement them effectively in large-scale general-population surveys remains largely unsettled.

In their study using the SHARE data across a dozen European countries, Voňková and Hullegie (2011) found that the effectiveness of the vignette method varied significantly by both the choice of health domain (particularly problematic for cognition and breathing but less so for mobility) and the choice of vignette. This finding is worrisome because the same health domains and the corresponding vignettes have been adopted in, among others, WHO-WHS, WHO-MCSS, HRS, and CHARLS without any modification other than literal language translation, and thus the same sensitivity issues may have been widespread if not exacerbated. Furthermore, it is usually impractical to collect data on multiple health domains in general-purpose surveys in which financial resources and interview time for measuring health are highly constrained and self-rated general health is likely a primary alternative. It is against this background that we have conducted this study as a contribution to survey methodology, through designing simple anchoring vignettes for self-rated general health in a general-purpose survey (the CFPS) and providing an evaluation of the usefulness of these simple vignettes in correcting for reporting heterogeneity bias as well as their external applicability to a different survey (such as the CHARLS).

Capitalizing on the new vignettes designed for the CFPS-2012 sample, we reach two significant conclusions in this study. First, reporting heterogeneity plays a significant role in biasing the measurement of health disparities among Chinese adults. In fact, our empirical findings suggest that reporting heterogeneity appears to be a common rather than exceptional phenomenon in SRH because most of the socioeconomic and demographic characteristics examined here induce cut-point shifts, either parallel or nonparallel. Second, anchoring vignettes appear to be a cost-effective method of ameliorating the effects of reporting bias in surveys of SRH.

We quantified the consequential effect of reporting bias in SRH, revealing that coefficients could be underestimated or overestimated by twice as much as those without adjustment (e.g., education and episodic memory), depending on whether the cut points are shifted upward or downward. Moreover, the significance levels changed for other covariates (e.g., political capital and residential and migration status) after adjustment.

We also quantified the magnitude of reporting heterogeneity and found that the probability of reporting excellent health when applying the cut points of college education is less than half of that when applying the cut points of no schooling while fixing the level of latent health. This result is in marked contrast to previous research that reported less than a 10 percent difference (Bago d'Uva et al. 2008b).

Our analyses reveal three significant features of vignettes methodology. First, adjustment for reporting heterogeneity in the full sample can be achieved by extrapolating anchoring points from a relatively small subsample. In the CFSP data, administering vignettes to about 20 percent to 30 percent of the full sample was as effective as adding more cases. Second, using a single vignette can provide some anchoring that is comparable with using more vignettes. However, in a sample such as the CFSP that has a large age range, and hence great health differentials, a vignette that describes a relatively poor health scenario may lend more discriminant power to the lower end of the health spectrum, at which the most striking gap occurs, compared with a vignette that describes a relatively good health scenario. Third, collecting vignette data in a large-scale nationally representative sample can benefit other surveys of the same population, as we have demonstrated that the cut-point shifts estimated from the CFPS-2012 vignette data can be effectively applied to correct for reporting heterogeneity in the CHARLS-2011 sample.

Taken together, our findings have important implications for future research and public health policy. Given that measures of SRH have strong predictive power for objective health status and low data collection costs, they are likely to remain in use for research on health disparities in developing countries such as China. On the other hand, the rapid social changes and the associated rising socioeconomic inequalities and social stratification in transition societies will increasingly complicate the pattern of health disparities. Reporting heterogeneity in health surveys may become more substantial as people of different social groups continue to diverge in their choice of reference group and the criteria they apply to gauge good versus poor health. If adjustment techniques to account for such heterogeneity, such as anchoring vignettes, become common practice, our research will yield better estimates of health disparities and provide higher quality information for policymakers.

7. LIMITATIONS AND FUTURE RESEARCH

Our study has several limitations that merit future research. First, the vignette equivalence assumption may not hold in reality. For example, high-SES respondents may value mental health as much as physical health, whereas low-SES respondents may not. Also, given the complex multidimensional nature of health, vignette descriptions are likely to be incomplete, and respondents may call upon their own experience to impute the missing information (van Soest et al. 2011). Similarly, the response consistency assumption may be violated when respondents report their own situation with a certain strategic consideration that is absent from vignette assessment (Bago d'Uva et al. 2011b). A prominent example is that respondents from welfare-state countries tend to apply lower thresholds when assessing their own disability status than they do when evaluating the vignettes because of the economic incentive to exaggerate personal health problems for disability benefits eligibility (Gupta, Kristensen, and Pozzoli 2010). Although it is hard to contemplate such strategic behavior in China given that social welfare and health insurance benefits are largely contingent on social institutions (e.g., the household registration system) and collective entities (e.g., work units) rather than an individual's self-rating, we should still consider the possibility of the invalid response consistency assumption for other reasons.

Rigorous tests of these assumptions require extra data such as valid and reliable objective health measures, which are often available only in *ad hoc* studies. The present study is merely a first step toward a better understanding of the effects of reporting heterogeneity and the utility of anchoring vignettes in survey data on the socioeconomic and demographic disparities in SRH. Nevertheless, we find that even with short vignettes that do not attempt to incorporate particular aspects of health or age-specific health conditions, this method is useful in detecting reporting heterogeneity by SES and demographic characteristics and enabling appropriate anchoring to identify true health disparities.

We also find that vignette data collected in one population-based sample can be used to anchor reporting behaviors in a different sample of the same population or subpopulation. These methodological findings suggest that researchers may do well in designing their own anchoring vignettes, however simple they may seem, to fit a specific context or population, instead of merely borrowing standard ones that are context blind. Future research is needed to improve vignette design while retaining its simplicity and cost-effectiveness with respect to survey operation and anchoring performance, especially with general-purpose surveys in which resources are highly limited.

Acknowledgments

Earlier versions of this article were presented at the methodology seminar at the Quantitative Methodology Program of the Survey Research Center at the University of Michigan and the 2014 conference on Chinese sociological and demographic methodology in Guangzhou, China. We thank seminar participants, conference discussant Jun Li, and session participants for useful comments.

Funding

The author(s) disclosed receipt of the following financial support for the research, authorship, and/or publication of this article: This study was supported by the National Institutes of Health under an investigator grant (R01-HD074603) to Yu Xie and a center grant to the Population Studies Center at the University of Michigan (R24 HD041028).

Notes

1. The scale in the standard ordered probit model is normalized to 1 (i.e., the error term is assumed to follow a standard normal distribution), while it is estimated in HOPIT models (i.e., σ^2 in equation 4).
2. The reference person is a married man of the sample average age, with 9 years of schooling (junior high school) and an episodic memory test score of 4 (rounded up to the sample mean), employed as a rural nonmigrant, and living in the poorest family income quartile.

References

Angelini, Viola, Danilo Cavapozzi, and Omar Paccagnella. 2011. "Dynamics of Reporting Work Disability in Europe." *Journal of the Royal Statistical Society, Series A: Statistics in Society* 174(3):621–38.

Bago d'Uva, Teresa, Maarten Lindeboom, Owen O'Donnell, and Eddy van Doorslaer. 2011a. "Education-related Inequity in Healthcare with Heterogeneous Reporting of Health." *Journal of the Royal Statistical Society, Series A: Statistics in Society* 174(3):639–64.

Bago d'Uva, Teresa, Maarten Lindeboom, Owen O'Donnell, and Eddy van Doorslaer. 2011b. "Slipping Anchor? Testing the Vignettes Approach to Identification and Correction of Reporting Heterogeneity." *Journal of Human Resources* 46(4): 875–906.

Bago d'Uva, Teresa, Owen O'Donnell, and Eddy van Doorslaer. 2008a. "Differential Health Reporting by Education Level and Its Impact on the Measurement of Health Inequalities among Older Europeans." *International Journal of Epidemiology* 37(6): 1375–83.

Bago d'Uva, Teresa, Eddy van Doorslaer, Maarten Lindeboom, and Owen O'Donnell. 2008b. "Does Reporting Heterogeneity Bias the Measurement of Health Disparities?" *Health Economics* 17(3):351–75.

Benjamins, Maureen Reindl, Robert A. Hummer, Isaac W. Eberstein, and Charles B. Nam. 2004. "Self-reported Health and Adult Mortality Risk: An Analysis of Cause-specific Mortality." *Social Science and Medicine* 59(6):1297–1306.

Chen, Feinian, Yang Yang, and Guangya Liu. 2010. "Social Change and Socioeconomic Disparities in Health over the Life Course in China: A Cohort Analysis." *American Sociological Review* 75(1):126–50.

Dowd, Jennifer Beam, and Megan Todd. 2011. "Does Self-reported Health Bias the Measurement of Health Inequalities in U.S. Adults? Evidence Using Anchoring Vignettes from the Health and Retirement Study." *Journals of Gerontology Series B: Psychological Sciences and Social Sciences* 66B(4):478–89.

Farley, Reynolds, Howard Schuman, Suzanne Bianchi, Diane Colasanto, and Shirley Hatchett. 1978. "'Chocolate City, Vanilla Suburbs': Will the Trend toward Racially Separate Communities Continue?" *Social Science Research* 7(4):319–44.

Ferraro, Kenneth F.1980. "Self-ratings of Health among the Old and the Old-old." *Journal of Health and Social Behavior* 21(4):377–83.

Ferraro, Kenneth F.1993. "Are Black Older Adults Health-pessimistic?" *Journal of Health and Social Behavior* 34(3):201–14.

Goldberg, P., A. Guéguen, A. Schmaus, J.-P. Nakache, and M. Goldberg. 2001. "Longitudinal Study of Associations between Perceived Health Status and Self Reported Diseases in the French Gazel Cohort." *Journal of Epidemiology and Community Health* 55(4):233–38.

Grol-Prokopczyk, Hanna, Jeremy Freese, and Robert M. Hauser. 2011. "Using Anchoring Vignettes to Assess Group Differences in General Self-rated Health." *Journal of Health and Social Behavior* 52(2):246–61.

Gupta, Nabanita Datta, Nicolai Kristensen, and Dario Pozzoli. 2010. "External Validation of the Use of Vignettes in Cross-country Health Studies." *Economic Modelling* 27(4):854–65.

Harzing, Anne-Wil. 2006. "Response Styles in Cross-national Survey Research: A 26-country Study." *International Journal of Cross Cultural Management* 6(2):243–66.

Herrnstein, Richard J., and Charles Murray. 1996. *The Bell Curve: Intelligence and Class Structure in American Life*. New York: Free Press.

House, James S., James M. Lepkowski, David R. Williams, Richard P. Mero, Paula M. Lantz, Stephanie A. Robert, and Jieming Chen. 2000. "Excess Mortality among Urban Residents: How Much, for Whom, and Why?" *American Journal of Public Health* 90(12):1898–1904.

Hu, Yuqing, Xiaoyan Lei, James P. Smith, and Yaohui Zhao. 2012. "Effects of Social Activities on Cognitive Functions: Evidence from CHARLS." Pp. 279–306 in *National Research Council (US) Panel on Policy Research and Data Needs to Meet the Challenge of Aging in Asia*, edited by J. P. Smith and M. Majmundar. Washington, DC: National Academies Press.

Huisman, Martijn, Anton E. Kunst, and Johan P. Mackenbach. 2003. "Socioeconomic Inequalities in Morbidity among the Elderly: A European Overview." *Social Science and Medicine* 57(5):861–73.

Idler, Ellen L., and Yael Benyamini. 1997. "Self-rated Health and Mortality: A Review of Twenty-seven Community Studies." *Journal of Health and Social Behavior* 38(1): 21–37.

Jones, Andrew M., Nigel Rice, Teresa Bago d'Uva, and Silvia Balia. 2007. *Applied Health Economics*. New York: Routledge.

Jürges, Hendrik. 2007. "True Health vs Response Styles: Exploring Cross-country Differences in Self-reported Health." *Health Economics* 16(2):163–78.

Kakwani, Nanak, Adam Wagstaff, and Eddy van Doorslaer. 1997. "Socioeconomic Inequalities in Health: Measurement, Computation, and Statistical Inference." *Journal of Econometrics* 77(1):87–103.

Kapteyn, Arie, James P. Smith, and Arthur van Soest. 2007. "Vignettes and Self-reports of Work Disability in the United States and the Netherlands." *American Economic Review* 97(1):461–73.

Kapteyn, Arie, James P. Smith, Arthur van Soest, and Hana Vonkova. 2011. "Anchoring Vignettes and Response Consistency." RAND Working Paper Series No. WR-840. Santa Monica, CA: RAND.

King, Gary, Christopher J. L. Murray, Joshua A. Salomon, and Ajay Tandon. 2004. "Enhancing the Validity and Cross-cultural Comparability of Measurement in Survey Research." *American Political Science Review* 98(1):191–207.

Knesebeck, Olaf von dem, Günther Lüschen, William C. Cockerham, and Johannes Siegrist. 2003. "Socioeconomic Status and Health among the Aged in the United States and Germany: A Comparative Cross-sectional Study." *Social Science and Medicine* 57(9):1643–52.

Luo, Ye, and Ming Wen. 2002. "Can We Afford Better Health? A Study of the Health Differentials in China." *Health: An Interdisciplinary Journal for the Social Study of Health, Illness and Medicine* 6(4):471–500.

Marks, Gary N. 2013. *Education, Social Background and Cognitive Functioning: The Decline of the Social*. New York: Routledge.

McArdle, John J., James P. Smith, and Robert J. Willis. 2011. "Cognition and Economic Outcomes in the Health and Retirement Survey." Pp. 209–36 in *Explorations in the Economics of Aging*, edited by D. A. Wise. Chicago: University of Chicago Press.

Mirowsky, John, and Catherine E. Ross. 2008. "Education and Self-rated Health: Cumulative Advantage and Its Rising Importance." *Research on Aging* 30(1): 93–122.

Murray, Christopher J. L., Emre Özaltin, Ajay Tandon, Joshua A. Salomon, and Somnath Chatterji. 2003. "Empirical Evaluation of the Anchoring Vignette Approach in Health Surveys." Pp. 369–99 in *Health Systems Performance*

Assessment: Debates, Methods and Empiricism, edited by C. J. L. Murrary and D. B. Evans. Geneva, Switzerland: World Health Organization.

Nosanchuk, T.A.1972. "The Vignette as an Experimental Approach to the Study of Social Status: An Exploratory Study." *Social Science Research* 1(1):107–20.

Paccagnella, Omar. 2011. "Anchoring Vignettes with Sample Selection Due to Non-response." *Journal of the Royal Statistical Society, Series A: Statistics in Society* 174(3):665–87.

Pei, X., and E. Rodriguez. 2006. "Provincial Income Inequality and Self-reported Health Status in China during 1991–97." *Journal of Epidemiology and Community Health* 60(12):1065–69.

Peracchi, Franco, and Claudio Rossetti. 2013. "The Heterogeneous Thresholds Ordered Response Model: Identification and Inference." *Journal of the Royal Statistical Society, Series A: Statistics in Society* 176(3):703–22.

Ross, Catherine E., and Chia-ling Wu. 1995. "The Links between Education and Health." *American Sociological Review* 60(5):719–45.

Salomon, Joshua A., Ajay Tandon, and Christopher J. L. Murrary, and World Health Survey Pilot Study Collaborating Group. 2004. "Comparability of Self Rated Health: Cross-sectional Multi-country Survey Using Anchoring Vignettes." *BMJ* 328:258.

Schnittker, Jason. 2005. "When Mental Health Becomes Health: Age and the Shifting Meaning of Self-evaluations of General Health." *Milbank Quarterly* 83(3):397–423.

Singh-Manoux, Archana, Pekka Martikainen, Jane Ferrie, Marie Zins, Michael Marmot, and Marcel Goldberg. 2006. "What Does Self Rated Health Measure? Results from the British Whitehall II and French Gazel Cohort Studies." *Journal of Epidemiology and Community Health* 60(4):364–72.

Tandon, Ajay, Christopher J. L. Murray, Joshua A Salomon, and Gary King. 2003. "Statistical Models for Enhancing Cross-population Comparability." Pp. 727–46 in *Health Systems Performance Assessment: Debates, Methods and Empiricism*, edited by C.J.L. Murrary and D. B. Evans. Geneva, Switzerland: World Health Organization.

Tandon, Ajay, Juzhong Zhuang, and Somnath Chatterji. 2006. "Inclusiveness of Economic Growth in the People's Republic of China: What Do Population Health Outcomes Tell Us?" *Asian Development Review* 23(2):53–69.

van Soest, Arthur, Liam Delaney, Colm Harmon, Arie Kapteyn, and James P. Smith. 2011. "Validating the Use of Anchoring Vignettes for the Correction of Response Scale Differences in Subjective Questions." *Journal of the Royal Statistical Society, Series A: Statistics in Society* 174(3):575–95.

Voňková, Hana, and Patrick Hullegie. 2011. "Is the Anchoring Vignette Method Sensitive to the Domain and Choice of the Vignette?" *Journal of the Royal Statistical Society, Series A: Statistics in Society* 174(3):597–620.

Wen, Ming, Jessie Fan, Lei Jin, and Guixin Wang. 2010. "Neighborhood Effects on Health among Migrants and Natives in Shanghai, China." *Health and Place* 16(3): 452–60.

Whyte, Martin King, and Zhongxin Sun. 2010. "The Impact of China's Market Reforms on the Health of Chinese Citizens: Examining Two Puzzles." *China: An International Journal* 8(1):1–32.

Willson, Andrea E., Kim M. Shuey, and Glen H. Elder. 2007. "Cumulative Advantage Processes as Mechanisms of Inequality in Life Course Health." *American Journal of Sociology* 112(6):1886–1924.

Xie, Yu. 2012. *The User's Manual of the China Family Panel Studies (2010)*. Beijing, China: Institute of Social Science Survey, Peking University.

Xie, Yu. 2013. "Population Heterogeneity and Causal Inference." *Proceedings of the National Academy of Sciences* 110(16):6262–68.

Xie, Yu, and Jingwei Hu. 2014. "An Introduction to the China Family Panel Studies (CFPS)." *Chinese Sociological Review* 47(1):3–29.

Zhang, Chunni, Qi Xu, Xiang Zhou, Xiaobo Zhang, and Yu Xie. 2012. "Comparing Poverty Rates from CFPS, CGSS, CHIP, and CHFS." 19th Report in the Series Technical Reports of the China Family Panel Studies, edited by Y. Xie. Beijing, China: Institute for Social Science Survey, Peking University.

Zhao, Yaohui, Yisong Hu, James P. Smith, John Strauss, and Gonghuan Yang. 2014. "Cohort Profile: The China Health and Retirement Longitudinal Study (CHARLS)." *International Journal of Epidemiology* 43(1):61–68.

Zimmer, Zachary, and Pattama Amornsirisomboon. 2001. "Socioeconomic Status and Health among Older Adults in Thailand: An Examination Using Multiple Indicators." *Social Science and Medicine* 52(8):1297–1311.

Zimmer, Zachary, and Julia Kwong. 2004. "Socioeconomic Status and Health among Older Adults in Rural and Urban China." *Journal of Aging and Health* 16(1):44–70.

Zimmer, Zachary, Josephina Nativida, Hui-Sheng Lin, and Napaporn Chayovan. 2000. "A Cross-national Examination of the Determinants of Self-assessed Health." *Journal of Health and Social Behavior* 41(4):465–81.

Author Biographies

Hongwei Xu is a research assistant professor at the Survey Research Center of the Institute for Social Research, University of Michigan. His substantive research areas are focused on health inequalities, child development, and residential segregation. His methodological interests include hierarchical modeling of spatial, multilevel, and longitudinal data, causal inference using observational data, and survey methodology. His recent work has appeared in *Demography, Population Studies, Sociological Methodology, Journal of Marriage and Family*, and *Journal of Epidemiology and Community Health*.

Yu Xie is Bert G. Kerstetter '66 University Professor of Sociology and Princeton Institute of International and Regional Studies, Princeton University, and Visiting Chair Professor at Peking University. His main areas of interest are social stratification, demography, statistical methods, Chinese studies, and sociology of science. His recently published works include *Marriage and Cohabitation* (University of Chicago Press, 2007), with Arland Thornton and William Axinn; *Statistical Methods for Categorical Data Analysis* (Emerald, 2008, 2nd ed.), with Daniel Powers; and *Is American Science in Decline?* (Harvard University Press, 2012), with Alexandra Killewald.

Sociological Methodology
2016, Vol. 46(1) 121–152
© American Sociological Association 2016
DOI: 10.1177/0081175016641714
http://sm.sagepub.com
⑤SAGE

ॐ 4 ॐ

THE KISS PRINCIPLE IN SURVEY DESIGN: QUESTION LENGTH AND DATA QUALITY

Duane F. Alwin*
Brett A. Beattie*

Abstract

Writings on the optimal length for survey questions are characterized by a variety of perspectives and very little empirical evidence. Where evidence exists, support seems to favor lengthy questions in some cases and shorter ones in others. However, on the basis of theories of the survey response process, the use of an excessive number of words may get in the way of the respondent's comprehension of the information requested, and because of the cognitive burden of longer questions, there may be increased measurement errors. Results are reported from a study of reliability estimates for 426 (exactly replicated) survey questions in face-to-face interviews in six large-scale panel surveys conducted by the University of Michigan's Survey Research Center. The findings suggest that, at least with respect to some types of survey questions, there are declining levels of reliability for questions with greater numbers of words and provide further support for the advice given to survey researchers that questions should be as short as possible, within constraints defined by survey objectives. Findings reinforce conclusions of previous studies that verbiage in survey questions— either in the question text or in the introduction to the question—has

---

*Pennsylvania State University, University Park, PA, USA

Corresponding Author:
Duane F. Alwin, Pennsylvania State University, 309 Pond Laboratory, University Park, PA 16802, USA
Email: dfa2@psu.edu

negative consequences for the quality of measurement, thus supporting the KISS principle ("keep it simple, stupid") concerning simplicity and parsimony.

Keywords

survey design, measurement, reliability, question length

One of the things I've learned as a reporter is that you get the best answers, not when you ask long questions, but when you ask short ones.
—Bob Schieffer, broadcast journalist, CBS News, August 31, 2004

1. INTRODUCTION

There is no lack of expert opinion among survey researchers on how to write good questions and develop good questionnaires. Over the past century, vast amounts have been written on the topic of what constitutes a good question, from the earliest uses of surveys down to the present (e.g., see Belson 1981; Converse and Presser 1986; Galton 1893; Krosnick and Fabrigar 1997; Krosnick and Presser 2010; Ruckmick 1930; Saris and Gallhofer 2007; Schaeffer and Presser 2003). On some issues addressed in this literature, there is little consensus on the specifics of question design, which leads some to view the development of survey questions as more of an art than a science. Still, efforts have been made to codify what attributes are possessed by "good questions" and/or "good questionnaires," espousing principles based on a more scientific approach (e.g., Schaeffer and Dykema 2011; Schuman and Presser 1981; Sudman and Bradburn 1974, 1982; Tanur 1992).

With regard to question length, Payne's (1951:136) early writings on surveys, for example, suggested that questions should rarely number more than 20 words. In this tradition, a general rule for formulating questions and designing questionnaires is that questions should be short and simple (e.g., Brislin 1986; Fowler 1992; Sudman and Bradburn 1982; van der Zouwen 1999). Other experts suggest that lengthy questions may work well in some circumstances and have concluded, for example, that longer questions may lead to more accurate reports in some behavioral assessments (see Cannell, Marquis, and Laurent 1977; Marquis, Cannell, and Laurent 1972; Bradburn, Sudman, and Associates 1979). Advice to researchers on question length has therefore been somewhat mixed, and although the testimony of broadcast journalist Bob Schieffer, quoted above, may not be directly relevant to the goals of large-scale survey data collection, there is a common thread there that resonates with many survey researchers, namely, "keep it short and simple."

2. THE KISS PRINCIPLE

The KISS principle (in which KISS is an acronym for "keep it simple, stupid") emphasizes simplicity of design. Other variants of this principle can be found in the vernacular of the day: "keep it short and simple" and "keep it short and sweet," and we are sure there are others. The key goal of the KISS principle is that unnecessary redundancy and complexity should be avoided, and the achievement of perfection depends on parsimony. The principle is applied in scientific reasoning, in which parsimony is revered (as in Occam's razor), as well as in art, in which perfection, it is often claimed, is "reached not when there is nothing left to add, but when there is nothing left to take away."[1] It is relevant to many domains of life, and the principle has been applied in a variety of fields, from software development to film animation.

In survey research, the KISS principle may be applied to the issue of the length of survey questions, both with respect to the length of the actual question and the length of the introduction to the question (e.g., Alwin 2007; Andrews 1984; Saris and Gallhofer 2007; Scherpenzeel and Saris 1997). By focusing on question length, we may be sidestepping another important issue, question comprehension, but we do not normally have independent assessments of comprehension (which ultimately resides in the mind of the individual), whereas question length can be objectively measured (by counting the number of words used in the question). Furthermore, it is possible to distinguish between the length of the introduction to a question, or the introduction to the series of questions in which a given question is embedded, and the length of the question itself. It is argued that either type of question length is fundamentally related to the overall burden felt by the respondent, but lengthy introductions may be viewed by the researcher as a helpful aid to answering the questions. For example, Converse and Presser (1986) suggested that "the best strategy is doubtless to use short questions when possible and slow interviewer delivery—always—so that respondents have time to think" (pp. 12–13). At the same time, they conceded that "in other cases, long questions or introductions may be necessary to communicate the nature of the task" (p. 12). Both respondents and interviewers, on the other hand, may find lengthy introductions time-consuming and distracting.

3. QUESTION LENGTH AND THE SURVEY PROCESS

One of the most basic elements of survey quality is the respondent's comprehension of the question, and the issue of question length is germane to this objective (see, e.g., Tourangeau, Rips, and Rasinski 2000:23–61). Comprehension may be dependent on question length in multiple, and possibly countervailing, ways. Longer questions may be less, rather than more, clear in their meaning, and more complex (longer) questions may reduce comprehension (see, e.g., Holbrook, Cho, and Johnson 2006; Knauper et al. 1997; Yan and Tourangeau 2008). If a question is ambiguous in its meaning, or if parts of the question can have more than one meaning, then the likelihood of measurement error will be increased. Thus, from the point of view of communication, too many words may get in the way of the respondent's comprehension. One study (Holbrook et al. 2006) found that question length was related to both comprehension problems and mapping problems, as measured by behavior coding. The investigators also found that measures of complexity, such as the reading level of the question, were also predictive of respondents understanding questions. In their research, however, the relationship between question length, complexity, and question problems was not always straightforward.[2]

On the other hand, some experts encourage redundancy in survey questions precisely to enhance comprehension (Brislin 1986). Noting the trade-offs between adding redundancy and question length, Converse and Presser (1986) wrote,

> One should consider the use of redundancy now and then to introduce new topics and also to flesh out single questions, but if one larded all questions with "filler" phrases, a questionnaire would soon be bloated with too few, too fat questions. (P. 12)

Although the key element here is probably not question length *per se* but question clarity (Converse and Presser 1986:12), the addition of question text needs to be evaluated in terms of its effects on measurement precision. The phrase "now and then" is vague, and this does not provide clear guidelines regarding the use of redundancy in the phrasing of questions. Additional arguments by experts for using longer questions with redundant information are mentioned by Cannell, Miller, and Oksenberg (1981:406). They mentioned three possible explanations for the finding that a greater number of relevant health events were reported when longer questions were used: longer questions (with redundant information) state the question twice, and this (1) improves the understanding of the question, (2) provides the respondent longer time to think, and (3) encourages (motivates) the respondent to answer by showing higher interest in the interview. Finally, Bradburn et al. (1979:73–74) briefly discussed the advantages of greater question length when requesting information about socially undesirable activities.

Despite all of this advice, there is little evidence on the issue of the relationship between question length and measurement errors, although it is clear that there are a variety of points of view. Where there is evidence, the support seems to favor longer questions. Using the multitrait, multimethod (MTMM) approach (see Saris and Gallhofer 2007), Andrews (1984) combined the consideration of question length and the length of introductions to questions. He found that "short introductions followed by short questions are not good . . . and neither are long introductions followed by long questions" (p. 431). Errors of measurement were lowest when "questions were preceded by a medium length introduction (defined as an introduction of 16 to 64 words)" followed by medium or long questions (defined by Andrews [1984:431] as "16–24 words and 25+ words, respectively"). By contrast, also using the MTMM approach and similar definitions of question and introduction length, Scherpenzeel and Saris (1997) found long questions with long introductions to be superior, but like Andrews (1984), they did not separate conceptually the two issues of battery/series introduction length and question length.[3] Similarly, using an MTMM approach, Saris and Gallhofer (2007) showed that the length of the question has no significant impact on the reliability of measures, but it does have a significant effect on the trait validity coefficients. On the other hand, in their research, the mean number of words per sentence has a significant effect on reliability but not on validity. Alwin's (2007:202–19) research provides an interesting counterpoint to these results, which shows that question length interacts with the questionnaire context of the question—his results indicating that for stand-alone questions and questions contained within a series of questions on a common topic (but not for questions in batteries), there are diminishing returns to longer questions in terms of the estimated reliability of questions. This research poses an interesting puzzle with respect to the possible interaction between question context, question length, and reliability of measurement.

To be clear, one of the key issues in this debate has to do with the role of comprehension in understanding the relationship between question length and reliability of measurement, but we cannot address this here. Comprehension is a process that occurs within the respondent, and although it cannot be examined in the present study, there are ways in which it can be possibly investigated using cognitive interviews in future research. The assumption is that improved comprehension translates into greater reliability, but what remains unclear is the role that question length contributes to greater comprehension. Greater question length can improve comprehension, but it may also contribute to cognitive burden and confusion, which may reduce accuracy of measurement.

In this article, we focus specifically on the role of survey content and its interaction with question length in the evaluation of the role of question length on reliability of measurement using a longitudinal approach (see Alwin 2007:122–27). There is a long-standing distinction in the survey methods literature between objective and subjective questions (e.g., Kalton and Schuman 1982; Schuman and Kalton 1985; Turner and Martin 1984). The distinction used here between "fact" and "nonfact" is derived from this early work, in which the former refers to information "directly accessible to the external observer" and the latter to phenomena that "can be directly known, if at all, only by persons themselves" (Schuman and Kalton 1985:643).[4] In the words of Schuman and Kalton (1985),

> the distinction is a useful one, since questions about age, sex, or education could conceivably be replaced or verified by use of records of observations, while food preferences, political attitudes, and personal values seem to depend ultimately on respondent self-reports. (P. 643)

From a practical point of view, there is a potential confounding of (1) question length driven by the choice of the variables for study and (2) question length driven by the design considerations mentioned in the above review of the literature. This confounding is problematic because (1) and (2) can lead to very different recommendations to practitioners. If question length driven by design considerations leads only to serious degradation of data quality, then we may need to consider fundamentally different approaches to capturing the information of interest. If length driven by (2) leads to serious degradation of data quality, then we may wish to continue an effort to capture the variable X through a survey, but with more restraint on the use of the aforementioned verbiage. Hence, there is the need to attempt to separate the effects of question content from question design (in this case question length) in analyzing their joint effects on data quality.

4. EVALUATING SURVEY QUESTION LENGTH

There are a number of different approaches to the evaluation of the attributes of questions that affect data quality (e.g., see Madans et al. 2011). Indeed, as already noted, there is a large literature that makes an effort to provide practical guidelines for the "best practices" of question and questionnaire design. Many of these approaches use subjective criteria, and rarely do they use rigorous methods for defining the desirable attributes of questions. Sudman and Bradburn (1974) were pioneers in their effort to quantify the "response effects" of various question forms. More recently, several efforts have been made to specify an empirical criteria of data quality—for example, using the MTMM approach to reliability and validity assessment, or the use of longitudinal methods of reliability assessment (see Alwin 1992, 2007; Alwin and Krosnick 1991; Andrews 1984; Saris and Andrews 1991; Saris and Gallhofer 2007; Saris and van Meurs 1990; Scherpenzeel 1995; Scherpenzeel and Saris 1997).

In this article we use the concept of *measurement reliability* as a criterion for evaluating the quality of survey data, and we reevaluate the question of the relationship between question length and reliability of measurement. Reliability refers to consistency of measurement—it is a *sine qua non* of scientific research (see Alwin 2005, 2010). It is typically conceived of as the absence of "errors of measurement" or the obverse of unreliability. Operationally, the psychometric concept of reliability refers to the correlational consistency "between two efforts to measure the same variable, using maximally similar measurements, and independent of any true change in the quantity being measured" (see Campbell and Fiske 1959; Lord and Novick 1968). This analysis builds on the previous analysis of this issue by Alwin (2007) and is based on a reexamination and reanalysis of reliability data assembled by that project.

The concept of reliability has been applied to survey measurement previously (e.g., Alwin 1989, 1992, 2007, 2010; Alwin and Krosnick

1991; Marquis and Marquis 1977), and it has proved useful as a measure of data quality (Biemer et al. 1991; Groves 1989); however, there is a reluctance on the part of many survey methods experts to evaluate questions in terms of their reliability (e.g., see Krosnick and Presser 2010; Schaeffer and Dykema 2011). In general, the psychometric approach defines the observed score as a function of a true score and an error score—that is, as $y = \tau + \varepsilon$, where $E(\varepsilon) = E(\tau\varepsilon) = 0$ and $E(\tau) = E(y)$. The idea of a true value may be difficult for some analysts to accept (see Lord and Novick 1968:27), but it follows from this classical true score theory (CTST) model that the sample estimate of *reliability* is the squared correlation between observed and true scores (i.e., $\rho_{y\tau}^2$), which equals the ratio of true variance to the observed variance (i.e., σ_τ^2/σ_y^2), or the proportion of the observed variance that is true variance. The challenge is to design surveys that will produce a valid estimate of this ratio (see Section 5.2).

5. RESEARCH METHODS

The purpose of this article is to present a more detailed analysis of the issue of the linkage between question length and reliability of measurement on the basis of a reanalysis of the data assembled in Alwin (2007:202–10), a project dealing with the relationships of various attributes of questions and the reliability of measurement. Here we provide a more thorough investigation of this topic, examining the relationship between question length and the reliability of measurement, controlling for question content, question context, and length of unit (series and battery) introductions. In this section we discuss the source of the data on which the present analysis is based, the methods we use to estimate the reliability of measurement, and our strategy for analyzing these data.

5.1. *Samples and Data*

Our study design requires the use of large-scale panel studies that are representative of known populations, with a minimum of three waves of measurement separated by two-year reinterview intervals. Questions were selected for use only if they were exactly replicated (exact wording, response categories, mode of interviewing, etc.) across the three waves and if the underlying variable measured was continuous (rather than categorical) in nature. Specifically, this research is based on six nationally (or regionally) representative panel surveys of the U.S. population, all involving probability samples and all using face-to-face interviews, as shown in Table 1. These data sets are as follows: (1) the 1956, 1958, and 1960 National Election Study (NES) panel; (2) the 1972, 1974, and 1976 NES panel; (3) the 1992, 1994, and 1996 NES panel; (4) the 1986, 1989, and 1994 Americans' Changing Lives panel study; (5) the Study of American Families (Detroit Area) panel study of mothers; and (6) the Study of American Families (Detroit Area) panel study of children (see Alwin 2007:119–22). These selection criteria yielded 426 self-report and proxy-report questions. Table 1 presents descriptive information on these six panel studies (see Alwin 2007:118–22). This table presents the total sample sizes of these studies, along with the number of cases with data present at all three waves of the panel (i.e., listwise cases).

One of the main advantages of the reinterview or panel design using long reinterview intervals is that under appropriate circumstances, it is possible to eliminate the confounding of the systematic and random error components. To address the question of stable components of error, the panel survey must deal with the problem of memory, because in the panel design, by definition, measurement is repeated. So, although this overcomes one limitation of cross-sectional surveys—namely, the failure to meet the assumption of the independence of errors—it presents problems if respondents can remember what they said in a previous interview and are motivated to provide consistent responses (Moser and Kalton 1972). Estimation of reliability from reinterview designs makes sense only if we can rule out memory as a factor in the covariance of measures over time, and thus the occasions of measurement must be separated by sufficient periods of time to rule out the operation of memory. In cases where the remeasurement interval is insufficiently large to permit appropriate estimation of the reliability of the data, the estimate of the amount of reliability will most likely be inflated (see Alwin 1989, 1992; Alwin and Krosnick 1991), and the results of these studies suggest that longer remeasurement intervals, such as those used here, are highly desirable.

As noted, we include survey measures of continuous variables only, and within this class of variables, we implement estimates of reliability that are independent of scale properties of the observed measures, which may be dichotomous, polytomous-ordinal, or interval. In each of

Table 1. Sources of Data for Question-specific Estimates of Reliability and Attributes of Questions ($n = 426$)

Panel Studies	Population Sampled	Acronym	Total Sample	Listwise Sample	Number of Measures
1956, 1958, and 1960 NES panel	National household population	NES60s	2,529	1,132	42
1972, 1974, and 1976 NES panel	National household population	NES70s	4,455	1,296	100
1992, 1994, and 1996 NES panel	National household population	NES90s	2,439	597	98
ACL panel	National household population	ACL	3,617	2,223	86
Study of American Families, mother panel	Detroit area, mothers of 1961 births	SAF-Mo	1,113	879	46
Study of American Families, children panel	Detroit area, 1961 births	SAF-Ch	1,113	875	54
Total					426

Source: Alwin (2007).

Note: ACL = Americans' Changing Lives; NES = National Election Study.

these cases, the analysis uses a different estimate of the covariance structure of the observed data, but the model for reliability is the same. That is, when the variables are dichotomies, the appropriate covariance structure used in reliability estimation is based on tetrachoric correlations (Jöreskog 1990, 1994; Muthén 1984); when the variables are polytomous-ordinal, the appropriate covariance structure is either the polychoric correlation matrix or the asymptotic covariance matrix based on polychoric correlations; and when the variables can be assumed to be interval, ordinary Pearson-based correlations and covariance structures for the observed data are used (Brown 1989; Jöreskog 1990, 1994; Lee, Poon, and Bentler 1990; Muthén 1984). As noted, all of these models assume that the latent variable is continuous.

5.2. Methods of Reliability Estimation

Following Campbell and Fiske's (1959) famous definition of reliability as the "agreement between two efforts to measure the same thing, using maximally similar methods" (p. 83), the concept of reliability is often conceptually defined in terms of the consistency of measurement. This is an appropriate characterization, indicating the extent to which "measurement remains constant as it is repeated under conditions taken to be constant" (see Kaplan 1964:200). The key idea here is expressed by Lord and Novick (1968) in their classical statement of true score theory, wherein they state that "the correlation between truly parallel measurements taken in such a way that the person's true score does not change between them is often called *the coefficient of precision*" (p. 134). In this case, the only source contributing to measurement error is the unreliability or imprecision of measurement. The assumption here, as is true in the case of cross-sectional designs, is that "if a measurement were taken twice and *if no practice, fatigue, memory, or other factor* [emphasis added] affected repeated measurements," the correlation between the measures reflects the precision, or reliability, of measurement (Lord and Novick 1968:134). In practical situations in which there are in fact practice effects, fatigue, memory, or other spurious factors contributing to the correlation between repeated measures, the simple idea of the correlation between Y_1 and Y_2 is not the appropriate estimate of reliability. Indeed, in survey interviews of the type commonly used it would be almost impossible to ask the same question twice without memory or other factors contributing to the correlation of repeated measures. Thus, in general, asking the same question twice within the same interview would be the incorrect design for estimating the reliability of measuring the trait T, because the two observations are likely not independent.

It can be argued that for purposes of assessing the reliability of survey data, longitudinal data provide an optimal design (see Alwin 2007). Indeed, the idea of replication of questions in panel studies as a way of getting at measurement consistency has been present in the literature for decades, the idea of "test-retest correlations" as an estimate of reliability being the principle example of a longitudinal approach. The limitations of the test-retest design are well known, but they can be overcome by incorporating three or more waves of data separated by lengthy periods of time (see Alwin 2007:96–116). The multiple-wave reinterview design discussed in this article goes well beyond the traditional test-retest design (see Moser and Kalton 1972:353–54), and specifically by using models that permit change in the underlying true score (using the quasi-Markov simplex approach) allows us to overcome one of the key limitations of the test-test design (see, e.g., Heise 1969; Wiley and Wiley 1970). The literature discussing the advantages of the quasi-Markov simplex approach for separating unreliability from true change is extensive (see Appendix A in the online journal).[5]

Through the use of design strategies with relatively distant reinterview intervals (e.g., two-year intervals), the problem of consistency due to retest effects or memory can be remedied, or at least minimized. There are two main advantages of the reinterview design for reliability estimation. First, the estimate of reliability obtained includes all reliable sources of variation in the measure, both common and specific variance. Second, under appropriate circumstances it is possible to eliminate the confounding of the systematic error component discussed earlier, if systematic components of error are not stable over time. To address the question of stable components of error, the panel survey must deal with the problem of memory, because in the panel design, by definition, measurement is repeated. So, although this overcomes one limitation of cross-sectional surveys, it presents problems if respondents can remember what they say and are motivated to provide consistent responses. If reinterviews are spread over months or years, this can help rule out sources of bias that occur in cross-sectional studies. Given the difficulty of estimating memory functions, estimation of reliability from reinterview designs makes sense only if we can rule out memory as a factor in the covariance of measures over time, and thus, the occasions of measurement must be separated by sufficient periods of time to rule out the operation of memory.

The model used here falls into a class of autoregressive or quasi-Markov simplex models that specifies two structural equations for a set of p over-time measures of a given variable Y (where $t = 1, 2, \ldots, p$) as follows:

$$Y_t = T_t + E_t \tag{1}$$

and

$$T = \beta_{t, t-1} T_{t-1} + Z_t. \tag{2}$$

Equation (1) represents a set of measurement assumptions indicating (1) that over-time measures are assumed to be τ-equivalent, except for true score change and (2) that measurement error is random (see Alwin 1989, 2007, 2011; Heise 1969; Jöreskog 1970; Wiley and Wiley 1970). Equation (2) specifies the causal processes involved in change of the latent variable over time. A formal statement of the model is provided in Appendix A in the online journal.

It is important to note that this model assumes that the latent variable will change over time and that it follows a Markovian process in which the distribution of the true variables at time t is dependent only on the distribution at time $t - 1$ and not directly dependent on distributions of the variable at earlier times. If these assumptions do not hold, then this type of simplex model may not be appropriate. In order to estimate such models, it is necessary to make some assumptions regarding the measurement error structures and the nature of the true change processes underlying the measures. All estimation strategies available for such three-wave data require a lag-1 assumption regarding the nature of the true change. This assumption in general seems a reasonable one, but erroneous results can result if it is violated. The various approaches differ in their assumptions about measurement error. One approach assumes equal reliabilities over occasions of measurement (Heise 1969). This is often a realistic and useful assumption, especially when the process is not in dynamic equilibrium, that is, when the observed variances vary with time. Another approach to estimating the parameters of the above model is to assume constant measurement error variances rather than constant reliabilities (Wiley and Wiley 1970). Where

$P = 3$, either model is just-identified, and where $P > 3$, both models are overidentified with degrees of freedom equal to $.5[P(P + 1)] - 2P$. The four-wave model has two degrees of freedom, which can be used to perform likelihood-ratio tests of the fit of the model.

Wiley and Wiley (1970) showed that by invoking the assumption that the measurement error variances are equal over occasions of measurement, the $P = 3$ model is just-identified, and parameter estimates can be defined. They suggested that measurement error variance is "best conceived as a property of the measuring instrument itself and not of the population to which it is administered" (p. 112). Following this reasoning, we might expect that the properties of our measuring instrument would be invariant over occasions of measurement and that such an assumption would be appropriate. Following the CTST model for reliability, the reliability for the observed score Y_t is the ratio of the observed variance—that is, $\text{VAR}(T_t)/\text{VAR}(Y_t)$—and this model permits the calculation of distinct reliabilities at each occasion of measurement using the above estimates of $\text{VAR}(E_t)$ and $\text{VAR}(T_t)$ (see Wiley and Wiley 1970; see also Alwin 2007:109).

To summarize, the three-wave model is just-identified—that is, there are no overidentifying restrictions that allow for the possibility of testing the model against a null model of interest. Where $P > 3$, the simplex model is overidentified, and a test of the model (with degrees of freedom equal to $.5[P(P + 1)] - 2P$) is possible. The four-wave model has two degrees of freedom, which can be used to perform likelihood-ratio tests of the fit of the model. We restricted the present analysis to only three waves because the panel studies on which we draw include only three waves. Note that because the models we use are just-identified, standard errors for the parameter estimates cannot be computed. Further testing of these models using more than three waves is essential for generalization of the findings reported here. On the other hand, although it is important to use multiwave panel data in this case, it is also true that more waves add complexities that must be dealt with. It is important to be able to deal with attrition, for example, and more waves add to the problems of missing data.

5.3. *Analysis Strategy*

In the following analysis we use several measures of the attributes of the questions and use these to predict the estimated reliability of the questions. First, our primary explanatory variable is question length (i.e., the number of words in a question), and we examine several aspects of question length, including the number of words in both the "prequestion text" and the actual question. In our previous discussion of the reliability of questions in series and batteries, we considered the role of the presence and length of subunit introductions on estimates of measurement reliability, finding that it had modest effects (see Alwin 2007:208). Here we consider the number of words appearing in the text for a given question following any prequestion text that might be included as an introduction. Our analysis focuses on both the effects of question length and introduction length as independent factors. In both cases, we express question and introduction length in units of 10 (or fractions thereof) and center this measure by expressing it as a deviation from its mean.

Second, we use a measure of question content, an important predictor of question reliability, to control for this source of variation. In this case, question content refers to whether the content measured is a fact (content that can potentially be verified by consulting other sources) versus a nonfact such as beliefs, values, attitudes, or self-descriptions—content that is primarily subjective and cannot be verified by consulting other sources.

Third, we use a three-category variable that we refer to as question context, which classifies questions according to the architecture of survey questionnaires: (1) stand-alone questions, which do not bear any particular topical relationship with adjacent questions; (2) questions that are a part of a series of questions that all focus on the same specific topical content; and (3) questions that appear in batteries focusing on the same or similar subject matter, and more specifically use the identical response format (for some examples, see Alwin 2007:205–207). (Appendix C in the online journal presents illustrative examples of these types of contexts from an actual survey questionnaire used here.) In our analyses of the total sample of questions, we use a set of dummy variables to represent this variable and omit the category of "questions in batteries" to deal with the redundancy.

Finally, in subclass regressions for survey context, we additionally consider as a separate factor the number of words in the introduction to a series or battery.

Table 2. Regression of Reliability Estimates on Length of Question and Attributes of Question Content and Question Context

Predictor	Model				
	1	2	3	4	5
Intercept	.668	.636	.625	.620	.600
QL	.000	−.001	−.006	−.004	−.006
Fact vs. nonfact		.173***		.117***	.092***
SA			.228***	.146***	.263***
In series			.098***	.060***	.156***
QL × SA					−.040***
QL × In Series					−.034***
R^2	.000	.158	.149	.200	.239
Number of cases	426	426	426	426	426

Note: QL = question length; SA = stand-alone.
***$p \leq .001$.

6. RESULTS

In Table 2 we present a set of regression models in which we predict the reliability of survey measures from several question characteristics using the entire database of 426 questions. These models use predictor variables for question length, a dummy variable representing question content (nonfact is the omitted category), a set of two dummy variables representing question context (questions in batteries form the omitted category), and two interaction terms expressing the interaction between question length and questions in the stand-alone category and questions in series.[6] These interaction terms are important, in that as specified they express the role of question length within the two categories of stand-alone questions and questions in series. There are two key differences between the following results and Alwin's (2007) presentation of these data. First, we control here for question content when assessing the combined effects of question length and reliability of measurement; second, we separate the effects of question length and the length of introductions to series and batteries.

In this table, model 1 contains question length as the sole predictor; this variable is statistically independent of question reliability in the total set of 426 measures (see also Figure 1). The second model adds question content (fact vs. nonfact), as described above, which is an important predictor of reliability; the coefficient of .173 indicates that

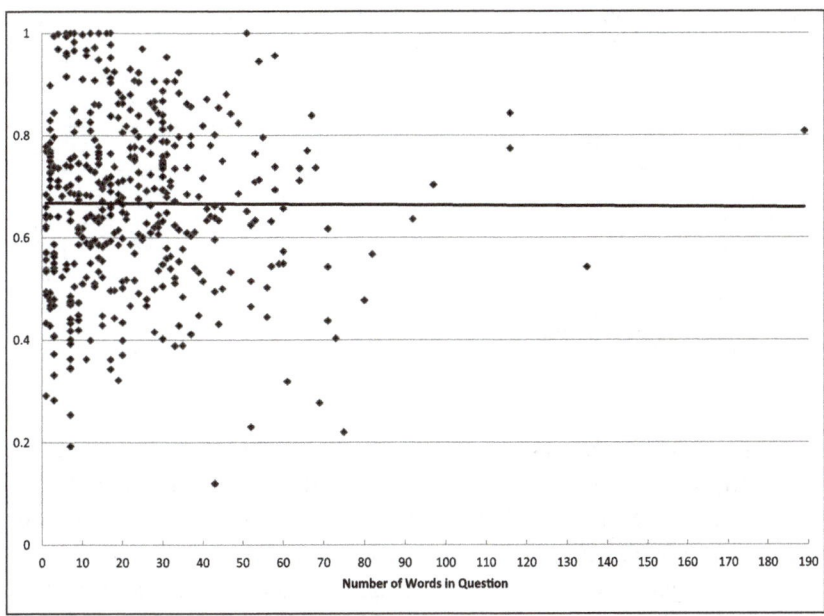

Figure 1. Regression of reliability estimates on question length for all factual and nonfactual questions (*n* = 426).

there is a predicted difference of this magnitude on average between questions that measure facts and those that measure nonfacts. This variable alone accounts for about 15 percent of the variation in reliability among survey questions. Model 3 adds our classification of survey context (removing the measure of question content), and model 4 includes both survey content and survey context. In model 3, the reference category (or omitted category) is the category of questions in batteries, and thus the regression coefficients for the dummy variables for "stand-alone" and "in series" refer to the differences in mean reliability for these categories relative to questions in batteries. The R^2 value for model 3 indicates that the survey context classification also by itself accounts for about 15 percent of the variation in reliability, when entered alone. Stand-alone questions are the most reliable, followed by questions in series; questions in batteries (which have an average reliability of .624) are the least reliable. We present the results of several alternative specifications of the model in the supplementary tables provided in Appendix B in the online journal.[7]

Table 3. Regression of Reliability Estimates on Length of Question and Question Content: Stand-alone Questions

Predictors	Model	
	1	2
Intercept	.956	.763
Question length	−.045***	−.023[†]
Fact vs. nonfact		.190**
R^2	.381	.556
Number of cases	30	30

[†]$p \leq .10.$ **$p \leq .01.$ ***$p \leq .001.$

Returning to the results in Table 2, survey content and survey context are clearly not independent, as indicated by the results of model 4: factual questions are more likely to be measured in a stand-alone format or in a series, whereas nonfacts are more likely to be placed in series and batteries. Together these two sets of factors account for about 20 percent of the variance, a rather remarkable result. In other words, by knowing only two things about survey questions—what they are measuring and the placement of the question in the organizational context of the questionnaire—we can account for roughly one fifth of the variation in reliability of measurement. The addition of the interaction terms between question length and question context, which allow the effects of question length to vary by question context, significantly improves the prediction of question reliability—to about 25 percent of the variation. These results indicate that although there is no relationship between question length and reliability among questions included in batteries (note that this effect is represented by the effects of question length in model 5), there is a significant decline in reliability linked to question length for both stand-alone questions and questions in series.[8]

To explore these relationships further, we examine the patterns of association between reliability and question length separately for stand-alone questions and questions in series. Table 3 presents results for the 30 stand-alone questions, and although this represents a relatively small pool of survey questions, these results are very revealing. As noted above, among stand-alone questions, there is a modest decline in reliability as the length of the question increases (see Figure 2). Very short questions are highly reliable, but as the length of the question increases,

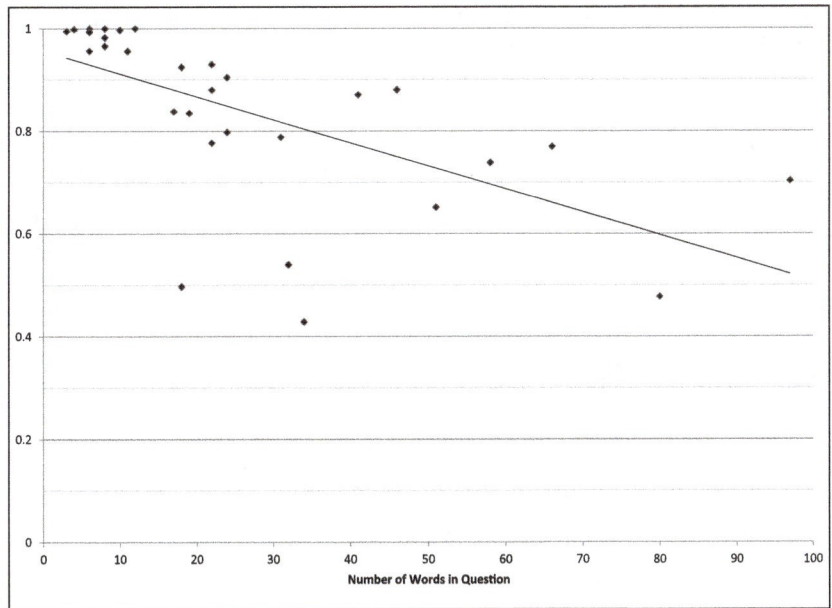

Figure 2. Regression of reliability estimates on question length for stand-alone questions ($n = 30$).

reliability of responses suffers. This relationship is, however, due almost entirely to the fact that stand-alone questions are much more likely to be measuring factual content: when this variable is added to the equation (see model 2 in Table 3), the relationship with question length is almost entirely removed. It is noteworthy, however, that there is a very slight negative decline in reliability due to question length. This pattern was stronger (statistically speaking) in Table 2, in which the overall test of this effect had the benefit of greater power due to the overall larger sample size.

When we examine these relationships separately for questions in series (see Table 4 and Figure 3), the same patterns emerge as those presented in Table 2—greater question length suppresses reliability of measurement. The effects of question length are not removed by controlling for survey content, and it is not possible to argue on the basis of these results that the effect of question length is due to its confounding with question content. As above, the major factor affecting measurement reliability is survey content, indicating that among questions in series,

Table 4. Regression of Reliability Estimates on Length of Question and Question Content: Questions in Series

Predictor	Model 1	2	3	4	5	6
Intercept	.799	.762	.759	.832	.738	.755
Question length	−.033***	−.029***	−.032***	−.037**	−.025*	−.024*
Fact vs. nonfact		.079***	.078**	.063***	.085***	.070***
First in series			.048[†]			
Series introduction length						−.009*
R^2	.098	.150	.166	.201	.144	.175
Number of cases	177	177	177	42	135	135

Note: Model 4 is for those cases first in series. Models 5 and 6 are for those cases second or later in series.

[†]$p \leq .10.$ *$p \leq .05.$ **$p \leq .01.$ ***$p \leq .001.$

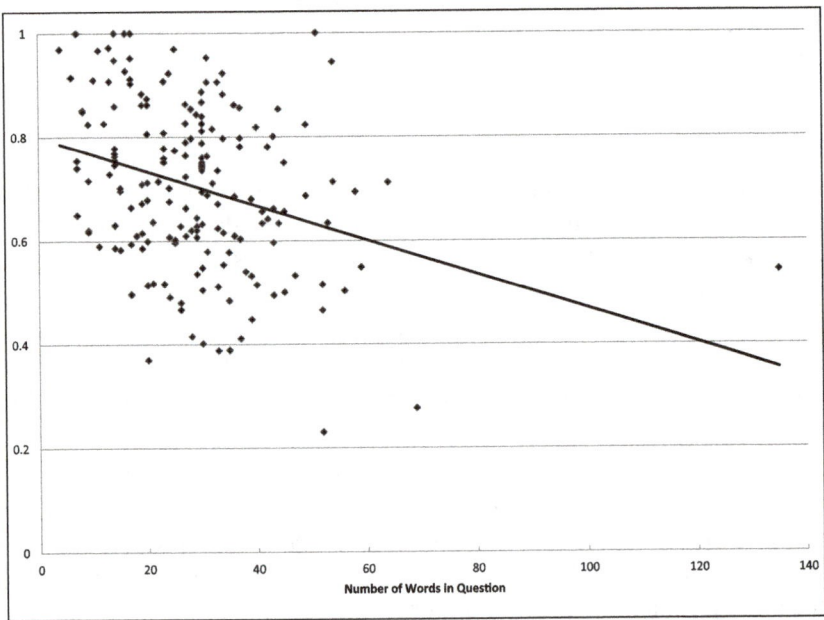

Figure 3. Regression of reliability estimates on question length for question in series ($n = 177$).

factual questions exceed nonfactual ones by about .07 to .08 in reliability, depending on which subset of questions we consider (compare model 2 in Table 3 and model 2 in Table 4). There is a slight advantage in reliability for the first questions in a series (see model 3 in Table 4), but this effect is marginally significant.

Note that in Table 4, the first three models apply to the full set of questions in the pool that were placed in series, whereas the later models apply to specific subsets. As already noted, among questions in series, there is a significant reduction in reliability associated with greater question length (see model 1), which is not removed when controlling for question content (see model 2). Model 3 tests for whether being the first question in a series improves reliability, and there is a marginally ($p < .10$) significant improvement. We tested for statistical interaction between being first in the series and question length, which was not significant ($p = .4326$). Note that model 4 is based only on those 42 questions that were the first items in the series. For this subsample of questions, the negative effect of question length is slightly enhanced in this subsample. Models 5 and 6 pertain to the subset of 135 questions that were second or later in the series, and there continues to be a significant effect of question length. The advantages that accrue to shorter questions and measures of facts do not depend on whether the question is the first one in the series or the second or later; compare models 4 and 5 in Table 4, for example. When we compare the effect of the length of the introductions with the questions in series (see model 6), we find a slight depressing effect on reliability of longer introductions; this effect, however, is quite small.

Finally, Table 5 presents the results of an analysis for questions in batteries. The first two models in this table were estimated on the basis of all of the questions in the pool that were placed with batteries; models 3 and 4 pertain to subsets of these questions in batteries. Model 3 was conducted on the questions that were the first questions in the batteries, whereas model 4 applies to the questions appearing second or later in the batteries. In these models, we have not included the survey content variable, in that virtually all questions in batteries measure nonfactual content. Consistent with prior results (Alwin 2007), there appears to be little evidence that question length for questions in batteries has any bearing on their reliability. We tested for statistical interaction on the basis of whether the question was first in a battery and question length in model 2, which was nonsignificant ($p = .3694$). Questions in batteries

Table 5. Regression of Reliability Estimates on Length of Question and Other Content: Questions in Batteries

	Model			
Predictor	1	2	3	4
Intercept	.600	.601	.647	.582
Question length	.007	.004	.000	.015
First in batteries		.026		
Batteries introduction length				.002
R^2	.012	.014	.000	.000
Number of cases	219	219	41	178

Note: Model 3 is for those cases first in batteries; model 4 is for those cases second or later in batteries.

are overall less reliable (e.g., compare the intercepts across the models in Table 5 with Table 4), but there is no association between question length and reliability. We consider the possible explanations for this in Section 7.

Question length is related to reliability for questions in series and to a lesser extent in stand-alone questions once content is controlled for. Given the importance of content, the question was raised about whether question length interacts with content—that is, are there differences in the relationship of question length and reliability by survey content (fact vs. nonfact)?[9] We explored this hypothesis and found that there was an interaction between content and question length, but it was driven entirely by question context. Fact-based questions are asked only in stand-alone questions or within a series of questions, and once question context is controlled, the relationship between content and question length disappears. The overall results suggest that content and context are clearly predictors of reliability, and question length adds to the prediction only within specific contexts, questions in series, and to some extent, stand-alone questions.

7. DISCUSSION

There is a growing literature that addresses the practical question of the desirable length of questions in surveys. There is no consensus on this issue and a mixture of opinions, few of which are grounded in empirical assessments. The present research builds on work that established an empirical regularity of a relationship between question length and data quality in some survey contexts, as assessed by measurement reliability. The focus of this research entertained the modest objective of the possible confounding in those results of question length with question content, while controlling for question context (i.e., the placement of the question within the organization of the questionnaire). There is no question that there may be substantial confounding between question length and the substantive focus of the question, wherein subjective content tends to be measured using longer questions. The present research has focused explicitly on the confounding of question length and survey content, net of survey context.

This research assesses question content using the long-standing focus in the survey methods literature on objective and subjective questions, operationalized here in terms of questions seeking factual versus nonfactual information. Prior research suggests that factual (or objective) material can be more precisely measured than content that is essentially subjective (Alwin 1989, 2007; Kalton and Schuman 1982; Schuman and Kalton 1985), although there is considerable overlap. Few survey questions are perfectly reliable—but the typical factual question can be shown to be substantially more reliably measured on average than the typical nonfactual question. Our more detailed examination of this issue in the present article confirms the strong effects of question content (fact vs. nonfact) and question context (stand-alone questions, questions in series, and questions in batteries) as important predictors of reliability, together accounting for some 20 percent of the variation in measurement reliability.

Within the constraints of the purpose of the survey, one element of survey question writing to which the majority of (but not all) researchers subscribe is that questions should be as short as possible, although there are opposing views. The question raised in the present research is whether the length of questions (and for questions in series and batteries, the length of introductory text) produces any significant decrement to reliability of measurement. Bear in mind our findings are conditional on our parameters for question inclusion (see Section 5.1 for qualifying criteria). Certainly, there may be limitations to any generalization concerning the practical advantages and disadvantages of any particular study, but it is important to address the practical conclusions of the present research. The major practical implications are that, exclusive of questions in batteries, other things being equal, shorter questions are more reliable. In the case of questions in batteries, the concept of question length, apart from the length of the introduction, is somewhat ambiguous. It will therefore be valuable for future research on the effects of question length to introduce further clarification of the types of questions we have considered to be part of questionnaire batteries (see Alwin 2007:205–207).

The overarching practical consideration in the case of batteries, then, is not the length of the questions but whether to use them at all. Consistent with prior research (e.g., Alwin 2007; Andrews 1984), our results provide strong support for the view that questions in what we referred to as a "topical series" are less reliable than "stand-alone questions" (at least among factual questions) and questions in "batteries" are less reliable than questions in series (among nonfactual questions) (Alwin 2007:171–72). Perhaps the best explanation of this phenomenon is that the same factors motivating the researcher to group questions together—contextual similarity—are the same factors that promote measurement errors (see Andrews 1984:431). Similarity of question content and response formats may actually distract the respondent from fully considering what information is being asked for, and this may reduce the respondent's attention to the specificity of questions (i.e., they may increase his or her tendency to "satisfice"; Krosnick and Alwin 1987). Thus, measurement errors may be generated in response to the "efficiency" features of the questionnaire, and unfortunately, as Andrews (1984:431) concluded, it appears that the respondents may also be more likely to "streamline" their answers when the investigator "streamlines" the questionnaire.

8. CONCLUSION

Our research concludes that the consideration of the length of questions adds to the understanding of levels of measurement error associated with question attributes, but the results must be understood in terms of the interaction of question length and question context. That is, as indicated above, the question length of stand-alone questions and questions in a topical series are found to have a negative effect on measurement reliability. The length of questions in batteries, however, reveals no relationship to reliability. This may be due in part to the fact that questions in batteries are quite a lot shorter; the typical nonfactual question in batteries has a question length of 12 words (see Alwin 2007:204).

This is because often the actual question stimulus is just a few words, given the existence of a lengthy introduction to the battery that explains how the respondent should use the rating scale. One of the important contributions of the present research is its emphasis on the separation of the length of questions from the length of introductions to series and batteries. Although it is true that the length of a question is not independent of the nature and length of the unit (series or battery) introduction, neither the length of the battery introduction nor the length of questions in batteries have any measurable effect on reliability of measurement. Indeed, the very existence of lengthy introductions in batteries of questions may promote greater reliability.

There are admitted limitations related to the present research. First, as noted at the outset, we focus on only one aspect of data quality, and given the limitations of the research design, we cannot examine aspects of data quality, such as bias and nonresponse (see Groves 1989). Similarly, there are other aspects of question complexity that are beyond the scope of the present article, which may limit its applicability to assessing question characteristics. Prior research has found that other measures of complexity, such as the reading level of the question, were predictive of the comprehension of the question. Ultimately, question comprehension is an attribute of the respondent and not necessarily an objective characteristic of the question. Clearly, the relationship between question length, complexity, and other aspects of questionnaires is an important set of issues, and the narrow focus on the issues of question length leaves other issues unaddressed. The assumption is that improved comprehension translates into greater reliability, but what is unclear is the role that question length contributes to greater comprehension. There are ways in which the issue of comprehension can be investigated in future research using cognitive interviews in laboratory settings.

Second, as noted, our estimates of reliability are limited to three-wave panels, which are just-identified and assume a lag-1 structure for the latent variables. Our models also indicate that the error variances are assumed to represent random error and that stable nonrandom sources of error are included in the underlying latent trait variable. In our defense, it should be noted that multiple-wave ($P > 3$) studies are generally unavailable, which limits the possibilities of developing overidentified models. Future research will need to address these issues using the expanding opportunities in longitudinal studies that meet the design requirements of this approach. Also as a minor but still important issue, we note that the present article addresses only questions used in the United States (and more specifically only questions from the University of Michigan surveys). One critical reviewer commented that the questions in the United States are in general "considerably longer than questions in Europe," which, if true, admittedly may influence the generalizability of the results of the present article. We therefore caution the reader to not overgeneralize the findings of the present study.

Finally, we should note we can rule out the broad features of question content as an explanation of the relationship between question length and reliability. As we stressed at the beginning of the article, the issue of question length may be sidestepping an important set of issues, namely, question clarity and comprehension (Converse and Presser 1986). Although it is an important practical concern of many survey methodologists, it is probably the case that some people would think that question length and introduction length *per se* are not the issue. However, clarity and comprehension ultimately reside in the mind of the individual respondent and are not specifically linked to the properties of the question itself. Question length can be measured objectively, and if a variable such as this is related to increased errors of measurement, we should attempt to understand why. One problem with long questions is that, if there are a lot of them, this may lead to satisficing on the part of the respondent (see Krosnick and Alwin 1987), and this may be likely to hurt reliability. But again this is beyond the potential question-level analysis that is possible here. Moreover, satisficing is an interpretative tool for respondent behavior, and as far as we know, no one has figured out a way of measuring it within the survey context. There are some avenues, however, that may be pursued in future studies as a way of explaining the effects of question length. For example, as we mentioned earlier, if a question gets to be too complex syntactically, it may lead to poor comprehension, which again is likely to reduce reliability. The relation between question length and question complexity is a topic that should be pursued in future work. On the other hand, a long question may be long because it includes a definition for a vague or unfamiliar term and that may actually improve comprehension (and reliability). Or, a long question may incorporate lots of memory cues, and that may increase reliability (or at least accuracy). Each of these hypotheses suggests interactions that could be incorporated into future studies that would try to account for the present findings. For example, if question length is associated with syntactic complexity, then it may have fewer (or different) effects on highly educated respondents. If length is associated with more memory cues for factual questions involving memory retrieval, then it may have positive effects for such questions. There are a number of ways in which hypotheses may be developed to afford better explanations for the findings presented here.

Acknowledgments

This article was submitted and reviewed during the editorship of Tim Futing Liao, editor-in-chief of *Sociological Methodology* through 2015. Decision making on this article was made entirely during his editorship. Tim Liao, previous reviewers, and Aaron Maitland provided valuable comments and suggestions on earlier versions. We acknowledge the research assistance of Mona Ostrowski, Audrey Jackson, Alyson Otto, and Judy Bowes.

Notes

1. This quotation is attributed to Antoine de Saint-Exupéry (1900–1944), noted French writer and aviator.
2. Although we believe that Holbrook et al.'s (2006) research helps account for the mechanisms by which question length may affect the quality of survey data, any exploration here of the complexity of questions, such as reading level and comprehensiveness, is beyond the scope of the present research.
3. Results reported by Andrews (1984) and Scherpenzeel and Saris (1997) are somewhat confusing because in the typical survey, questions by themselves do not have introductions. On the other hand, series of questions, or batteries, or entire sections, do typically have introductions (see below). We see the introduction to series and/or batteries of questions to be a separate topic conceptually from that of question length, and we therefore distinguish between question length and the length of introductions to organizational units larger than the question.
4. The distinction used here between "fact" and "nonfact" is derived from early work in survey methods (e.g., Kalton and Schuman 1982; Schuman and Kalton 1985; Turner and Martin 1984). The distinction was further used in Alwin's (1989, 2007) work, and it has been proved to be easily coded. Facts are defined as objective information regarding the respondent or members of the household, which can be verified against objective records—for example, information on the respondent's personal characteristics, such as the date of birth, amount of schooling, amount of family income, and the timing, duration, and frequencies of certain behaviors (Alwin 2007:123; see also Alwin 2007:157, Table 7.4). Nonfacts include beliefs, attitudes, and values that are a matter of personal judgment for which no objective information exists; nonfacts also include self-descriptions—that is, subjective assessments or evaluations of the state of the respondent within certain domains (see Alwin 2007:123–24, Table 6.1). In the present research, agreement was achieved among four investigators, and little ambiguity exists for the vast majority of

questions. It is worth noting that this dichotomous predictor may not always be adequate, and that the nonfact category is normally broken down further into attitudes, beliefs, values, self-appraisals, and self-evaluations (see Alwin 2007:153–62). In the present research, the fact-nonfact distinction was deemed adequate given the goals of the research.

5. Appendix A in the online journal presents an extended discussion of the simplex model as applied to multiwave panel data. In the three-wave case, the parameters of the model are just-identified and can easily be estimated by hand given the correlations among the variables. As noted in the text, for ordinal variables we base our estimates on polychoric correlations, and for continuous variables we use Pearson correlations. The reader is invited to contact us with any additional questions regarding the estimation models used to obtain the correlations involved.

6. The reader is referred to introductory statistics materials that cover the use of dummy variables and interaction terms in regression analysis (e.g., Hardy 1993).

7. See Appendix B in the online journal for a detailed discussion of these alternative models.

8. We concede that statistical tests on differences in attributes of questions measured by the predictor variables are not technically appropriate, because the questions have neither been randomly selected from some known universe of questions, nor are they independent in a sampling sense. Nonetheless, we present information from statistical tests as a qualitative measure of the relative magnitude of a particular relationship, not as a basis for generalizing to some known universe of questions.

9. One reviewer made the case for considering the interaction of content with question length. The net result is that there are no interactions of question length with survey content (fact vs. nonfact) in predicting reliability.

References

Achen, Christopher H. 1975. "Mass Political Attitudes and the Survey Response." *American Political Science Review* 69:1218–31.

Alwin, Duane F. 1989. "Problems in the Estimation and Interpretation of the Reliability of Survey Data." *Quality and Quantity* 23(3):277–331.

Alwin, Duane F. 1992. "Information Transmission in the Survey Interview: Number of Response Categories and the Reliability of Attitude Measurement." *Sociological Methodology* 22:83–118.

Alwin, Duane F. 2005. "Reliability." Pp. 351–59 in *Encyclopedia of Social Measurement*, edited by K. Kempf-Leonard. New York: Academic Press.

Alwin, Duane F. 2007. *Margins of Error—A Study of Reliability in Survey Measurement*. Hoboken, NJ: John Wiley.

Alwin, Duane F. 2010. "How Good Is Survey Measurement? Assessing the Reliability and Validity of Survey Measures." Pp. 405–34 in *Handbook of Survey Research*, edited by Peter V. Marsden and James D. Wright. Bingley, UK: Emerald Group.

Alwin, Duane F. 2011. "Evaluating the Reliability and Validity of Survey Interview Data Using the MTMM Approach." Pp. 265–93 in *Question Evaluation Methods: Contributing to the Science of Data Quality*, edited by Jennifer Madans, Kristen Miller, Aaron Maitland, and Gordon Willis. Hoboken, NJ: John Wiley.

Alwin, Duane F., and Jon A. Krosnick. 1991. "The Reliability of Survey Attitude Measurement: The Influence of Question and Respondent Attributes." *Sociological Methods and Research* 20(1):139–81.

Andrews, Frank M. 1984. "Construct Validity and Error Components of Survey Measures: A Structural Modeling Approach." *Public Opinion Quarterly* 48(2): 409–42.

Belson, William A. 1981. *The Design and Understanding of Survey Questions.* Aldershot, UK: Gower.

Biemer, Paul P., Robert M. Groves, Lars E. Lyberg, Nancy A. Mathiowetz, and Seymour Sudman, eds. 1991. *Measurement Errors in Surveys.* New York: Wiley.

Bradburn, Norman M., and S. Sudman, and Associates. 1979. *Improving Interviewing Methods and Questionnaire Design: Response Effects to Threatening Questions in Survey Research.* San Francisco, CA: Jossey-Bass.

Brislin, Richard W. 1986. "The Wording and Translation of Research Instruments." Pp. 137–64 in *Field Methods in Cross-cultural Research*, edited by W. J. Lonner and J. W. Berry. Newbury Park, CA: Sage.

Brown, R. L. 1989. "Using Covariance Modeling for Estimating Reliability on Scales with Ordered Polytomous Variables." *Educational and Psychological Measurement* 49(2):385–98.

Campbell, Donald T., and Donald W. Fiske. 1959. "Convergent and Discriminant Validation by the Multitrait-Multimethod Matrix." *Psychological Bulletin* 6(1): 81–105.

Cannell, Charles F., Kent H. Marquis, and Andre Laurent. 1977. "A Summary of Studies of Interviewing Methodology." *Vital and Health Statistics.* Series 2, No. 69, March.

Cannell, Charles F., Peter V. Miller, and Lois Oksenberg. 1981. "Research on Interviewing Techniques." *Sociological Methodology* 12:389–437.

Converse, Jean M., and Stanley Presser. 1986. *Survey Questions: Handcrafting the Standardized Questionnaire.* Beverly Hills, CA: Sage.

Fowler, Floyd J. 1992. "How Unclear Terms Affect Survey Data." *Public Opinion Quarterly* 56(2):218–31.

Galton, Francis. 1893. *Inquiries into the Human Faculty and Its Development.* London: Macmillan.

Groves, Robert M. 1989. *Survey Errors and Survey Costs.* New York: John Wiley.

Hardy, Melissa A. 1993. *Regression with Dummy Variables: Quantitative Applications in the Social Sciences.* Newbury Park, CA: Sage.

Heise, David R. 1969. "Separating Reliability and Stability in Test-retest Correlation." *American Sociological Review* 34(1):93–191.

Holbrook, Allyson, Young Ik-Cho, and Timothy Johnson. 2006. "The Impact of Question and Respondent Characteristics on Comprehension and Mapping Difficulties." *Public Opinion Quarterly* 70(4):565–95.

Jöreskog, Karl G. 1970. "Estimating and Testing of Simplex Models." *British Journal of Mathematical and Statistical Psychology* 23:121–45.

Jöreskog, Karl G. 1990. "New Developments in LISREL: Analysis of Ordinal Variables Using Polychoric Correlations and Weighted Least Squares." *Quality and Quantity* 24(4):387–404.

Jöreskog, Karl G. 1994. "On the Estimation of Polychoric Correlations and Their Asymptotic Covariance Matrix." *Psychometrika* 59(3):381–89.

Kalton, Graham, and Howard Schuman. 1982. "The Effect of the Question on Survey Response: A Review." *Journal of the Royal Statistical Society* 145(1):42–73.

Kaplan, Abraham .1964. *The Conduct of Inquiry*. San Francisco, CA: Chandler.

Knauper, Barbel, Robert F. Belli, Daniel H. Hill, and A. Regula Herzog. 1997. "Question Difficulty and Respondents' Cognitive Ability: The Effect on Data Quality." *Journal of Official Statistics* 13(2):181–99.

Krosnick, Jon A., and Duane F. Alwin. 1987. "Satisficing: A Strategy for Dealing with the Demands of Survey Questions." General Social Survey Technical Report No. 74; Methodological Report No. 46. Chicago: National Opinion Research Center.

Krosnick, Jon A., and Leandre R. Fabrigar. 1997. "Designing Rating Scales for Effective Measurement in Surveys." Pp. 141–64 in *Survey Measurement and Process Quality*, edited by Lars E. Lyberg, Paul P. Biemer, Martin Collins, Edith D. de Leeuw, Cathryn Dippo, Norbert Schwarz, and Dennis Trewin. New York: John Wiley.

Krosnick, Jon A., and Stanley Presser. 2010. "Question and Questionnaire Design" Pp. 263–313 in *Handbook of Survey Research*, edited by Peter V. Marsden and James D. Wright. Bingley, UK: Emerald Group.

Lee, Sik-Yum, Wai-Yin Poon, and Peter M. Bentler. 1990. "A Three-stage Estimation Procedure for Structural Equation Models with Polytomous Variables." *Psychometrika* 55(1):45–51.

Lord, Frederick M., and Melvin L. Novick. 1968. *Statistical Theories of Mental Test Scores*. Reading, MA: Addison-Wesley.

Madans, Jennifer, Kristen Miller, Aaron Maitland, and Gordon Willis, eds. 2011. *Question Evaluation Methods: Contributing to the Science of Data Quality*. Hoboken, NJ: John Wiley.

Marquis, Kent H., Charles F. Cannell, and Andre Laurent. 1972. "Reporting Health Events in Household Interviews: Effects of Reinforcement, Question Length, and Reinterviews." *Vital and Health Statistics*. Series 2, No. 45.

Marquis, M. Susan, and Kent H. Marquis. 1977. *Survey Measurement Design and Evaluation Using Reliability Theory*. Santa Monica, CA: RAND.

Moser, Claus A., and Graham Kalton. 1972. *Survey Methods in Social Investigation*. 2nd ed. New York: Basic Books.

Muthén, Bengt O. 1984. "A General Structural Equation Model with Dichotomous, Ordered Categorical, and Continuous Latent Variable Indicators." *Psychometrika* 49(1):115–32.

Payne, Stanley L. 1951. *The Art of Asking Questions*. Princeton, NJ: Princeton University Press.

Ruckmick, Christian A. 1930. "The Uses and Abuses of the Questionnaire Procedure." *Journal of Applied Psychology* 14(1):32–41.

Saris, Willem E., and Frank M. Andrews. 1991. "Evaluation of Measurement Instruments Using a Structural Modeling Approach." Pp. 575–97 in *Measurement Errors in Surveys*, edited by Paul B. Biemer, Robert M. Groves, Lars E. Lyberg, Nancy A. Mathiowetz, and Seymour Sudman. New York: John Wiley.

Saris, Willem E., and Irmtraud N. Gallhofer. 2007. *Design, Evaluation, and Analysis of Questionnaires for Survey Research*. New York: John Wiley.

Saris, Willem E., and A. van Meurs. 1990. *Evaluation of Measurement Instruments by Meta-analysis of Multitrait Multimethod Studies*. Amsterdam, the Netherlands: North-Holland.

Schaeffer, Nora Cate, and Jennifer Dykema. 2011. "Questions for Surveys: Current Trends and Future Directions." *Public Opinion Quarterly* 75(5):909–61.

Schaeffer, Nora Cate, and Stanley Presser. 2003. "The Science of Asking Questions." *Annual Review of Sociology* 29:65–88.

Scherpenzeel, Annette C. 1995. "A Question of Quality: Evaluating Survey Questions by Multitrait-multimethod Studies." PhD dissertation, Department of Methodology, University of Amsterdam.

Scherpenzeel, Annette C., and Willem E. Saris. 1997. "The Validity and Reliability of Survey Questions: A Meta-analysis of MTMM Studies." *Sociological Methods and Research* 25(3):341–83.

Schuman, Howard, and Graham Kalton. 1985. "Survey Methods." Pp. 634–97 in *Handbook of Social Psychology*, 3rd ed., edited by Gardner Lindzey and Eliott Aronson. New York: Random House.

Schuman, Howard, and Stanley Presser. 1981. *Questions and Answers: Experiments in Question Wording, Form, and Context*. New York: Academic Press.

Sudman, Seymour, and Norman M. Bradburn. 1974. *Response Effects in Surveys*. Chicago: Aldine.

Sudman, Seymour, and Norman N. Bradburn. 1982. *Asking Questions: A Practical Guide to Questionnaire Design*. San Francisco, CA: Jossey-Bass.

Sudman, Seymour, Norman M. Bradburn, and Norbert Schwarz. 1996. *Thinking about Answers: The Application of Cognitive Processes to Survey Methodology*. San Francisco, CA: Jossey-Bass.

Tanur, Judith M., ed. 1992. *Questions about Questions—Inquiries into the Cognitive Bases of Surveys*. New York: Russell Sage.

Tourangeau, Roger, Lance J. Rips, and Kenneth Rasinski. 2000. *The Psychology of Survey Response*. Cambridge, UK: Cambridge University Press.

Turner, Charles F., and Elizabeth Martin. 1984. *Surveying Subjective Phenomena*. Vol. 1. New York: Russell Sage.

van der Zouwen, Johannes. 1999. "An Assessment of the Difficulty of Questions Used in the ISSP-Questionnaires, the Clarity of Their Wording, and the Comparability of the Responses." *ZA-Information* 46(1):96–114.

Wiley, David E., and James A. Wiley. 1970. "The Estimation of Measurement Error in Panel Data." *American Sociological Review* 35(1):112–17.

Yan, Ting, and Roger Tourangeau. 2008. "Fast Times and Easy Questions: The Effects of Age, Experience, and Question Complexity on Web Survey Response Times." *Applied Cognitive Psychology* 22(1):51–68.

Author Biographies

Duane F. Alwin is the inaugural holder of the Tracy Winfree and Ted H. McCourtney Professorship in Sociology and Demography at Pennsylvania State University, where he directs the Center for Life Course and Longitudinal Studies. He is also emeritus research professor at the Survey Research Center, Institute for Social Research, and emeritus professor of sociology, University of Michigan, Ann Arbor. In addition to survey methodology, his research interests focus on the integration of demographic and developmental perspectives in the study of human lives. His current research work includes the study of socioeconomic inequality and health, parental child-rearing values, children's use of time, and social factors in cognitive aging. He has published extensively on these and related topics and is the recipient of numerous prestigious awards, grants, and special university honors.

Brett A. Beattie is a 2015 PhD recipient from the Department of Sociology and Criminology, Pennsylvania State University, specializing in family demography and quantitative methodology.

Sociological Methodology
2016, Vol. 46(1) 153–186
© American Sociological Association 2016
DOI: 10.1177/0081175016665425
http://sm.sagepub.com

❧ 5 ☙

GENERALIZING THE NETWORK SCALE-UP METHOD: A NEW ESTIMATOR FOR THE SIZE OF HIDDEN POPULATIONS

Dennis M. Feehan*
Matthew J. Salganik[†]

Abstract

The network scale-up method enables researchers to estimate the sizes of hidden populations, such as drug injectors and sex workers, using sampled social network data. The basic scale-up estimator offers advantages over other size estimation techniques, but it depends on problematic modeling assumptions. The authors propose a new generalized scale-up estimator that can be used in settings with nonrandom social mixing and imperfect awareness about membership in the hidden population. In addition, the new estimator can be used when data are collected via complex sample designs and from incomplete sampling frames. However, the generalized scale-up estimator also requires data from two samples: one from the frame population and one from the hidden population. In some situations these data from the hidden population can be collected by adding a small number of questions to already planned studies. For other situations, the authors develop interpretable adjustment factors that can be applied to the basic scale-up estimator.

*University of California, Berkeley, CA, USA
[†]Princeton Univerity, Princeton, NJ, USA

Corresponding Author:
Dennis M. Feehan, University of California, Berkeley, Department of Demography, 2232 Piedmont Avenue, Berkeley, CA 94720, USA
Email: feehan@berkeley.edu

The authors conclude with practical recommendations for the design and analysis of future studies.

Keywords

hidden populations, social networks, sampling, network scale-up method

1. INTRODUCTION

Many important problems in social science, public health, and public policy require estimates of the sizes of hidden populations. For example, in HIV/AIDS research, estimates of the size of the most at-risk populations—drug injectors, female sex workers, and men who have sex with men—are critical for understanding and controlling the spread of the epidemic. However, researchers and policymakers are unsatisfied with the ability of current statistical methods to provide these estimates (Joint United Nations Programme on HIV/AIDS 2010). We address this problem by improving the network scale-up method, a promising approach to size estimation. Our results are immediately applicable in many substantive domains in which size estimation is challenging, and the framework we develop advances the understanding of sampling in networks more generally.

The core insight behind the network scale-up method is that ordinary people have embedded within their personal networks information that can be used to estimate the sizes of hidden populations, if that information can be properly collected, aggregated, and adjusted (Bernard et al. 1989, 2010). In a typical scale-up survey, randomly sampled adults are asked about the number of connections they have to people in a hidden population (e.g., "How many people do you know who inject drugs?") and a series of similar questions about groups of known size (e.g., "How many widowers do you know?" "How many doctors do you know?"). Responses to these questions are called *aggregate relational data* (McCormick et al. 2012).

To produce size estimates from aggregate relational data, previous researchers have begun with the *basic scale-up model*, which makes three important assumptions: (1) Social ties are formed completely at random (i.e., random mixing), (2) respondents are perfectly aware of the characteristics of their alters, and (3) respondents are able to provide accurate answers to survey questions about their personal networks. From the basic scale-up model, Killworth, McCarty, et al. (1998)

derived the *basic scale-up estimator*. This estimator, which is widely used in practice, has two main components. For the first component, the aggregate relational data about the hidden population are used to estimate the number of connections that respondents have to the hidden population. For the second component, the aggregate relational data about the groups of known size are used to estimate the number of connections that respondents have in total. For example, a researcher might estimate that members of her sample have 5,000 connections to people who inject drugs and 100,000 connections in total. The basic scale-up estimator combines these pieces of information to estimate that 5 percent $(5,000/100,000)$ of the population injects drugs. This estimate is a sample proportion, but rather than being taken over the respondents, as would be typical in survey research, the proportion is taken over the respondents' alters. Researchers who desire absolute size estimates multiply the alter sample proportion by the size of the entire population, which is assumed to be known (or estimated using some other method).

Unfortunately, the three assumptions underlying the basic scale-up model have all been shown to be problematic. Scale-up researchers call violations of the random mixing assumption *barrier effects* (Killworth et al. 2006; Maltiel et al. 2015; Zheng, Salganik, and Gelman 2006), they call violations of the perfect awareness assumption *transmission error* (Killworth et al. 2006; Maltiel et al. 2015; Salganik, Mello, et al. 2011; Shelley et al. 1995, 2006), and they call violations of the respondent accuracy assumption *recall error* (Killworth et al. 2003, 2006; Maltiel et al. 2015; McCormick and Zheng 2007).

In this paper, we develop a new approach to producing size estimates from aggregate relational data. Rather than depending on the basic scale-up model or its variants (e.g., Maltiel et al. 2015), we use a simple identity to derive a series of new estimators. Our new approach reveals that one of the two main components of the basic scale-up estimator is problematic. Therefore, we propose a new estimator—the *generalized scale-up estimator*—that combines the aggregate relational data traditionally used in scale-up studies with similar data collected from the hidden population. Collecting data from the hidden population is a major departure from current scale-up practice, but we believe that it enables a more principled approach to estimation. For researchers who are not able to collect data from the hidden population, we propose a series of adjustment factors that highlight the possible biases of the basic scale-up estimator. Ultimately, researchers must balance the trade-offs between the basic scale-up estimator, generalized scale-up estimator, and other size estimation techniques on the basis of the specific features of their research setting.

In Section 2, we derive the generalized scale-up estimator, and we describe the data collection procedures needed to use it. In Section 3, we compare the generalized and basic scale-up approaches analytically and with simulations; our comparison leads us to propose a decomposition that separates the difference between the two approaches into three measurable and substantively meaningful factors (equation 15). In Section 4, we make practical recommendations for the design and analysis of future scale-up studies, and in Section 5, we conclude with a discussion of the steps that follow. Appendices A to G in the online journal provide technical details and supporting arguments.

2. THE GENERALIZED SCALE-UP ESTIMATOR

The generalized scale-up estimator can be derived from a simple accounting identity that requires no assumptions about the underlying social network structure in the population. Figure 1 helps illustrate the derivation, which was inspired by earlier research on multiplicity estimation (Sirken 1970) and indirect sampling (Lavallée 2007). Consider a population of seven people, two of whom are drug injectors (Figure 1a). In this population, two people are connected by a directed edge $i \rightarrow j$ if person i would count person j as a drug injector when answering the question "How many drug injectors do you know?" Whenever $i \rightarrow j$, we say that i makes an *out-report* about j and that j receives an *in-report* from i.[1]

Each person can be viewed as both a source of out-reports and a recipient of in-reports, and in order to emphasize this point, Figure 1b shows the population with each person represented twice: on the left as a sender of out-reports and on the right as a receiver of in-reports. This visual representation highlights the following identity:

$$\text{total out-reports} = \text{total in-reports}. \qquad (1)$$

Despite its simplicity, the identity in equation (1) turns out to be very useful because it leads directly to the new estimator that we propose.

In order to derive an estimator from equation (1), we must define some notation. Let U be the entire population, and let $H \subset U$ be the hidden population. Furthermore, let $y_{i,H}$ be the total number of

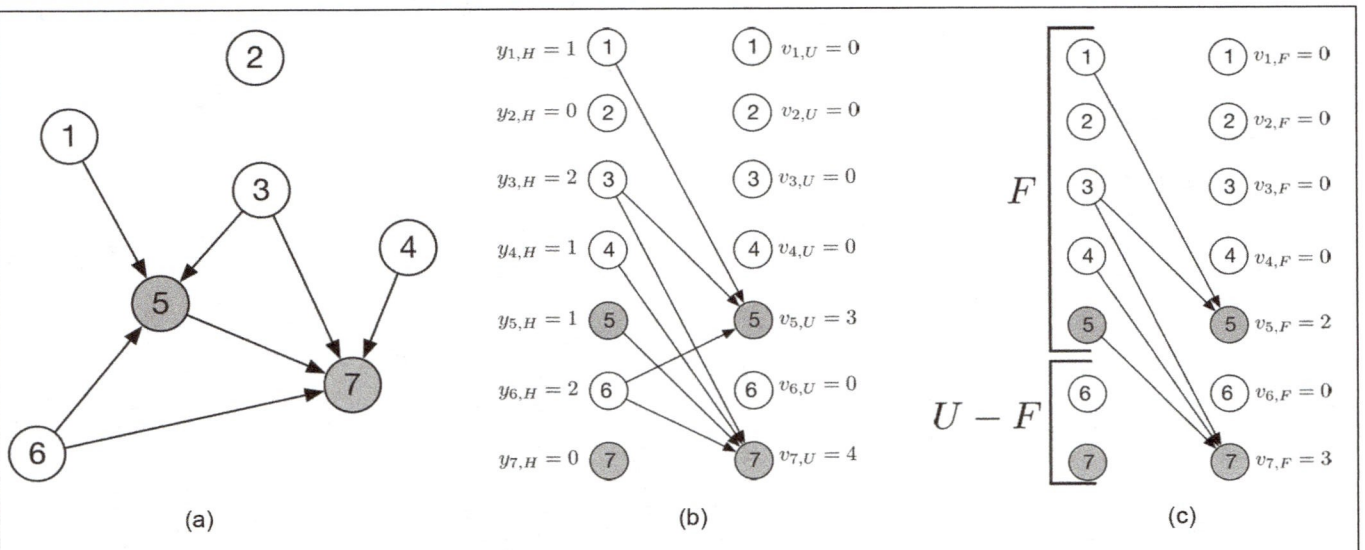

Figure 1. Illustration of the derivation of the generalized scale-up estimator. (a) A population of seven people, two of whom are drug injectors (shown in gray). A directed edge $i \rightarrow j$ indicates that i counts j as a drug injector when answering the question "How many drug injectors do you know?" (b) The same population, redrawn so that each person now appears twice: as a source of out-reports, on the left, and as a recipient of in-reports, on the right. This arrangement shows that total out-reports and total in-reports must be equal. (c) The same population again, but now some of the people are in the frame population F and some are not. In real scale-up studies, we can only learn about out-reports from the frame population.

out-reports from person i (i.e., person i's answer to the question "How many drug injectors do you know?"). For example, Figure 1b shows that person 5 would report knowing one drug injector, so $y_{5,H} = 1$. Let $v_{i,U}$ be the total number of in-reports to i if everyone in U is interviewed; that is, $v_{i,U}$ is the *visibility* of person i to people in U. For example, Figure 1b shows that person 5 would be reported as a drug injector by three people so $v_{5,U} = 3$. Because total out-reports must equal total in-reports, it must be the case that

$$y_{U,H} = v_{U,U}, \tag{2}$$

where $y_{U,H} = \sum_{i \in U} y_{i,H}$ and $v_{U,U} = \sum_{i \in U} v_{i,U}$. Multiplying both sides of equation (2) by N_H, the number of people in the hidden population, and then rearranging terms, we get

$$N_H = \frac{y_{U,H}}{v_{U,U}/N_H}. \tag{3}$$

Equation (3) is an expression for the size of the hidden population that does not depend on any assumptions about network structure or reporting accuracy; it is just a different way of expressing the identity that the total number of out-reports must equal the total number of in-reports. If we could estimate the two terms on the right side of equation (3)—one term related to out-reports ($y_{U,H}$) and one term related to in-reports ($v_{U,U}/N_H$)—then we could estimate N_H.

However, in order to make the identity in equation (3) useful in practice we need to modify it to account for an important logistical requirement of survey research. In real scale-up studies, researchers do not sample from the entire population U, but instead they sample from a subset of U called the frame population, F. For example, in almost all scale-up studies the frame population has been adults (but note that our mathematical results hold for any frame population). In standard survey research, restricting interviews to a frame population does not cause problems because inference is being made about the frame population. In other words, when respondents report about themselves, it is clear to which group inferences apply. However, with the scale-up method, respondents report about others, so the group that inferences are being made about is not necessarily the same as the group that is being interviewed. As we show in Section 4.2, failure to consider this fact requires the introduction of an awkward adjustment factor that had previously gone unnoticed. Here, we avoid this awkward adjustment factor by deriving an identity explicitly in terms of the frame population. Restricting our attention to out-reports coming from people in the frame population, it must be the case that

$$N_H = \frac{y_{F,H}}{v_{U,F}/N_H},$$ (4)

where $y_{F,H} = \sum_{i \in F} y_{i,H}$ and $v_{U,F} = \sum_{i \in U} v_{i,F}$. The only difference between equation (3) and equation (4) is that equation (4) restricts out-reports and in-reports to come from people in the frame population (Figure 1c). The identity in equation (4) is extremely general: it does not depend on any assumptions about the relationship between the entire population U, the frame population F, and the hidden population H. For example, it holds if no members of the hidden population are in the frame population, if there are barrier effects, and if there are transmission errors. Thus, if we could estimate the two terms on the right side of equation (4)—one term related to out-reports ($y_{F,H}$) and one term related to in-reports ($v_{U,F}/N_H$)—then we could estimate N_H under very general conditions.

Unfortunately, despite repeated attempts, we were unable to develop a practical method for estimating the term related to in-reports ($v_{U,F}/N_H$). However, if we make an assumption about respondents' reporting behavior, then we can re-express equation (4) as an identity made up of quantities that we can actually estimate. Specifically, if we assume that the out-reports from people in the frame population only include people in the hidden population, then it must be the case that the visibility of everyone not in the hidden population is 0: $v_{i,F} = 0$ for all $i \notin H$. In this case, we can rewrite equation (4) as

$$N_H = \frac{y_{F,H}}{v_{H,F}/N_H} = \frac{y_{F,H}}{\bar{v}_{H,F}} \quad \text{if } v_{i,F} = 0 \text{ for all } i \notin H,$$ (5)

where $\bar{v}_{H,F} = v_{H,F}/N_H$.

To understand the reporting assumption substantively, consider the two possible types of reporting errors: false positives and false negatives. Previous scale-up research on transmission error focused on the problem of false negatives, where a respondent is connected to a member of the hidden population but does not report this, possibly because she is not aware that the person she is connected to is in the hidden population (Bernard et al. 2010). Because hidden populations such as drug

injectors are often stigmatized, it is reasonable to suspect that false negatives will be a serious problem for the scale-up method. Fortunately,
equation (5) holds even if there are false negative reporting errors.
However, false positives—which do not seem to have been considered
previously in the scale-up literature—are also possible. For example, a
respondent who is not connected to any drug injectors might report that
one of her acquaintances is a drug injector. These false positive reports
are not accounted for in the identity in equation (5) and the estimators
that we derive subsequently. If false positive reports exist, they will introduce a positive bias into estimates from the generalized scale-up estimator. Therefore, in Appendix A in the online journal we (1) formally
define an interpretable measure of false positive reports, the *precision of
out-reports*; (2) analytically show the bias in size estimates as a function
of the precisions of out-reports; and (3) discuss two research designs that
could enable researchers to estimate the precision of out-reports.

2.1. Estimating N_H from Sampled Data

Equation (5) relates our quantity of interest, the size of the hidden population (N_H), to two other quantities: the total number of out-reports from
the frame population ($y_{F,H}$) and the average number of in-reports in the
hidden population ($\bar{v}_{H,F}$). We now show how to estimate $y_{F,H}$ with a
probability sample from the frame population and $\bar{v}_{H,F}$ with a relative
probability sample from the hidden population.

The total number of out-reports ($y_{F,H}$) can be estimated from respondents' reported number of connections to the hidden population,

$$\widehat{y}_{F,H} = \sum_{i \in s_F} \frac{y_{i,H}}{\pi_i}, \qquad (6)$$

where s_F denotes the sample, $y_{i,H}$ denotes the reported number of connections between i and H, and π_i is i's probability of inclusion from a
conventional probability sampling design from the frame population.
Because $\widehat{y}_{F,H}$ is a standard Horvitz-Thompson estimator, it is consistent
and unbiased as long as all members of F have a positive probability of
inclusion under the sampling design (Sarndal, Swensson, and Wretman
1992); for a more formal statement, see Result B.1. This estimator
depends only on an assumption about the sampling design for the frame
population, and in Table D.2, we show the sensitivity of our estimator
to violations of this assumption.

Estimating the average number of in-reports for the hidden population ($\bar{v}_{H,F}$) is more complicated. First, it will usually be impossible to obtain a conventional probability sample from the hidden population. As we show below, however, estimating $\bar{v}_{H,F}$ requires only a relative probability sampling design in which hidden population members have a nonzero probability of inclusion and respondents' probabilities of inclusion are known up to a constant of proportionality, $c\pi_i$ (see Appendix C.1 in the online journal for a more precise definition). Of course, even selecting a relative probability sample from a hidden population can be difficult.

A second problem arises because we do not expect respondents to be able to easily and accurately answer direct questions about their visibility ($v_{i,F}$). That is, we do not expect respondents to be able to answer questions such as "How many people on the sampling frame would include you when reporting a count of the number of drug injectors that they know?" Instead, we propose asking hidden population members a series of questions about their connections to certain groups and their visibility to those groups. For example, each sampled hidden population respondent could be asked "How many widowers do you know?" and then "How many of these widowers are aware that you inject drugs?" This question pattern can be repeated for many groups (e.g., widowers, doctors, bus drivers). We call data with this structure *enriched aggregate relational data* to emphasize its similarity to the aggregate relational data that is familiar to scale-up researchers. An interviewing procedure called the *game of contacts* enables researchers to collect enriched aggregated relational data, even in realistic field settings (Salganik, Mello, et al. 2011; Maghsoudi et al. 2014).

Given a relative probability sampling design and enriched aggregate relational data, we can now formalize our proposed estimator for $\bar{v}_{H,F}$. Let $A_1, A_2, \ldots, A_J$, be the set of groups about which we collect enriched aggregate relational data (e.g., widowers, doctors). Here, to keep the notation simple, we assume that these groups are all contained in the frame population, so that $A_j \subset F$ for all j; in Appendix C.4 in the online journal, we extend the results to groups that do not meet this criterion. Let $\mathcal{A}$ be the concatenation of these groups, which we call the *probe alters*. For example, if A_1 is widowers and A_2 is doctors, then the probe alters $\mathcal{A}$ is the collection of all widowers and all doctors, with doctors who are widowers included twice. Also, let $\tilde{v}_{i,A_j}$ be respondent i's report about her visibility to people in A_j and let v_{i,A_j} be respondents i's actual visibility to people in A_j (i.e., the number of times that this respondent would be reported about if everyone in A_j was asked about their connections to the hidden population).

The estimator for $\bar{v}_{H,F}$ is

$$\widehat{\bar{v}}_{H,F} = \frac{N_F}{N_A} \frac{\sum_{i \in s_H} \sum_j \tilde{v}_i, A_j / (c\pi_i)}{\sum_{i \in s_H} 1/(c\pi_i)}, \tag{7}$$

where N_A is the number of probe alters, c is the constant of proportionality from the relative probability sample, and s_H is a relative probability sample of the hidden population. Equation (7) is a standard weighted sample mean (Sarndal et al. 1992, Section 5.7) multiplied by a constant, N_F/N_A. Result C.2 shows that this estimator is consistent and *essentially unbiased*,[2] when three conditions are satisfied: one about the design of the survey, one about reporting behavior, and one about sampling from the hidden population.

The first condition underlying the estimator in equation (7) is related to the design of the survey, and we call it the *probe alter condition*. This condition describes the required relationship between the visibility of the hidden population to the probe alters and the visibility of the hidden population to the frame population:

$$\frac{v_{H,A}}{N_A} = \frac{v_{H,F}}{N_F}, \tag{8}$$

where $v_{H,A}$ is the total visibility of the hidden population to the probe alters, $v_{H,F}$ is the total visibility of the hidden population to the frame population, N_A is the number of probe alters, and N_F is the number of people in the frame population. In words, equation (8) says that the rate at which the hidden population is visible to the probe alters must be the same as the rate at which the hidden population is visible to the frame population. For example, in a study to estimate the number of drug injectors in a city, drug treatment counselors would be a poor choice for membership in the probe alters because drug injectors are probably more visible to drug treatment counselors than to typical members of the frame population. On the other hand, postal workers would probably be a reasonable choice for membership in the probe alters because drug injectors are probably about as visible to postal workers as they are to typical members of the frame population. Additional results about the probe alter condition are presented in the online appendices: (1) Result C.3 presents three other algebraically equivalent formulations of probe alter condition, some of which offer additional intuition; (2) Result C.4 provides a method to empirically test the probe alter condition; and (3) Table D.1 quantifies the bias introduced when the probe alter condition is not satisfied.

The second condition underlying the estimator $\widehat{v}_{H,F}$ (equation 7) is related to reporting behavior, and we call it *accurate aggregate reports about visibility*:

$$\tilde{v}_{H,\mathcal{A}} = v_{H,\mathcal{A}}, \tag{9}$$

where $\tilde{v}_{H,\mathcal{A}}$ is the total reported visibility of members of the hidden population to the probe alters ($\sum_{i \in H} \sum_{A_j \in \mathcal{A}} \tilde{v}_{i,A_j}$) and $v_{H,\mathcal{A}}$ is the total actual visibility of members of the hidden population to the probe alters ($\sum_{i \in H} \sum_{A_j \in \mathcal{A}} v_{i,A_j}$). In words, equation (9) says that hidden population members must be correct in their reports about their visibility to probe alters in aggregate, but equation (9) does not require the stronger condition that each individual report be accurate. In practice, we expect that there are two main ways that there might not be accurate aggregate reports about visibility. First, hidden population members might not be accurate in their assessments of what others know about them. For example, research on the "illusion of transparency" suggests that people tend to overestimate how much others know about them (Gilovich, Savitsky, and Medvec 1998). Second, although we propose asking hidden population members what other people know about them (e.g., "How many of these widowers know that you are a drug injector?") what actually matters for the estimator is what other people would report about them (e.g., "How many of these widowers would include you when reporting a count of the number of drug injectors that they know?"). In cases in which the hidden population is extremely stigmatized, some respondents to the scale-up survey might conceal the fact that they are connected to people whom they know to be in the hidden population, and if this were to occur, it would lead to a difference between the information that we collect ($\tilde{v}_{i,\mathcal{A}}$) and the information that we want ($v_{i,\mathcal{A}}$). Unfortunately, there is currently no empirical evidence about the possible magnitude of these two problems in the context of scale-up studies. However, Table D.1 quantifies the bias introduced into estimates if the accurate aggregate reports about visibility condition is not satisfied.

Finally, the third condition underlying the estimator $\widehat{v}_{H,F}$ (equation 7) is that researchers have a relative probability sample from the hidden population. Currently the most widely used method for drawing relative probability samples from hidden populations is respondent-driven sampling (Heckathorn 1997); see Volz and Heckathorn (2008) for a set of conditions under which respondent-driven sampling leads to a relative probability sample. Although respondent-driven sampling has been used in hundreds of studies around the world (White et al. 2015), there is active debate about the characteristics of samples that it yields (Bengtsson and Thorson 2010; Gile and Handcock 2010, 2015; Gile, Johnston, and Salganik 2015; Goel and Salganik 2010; Heimer 2005; Li and Rohe 2015; McCreesh et al. 2012; Mills et al. 2012; Rohe 2015; Rudolph et al. 2013; Salganik 2012; Scott 2008; Yamanis et al. 2013). If other methods for sampling from hidden populations are demonstrated to be better than respondent-driven sampling (e.g., see Karon and Wejnert 2012; Kurant, Markopoulou, and Thiran 2011; Mouw and Verdery 2012), then researchers should consider these methods when using the generalized scale-up estimator. Furthermore, researchers can use Table D.2 to quantify the bias that results if the condition requiring a relative probability sample is not satisfied.

To recap, using two different data collection procedures—one with the frame population and one with the hidden population—we can estimate the two components of the expression for N_H given in equation (5). The estimator for the numerator $(\widehat{y}_{F,H})$ depends on an assumption about the ability to select a probability sample from the frame population (see Result B.1), and the estimator for the denominator $(\widehat{v}_{H,F})$ depends on assumptions about survey construction, reporting behavior, and the ability to select a relative probability sample from the hidden population (see Result C.2).

We can combine these component estimators to form the *generalized scale-up estimator*:

$$\widehat{N}_H = \frac{\widehat{y}_{F,H}}{\widehat{v}_{H,F}}. \tag{10}$$

Result C.8 proves that the generalized scale-up estimator will be consistent and essentially unbiased if (1) the estimator for the numerator $(\widehat{y}_{F,H})$ is consistent and essentially unbiased, (2) the estimator for the denominator $(\widehat{\overline{v}}_{H,F})$ is consistent and essentially unbiased, and (3) there are no false positive reports.

One attractive feature of the generalized scale-up estimator (equation 10) is that it is a combination of standard survey estimators. This structure enabled us to derive very general sensitivity results about the impact of violations of assumptions, either individually or jointly. We return to the issue of assumptions and sensitivity analysis when discussing recommendations for practice (Section 4).

3. COMPARISON BETWEEN THE GENERALIZED AND BASIC SCALE-UP APPROACHES

In Section 2, we derived the generalized network scale-up estimator by using an identity relating in-reports and out-reports as the basis for a design-based estimator. The approach we followed differs from previous scale-up studies, which have posited the basic scale-up model and derived estimators conditional on that model. In this section, we compare these two different approaches from a design-based perspective.

We begin our comparison by reviewing the basic scale-up model, which was used in most of the studies listed in Table 1. To review this model, we need to define another quantity: we call $d_{i,U}$ person i's degree, the number of undirected network connections she has to everyone in U.

The basic scale-up model assumes that each person's connections are formed independently, that reporting is perfect, and that visibility is perfect (Killworth, McCarty, et al. 1998). Together, these three assumptions lead to the probabilistic model:

$$y_{i,A_j} = d_{i,A_j} \sim \text{Binomial}\left(d_{i,U}, \frac{N_{Aj}}{N}\right), \tag{11}$$

for all i in U and for any group A_j. In words, this model suggests that the number of connections from a person i to members of a group A_j is the result of a series of $d_{i,U}$ independent random draws, where the probability of each edge being connected to A_j is $\frac{N_{Aj}}{N}$.

The basic scale-up model leads to what we call the basic scale-up estimator:

$$\widehat{N}_H = \frac{\sum_{i \in s_F} y_{i,H}}{\sum_{i \in s_F} \widehat{d}_{i,U}} \times N, \tag{12}$$

Table 1. Network Scale-up Studies That Have Been Completed

Hidden Population(s)	Location	Citation
Mortality in earthquake	Mexico City, Mexico	Bernard et al. (1989)
Rape victims	Mexico City, Mexico	Bernard et al. (1991)
HIV prevalence, rape, and homelessness	United States	Killworth, McCarty, et al. (1998)
Heroin use	14 U.S. cities	Kadushin et al. (2006)
Choking incidents in children	Italy	Snidero et al. (2007, 2009, 2012)
Groups most at risk for HIV/AIDS	Ukraine	Paniotto et al. (2009)
Heavy drug users	Curitiba, Brazil	Salganik, Fazito, et al. (2011)
Groups most at risk for HIV/AIDS	Kerman, Iran	Shokoohi, Baneshi, and Haghdoost (2012)
Men who have sex with men	Japan	Ezoe et al. (2012)
Groups most at risk for HIV/AIDS	Almaty, Kazakhstan	Scutelniciuc (2012a)
Groups most at risk for HIV/AIDS	Moldova	Scutelniciuc (2012b)
Groups most at risk for HIV/AIDS	Thailand	Aramrattan and Kanato (2012)
Groups most at risk for HIV/AIDS	Rwanda	Rwanda Biomedical Center (2012)
Groups most at risk for HIV/AIDS	Chongqing, China	Guo et al. (2013)
Groups most at risk for HIV/AIDS	Tabriz, Iran	Khounigh et al. (2014)
Men who have sex with men	Taiyuan, China	Jing et al. (2014)
Drug and alcohol users	Kerman, Iran	Sheikhzadeh et al. (2014)
Men who have sex with men	Shanghai, China	Wang et al. (2015)
Men who have sex with men	Tbilisi, Georgia	Sulaberidze et al. (2016)
Illicit drug users	Iran	Nikfarjam et al. (2016)

where $\widehat{d}_{i,U}$ is the estimated degree of respondent i from the known population method (Killworth, Johnsen, et al. 1998). Killworth, McCarty, et al. (1998) showed that equation (12) is the maximum likelihood estimator for N_H under the basic scale-up model, conditional on the additional assumption that $d_{i,U}$ is known for each $i \in s_F$.

Given this background, we can now compare the basic and generalized scale-up approaches by comparing their estimands; that is, we compare the quantities that they produce in the case of a census with perfectly observed degrees. The basic scale-up estimand can be written

$$\widehat{N}_H = \frac{y_{F,H}}{d_{F,U}} \times N = \frac{y_{F,H}}{\bar{d}_{U,F}}, \tag{13}$$

where $d_{F,U} = \sum_{i \in F} d_{i,U}$ and $\bar{d}_{U,F} = d_{U,F}/N = d_{F,U}/N$. Furthermore, as shown in Section 2, the generalized scale-up estimand is

$$\widehat{N}_H = \frac{y_{F,H}}{\bar{v}_{H,F}}. \tag{14}$$

Comparing equations (13) and (14) reveals that both estimands have the same numerator, but they have different denominators. The network reporting identity from Section 2 (total out-reports = total in-reports) shows that the appropriate way to adjust the out-reports is based on in-reports, as in the generalized scale-up approach. However, the basic scale-up approach instead adjusts out-reports with the degree of respondents. Although using the degree of respondents cleverly avoids any data collection from the hidden population, our results reveal that it will be correct only under a very specific special case ($\bar{d}_{U,F} = \bar{v}_{H,F}$).

To further clarify the relationship between the basic and generalized scale-up approaches, we propose a decomposition that separates the difference between the two estimands into three measurable and substantively meaningful *adjustment factors*:

$$N_H = \underbrace{\left(\frac{y_{F,H}}{\bar{d}_{U,F}}\right)}_{\substack{\text{basic} \\ \text{scale-up}}} \times \underbrace{\frac{1}{\bar{d}_{F,F}/\bar{d}_{U,F}}}_{\substack{\text{frame ratio} \\ \phi_F}} \times \underbrace{\frac{1}{\bar{d}_{H,F}/\bar{d}_{F,F}}}_{\substack{\text{degree ratio} \\ \delta_F}} \times \underbrace{\frac{1}{\bar{v}_{H,F}/\bar{d}_{H,F}}}_{\substack{\text{true positive rate} \\ \tau_F}} = \underbrace{\left(\frac{y_{F,H}}{\bar{v}_{H,F}}\right)}_{\substack{\text{generalized} \\ \text{scale-up}}}. \tag{15}$$

$$\underbrace{}_{\text{adjustment factors}}$$

The decomposition shows that when the product of the adjustment factors is 1, the two estimands are both correct. However, when the product of the adjustment factors is not 1, then the generalized scale-up estimand is correct but the basic scale-up estimand is incorrect. We now describe each of the three adjustment factors in turn.

First, we define the frame ratio, ϕ_F, to be

$$\phi_F = \frac{\text{avg \# connections from a member of } F \text{ to the rest of } F}{\text{avg \# connections from a member of } U \text{ to } F} = \frac{\bar{d}_{F,F}}{\bar{d}_{U,F}}. \quad (16)$$

ϕ_F can range from 0 to infinity, and in most practical situations we expect that ϕ_F will be greater than 1. Result B.6 shows that we can make consistent and essentially unbiased estimates of ϕ_F from a sample of F.[3]

Next, we define the degree ratio δ_F to be

$$\delta_F = \frac{\text{avg \# connections from a member of } H \text{ to } F}{\text{avg \# connections from a member of } F \text{ to the rest of } F} = \frac{\bar{d}_{H,F}}{\bar{d}_{F,F}}. \quad (17)$$

δ_F ranges from 0 to infinity, and it is less than 1 when the hidden population members have, on average, fewer connections to the frame population than frame population members. Result C.6 shows that we can make consistent and essentially unbiased estimates of δ_F from samples of F and H.

Finally, we define the true positive rate, τ_F, to be

$$\tau_F = \frac{\text{\# in-reports to } H \text{ from } F}{\text{\# edges connecting } H \text{ and } F} = \frac{v_{H,F}}{d_{H,F}} = \frac{\bar{v}_{H,F}}{\bar{d}_{H,F}}. \quad (18)$$

τ_F relates network degree to network reports.[4] τ_F ranges from 0, if none of the edges are correctly reported, to 1 if all of the edges are reported. Substantively, the more stigmatized the hidden population, the closer we would expect τ_F to be to 0. Result C.7 shows that we can make consistent and essentially unbiased estimates of τ_F from a sample of H.

Furthermore, the decomposition in equation (15) can be used to derive an expression for the bias in the basic scale-up estimator when we have a census and degrees are known:

$$\text{bias}\left(\widehat{N}_H^{\text{basic}}\right) \equiv \widehat{N}_H^{\text{basic}} - N_H \quad (19)$$

$$= \widehat{N}_H^{\text{basic}}\left[1 - \frac{1}{\phi_F \delta_F \tau_F}\right]. \quad (20)$$

The comparison between the basic and generalized scale-up approaches leads to two main conclusions. First, the estimand of the basic scale-up approach is correct only in one particular situation: when the product of the three adjustment factors is 1. The estimand of generalized scale-up approach, in contrast, is correct more generally. Second, as equation 15 shows, if the adjustment factors are known (or have been estimated), then they can be used to improve basic scale-up estimates.

3.1. *Illustrative Simulation*

To illustrate our comparison between the basic and generalized scale-up approaches, we conducted a series of simulation studies. The simulations were not meant to be a realistic model of a scale-up study, but rather, they were designed to clearly illustrate our analytic results. More specifically, the simulation investigated the performance of the estimators as three important quantities vary: (1) the size of the frame population F, relative to the size of the entire population U; (2) the extent to which people's network connections are not formed completely at random; and (3) the accuracy of reporting, as captured by the true positive rate τ_F (see equation 18).[5]

As described in detail in Appendix G in the online journal, we created populations of $5,000$ people with different proportions of the population on the sampling frame (p_F). Next, we connected the people with a social network created by a stochastic block model (Wasserman and Faust 1994; White, Boorman, and Breiger 1976) in which the randomness of the mixing was controlled by a parameter ρ such that $\rho = 1$ is equivalent to random mixing (i.e., an Erdos-Reyni random graph) and the mixing becomes more nonrandom as $\rho \rightarrow 0$. Then, for each combination of parameters, we drew 10 populations, and within each of these populations, we simulated 500 surveys. For each survey, we drew a probability sample of 500 people from the frame population, a relative probability sample of 30 people from the hidden population, and simulated responses with a specific level of reporting accuracy (τ_F). Finally, we used these reports and the appropriate sampling weights to calculate the basic and generalized scale-up estimates.

Figure 2 shows that the simulations support our analytic results. First, the simulations show that the generalized scale-up estimator is unbiased even in the presence of incomplete sampling frames, nonrandom mixing, and imperfect reporting. Second, they show that the basic scale-up

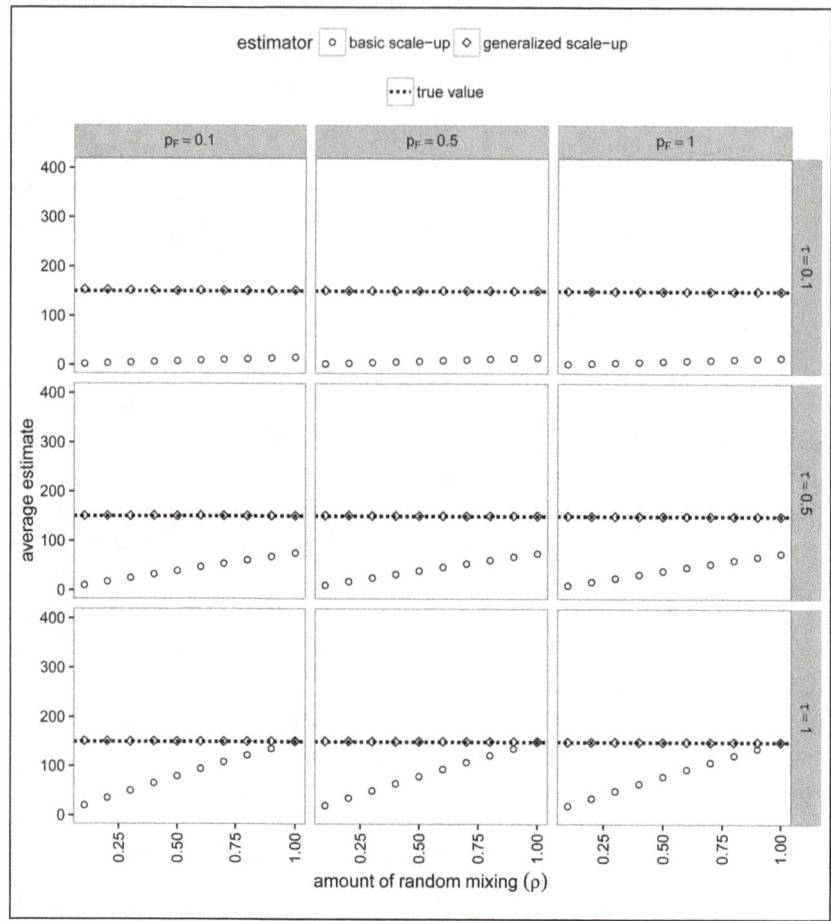

Figure 2. Estimated size of the hidden population for the generalized and basic scale-up estimators. Each panel shows how the two estimators change as the amount of random mixing is varied from low ($\rho = 0.1$; members of the hidden population are relatively unlikely to form contacts with nonmembers) to high ($\rho = 1$; members of the hidden population form contacts independent of other people's hidden population membership). The columns show results for different sizes of the frame population, from small (left column, $p_F = 0.1$), to large (right column, $p_F = 1$). The rows show results for different levels of reporting accuracy, from a small amount of true positives (top row, $\tau_F = 0.1$), to perfect reporting (bottom row, $\tau_F = 1$). For example, looking at the middle of the center panel, when $p_F = 0.5$, $\tau_F = 0.5$, and $\rho = 0.5$, we see that the average basic scale-up estimate is about 50, while the average generalized scale-up estimate is 150 (the true value). The generalized scale-up estimator is unbiased for all parameter combinations, while the basic scale-up estimator is only unbiased for certain special cases (e.g., when $\rho = 1$, $\tau_F = 1$, and $p_F = 1$). Full details of the simulation are presented in Appendix G in the online journal.

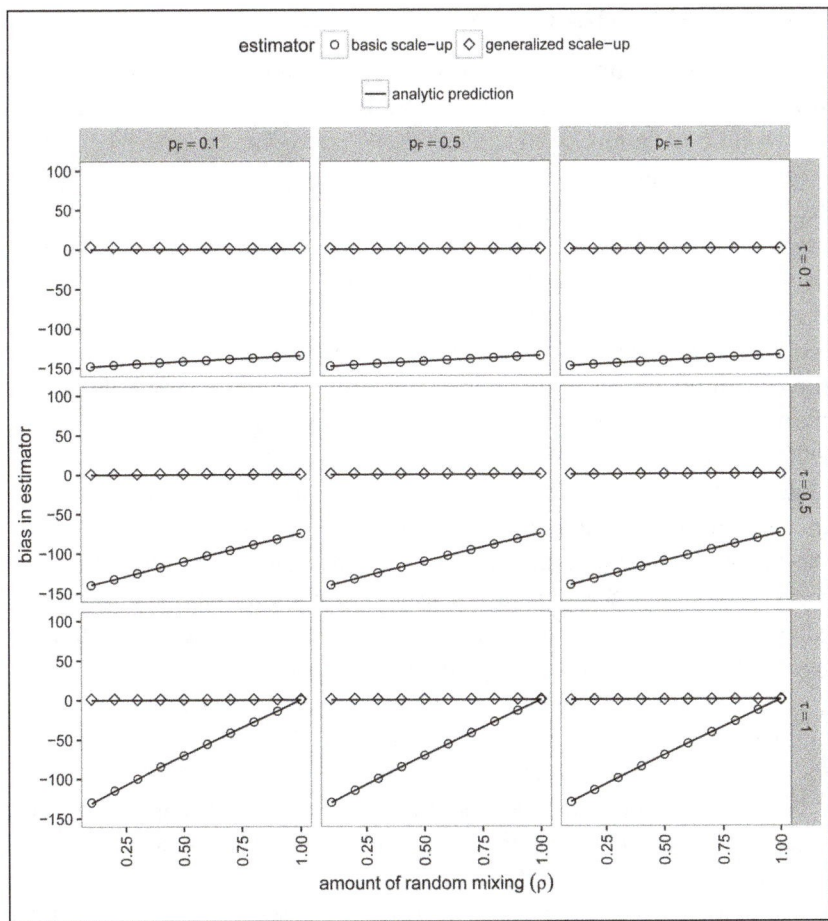

Figure 3. Bias (open circles and diamonds) and predicted bias (solid lines) in the basic scale-up estimates and generalized scale-up estimates for the same parameter configurations depicted in Figure 2. Our analytical results (equation 20) accurately predict the bias observed in our simulation study.

estimator is unbiased in a much smaller set of situations. More concretely, the basic scale-up estimator is unbiased in situations in which the basic scale-up model holds—when everyone is in the frame population ($p_F = 1$), there is random mixing ($\rho = 1$), and respondents' reports are perfect ($\tau_F = 1$).[6] Furthermore, Figure 3 illustrates that our analytic approach (equation 3) can correctly predict the bias of the basic scale-up estimator.

4. RECOMMENDATIONS FOR PRACTICE

The results in Sections 2 and 3 lead us to recommend a major departure from current scale-up practice. In addition to collecting a sample from the frame population, we recommend that researchers consider collecting a sample from the hidden population so that they can use the generalized scale-up estimator. As our results clarify, researchers using the scale-up method face a decision: they can collect data from the hidden population or they can make assumptions about the adjustment factors described in Section 3. The appropriate decision depends on a number of factors, but we think that two are most important: (1) the difficulty of sampling from the hidden population and (2) the availability of high-quality estimates of the adjustment factors in Section 3. For example, if it is particularly difficult to sample from a specific hidden population and high-quality estimates of the adjustment factors are already available, then a basic scale-up estimator may be appropriate. If, however, it is possible to sample from the hidden population and there are no high-quality estimates of adjustment factors, then the generalized scale-up estimator may be appropriate. Many realistic situations will be somewhere between these two extremes, and the trade-offs must be weighed on a case-by-case basis.

To aid researchers deciding between basic and generalized scale-up approaches, we collected the conditions needed for consistent and essentially unbiased estimates into Table 2; formal proofs of these results are presented in Online Appendices B and C. We find it helpful to group these conditions into four broad categories: sampling, survey construction, network structure, and reporting behavior.

A review of the conditions in Table 2 necessarily raises practical concerns. In situations in which researchers are trying to make estimates about real hidden populations, they probably will not know how close they are to meeting these conditions. Therefore, researchers may wonder how their estimates will be affected by violations of these assumptions, both individually (e.g., "How would my estimates be affected if there was a problem with the survey construction?") and jointly (e.g., "How would my estimate be affected if there was a problem with my survey construction and reporting behavior?"). To address this concern, in Appendix D in the online journal, we develop a framework for sensitivity analysis that shows researchers exactly how estimates will be affected by violations of all assumptions, either individually or jointly. Table 3 summarizes the results of our sensitivity framework.

Table 2. Summary of the Conditions Needed for the Generalized and Modified Basic Network Scale-up Estimators, and Their Components, to Produce Estimates That Are Consistent and Essentially Unbiased

Quantity	Conditions Required	Condition Type	Result
Reported connections to $H\left(\widehat{y}_{F,H}\right)$	1. Probability sample from F	Sampling	B1
Average personal network size of $F\left(\widehat{d}_{F,F}\right)$	1. Probability sample from F 2. Groups of known size total is accurate N_A 3. Probe alter condition $\left(\overline{d}_{A,F}=\overline{d}_{F,F}\right)$ 4. Accurate reporting condition $\left(y_{F,A}=d_{F,A}\right)$	Sampling Survey construction Survey construction Reporting behavior	B3
Average visibility of $H\left(\widehat{v}_{H,F}\right)$	1. Relative probability sample from H 2. Groups of known size total is accurate $N_{A_H\cap F}$ 3. Probe alter condition $\left(\dfrac{v_{H,A\cap F}}{N_{A\cap F}}=\dfrac{v_{H,F}}{N_F}\right)$ 4. Accurate aggregate reports about visibility $\left(\overline{v}_{H,A_H\cap F}=v_{H,A_H\cap F}\right)$	Sampling Survey construction Survey construction Reporting behavior	C2
Generalized scale-up $\left(\widehat{N}_H=\dfrac{\widehat{y}_{F,H}}{\widehat{\overline{v}}_{H,F}}\right)$	1. Conditions needed for $\widehat{y}_{F,H}$ 2. Conditions needed for $\widehat{\overline{v}}_{H,F}$ 3. No false positive reports about connections to $H(\eta_F=1)$	Sampling Sampling, survey construction, Reporting behavior Reporting behavior	C8
Modified basic scale-up $\left(\widehat{N}_H=\dfrac{\widehat{y}_{F,H}}{\widehat{\overline{d}}_{F,F}}\right)$	1. Conditions needed for $\widehat{y}_{F,H}$ 2. Condition needed for $\widehat{\overline{d}}_{F,F}$ 3. No false positive reports about connections to $H(\eta_F=1)$ 4. Members of H and members of F have same average personal network size $(\delta_F=1)$ 5. No false negative reports about connections to $H(\tau_F=1)$	Sampling Sampling, survey construction Reporting behavior Reporting behavior Network structure Reporting behavior	Sections 2 and 3

Note: This table uses the version of the basic scale-up estimator we recommend in Section 4.2.

Table 3. Analytical Expressions Researchers Can Use to Perform Sensitivity Analysis for Estimates Made Using Scale-up Estimators

Quantity	Conditions Required	Adjusted Estimand for Sensitivity Analysis
Generalized scale-up $\left(\widehat{N}_H = \dfrac{\widehat{\overline{y}}_{F,H}}{\widehat{\overline{v}}_{H,F}}\right)$	1. Probability sample from F with accurate weights ($K_{F_2}=0$ and $\overline{\in}_F=1$) 2. Relative probability sample from H with accurate weights ($K_H=0$) 3. Conditions needed for $\widehat{\overline{v}}_{H,F}\left(\widehat{\overline{v}}_{F,H}=k\overline{v}_{F,H}\right)$ 4. No false positive reports about connections to $H(\eta_F=1)$	$\widehat{N}_H \cdot \dfrac{(1+K_H)}{\overline{\in}_F(1+K_{F_2})} \cdot k \cdot \eta_F \rightsquigarrow N_H$
Modified basic scale-up $\left(\widehat{N}_H = \dfrac{\widehat{\overline{y}}_{F,H}}{\widehat{\overline{d}}_{F,F}}\right)$	1. Probability sample from F with accurate weights for $y_{F,H}(K_{F_2}=0)$ 2. Probability sample from F with accurate weights for $y_{F,A}(K_{F_1}=0)$ 3. Condition needed for $\widehat{\overline{d}}_{F,F}\left(\widehat{\overline{d}}_{F,F}=k\overline{d}_{,F}\right)$ 4. No false positive reports about connections to $H(\eta_F=1)$ 5. Members of H and members of F have same average personal network size ($\delta_F=1$) 6. No false negative reports about connections to $H(\tau_F=1)$	$\widehat{N}_H \cdot \dfrac{(1+K_{F_1})}{(1+K_{F_2})} \cdot k \cdot \dfrac{\eta_F}{\delta_F\tau_F} \rightsquigarrow N_H$

Note: See Appendix D in the online journal for more details. K_{F_1}, K_{F_2}, and K_H are indices that reflect how imperfect the sampling weights researchers use to make estimates are; when these K values are 0, the weights are exactly correct; the farther they are from 0, the more imperfect the weights are. (We use the symbol $\rightsquigarrow$ as a shorthand for "is consistent and essentially unbiased for.")

Another problem researchers face in practice is putting appropriate confidence intervals around estimates. The procedure currently used in scale-up studies was proposed by Killworth, McCarty, et al. (1998), but it has a number of conceptual problems, and in practice, it produces intervals that are anticonservative (i.e., the actual coverage rate is lower than the desired coverage rate). Both of these problems—theoretical and empirical—do not seem to be widely appreciated in the scale-up literature. Therefore, instead of the current procedure, we recommend that researchers use the rescaled bootstrap procedure (Rao, Wu, and Yue 1992; Rao and Wu 1988; Rust and Rao 1996), which has strong theoretical foundations, does not depend on the basic scale-up model, can handle both simple and complex sample designs, and can be used for both the basic scale-up estimator and the generalized scale-up estimator. In Appendix F in the online journal, we review the current scale-up confidence interval procedure and the rescaled bootstrap, highlighting the conceptual advantages of the rescaled bootstrap. Furthermore, we show that the rescaled bootstrap produces slightly better confidence intervals in three real scale-up data sets: one collected via simple random sampling (McCarty et al. 2001) and two collected via complex sample designs (Salganik, Fazito, et al. 2011; Rwanda Biomedical Center 2012). Finally, and somewhat disappointingly, our results show that none of the confidence interval procedures work very well in an absolute sense, a finding that highlights an important problem for future research.

We now provide more specific guidance for researchers based on the data they decide to collect. In Section 4.1 we present recommendations for researchers who collect a sample from both the frame population, F, and the hidden population, H; in Section 4.2, we present recommendations for researchers who only select a sample from the frame population.

4.1. Estimation with Samples from F and H

We recommend that researchers who have samples from F and H use a generalized scale-up estimator to produce estimates of N_H (see Section 2):

$$\widehat{N}_H = \frac{\widehat{y}_{F,H}}{\widehat{v}_{H,F}}. \tag{21}$$

For researchers using the generalized scale-up estimator, we have three specific recommendations. Of all the conditions needed for consistent and essentially unbiased estimation, the ones most under the control of the researcher are those related to survey construction, so we recommend that researchers focus on these during the study design phase. In particular, we recommend that the probe alters be designed so that the rate at which the hidden population is visible to the probe alters is the same as the rate at which the hidden population is visible to the frame population (see Result C.2 for a more formal statement, and see Section C5 for more advice about choosing probe alters). Second, when presenting estimates, we recommend that researchers use the results in Table 3 to also present sensitivity analyses highlighting how the estimates may be affefcted by assumptions that are particularly problematic in their setting. Finally, we recommend that researchers produce confidence intervals around their estimate using the rescaled bootstrap procedure, keeping in mind that this will likely produce intervals that are anticonservative.

We also have three additional recommendations that will facilitate the cumulation of knowledge about the scale-up method. First, although the generalized scale-up estimator does not require aggregate relational data from the frame population about groups of known size, we recommend that researchers collect these data so that the basic and generalized estimators can be compared. Second, we recommend that researchers publish estimates of δ_F and τ_F, although these quantities play no role in the generalized scale-up estimator (Figure 4). As a body of evidence about these adjustment factors accumulates (e.g., Salganik, Fazito, et al. 2011; Maghsoudi et al. 2014), studies that are not able to collect a sample from the hidden population will have an empirical foundation for adjusting basic scale-up estimates, either by borrowing values directly from the literature or by using published values as the basis for priors in a Bayesian model. Finally, we recommend that researchers design their data collections—both from the frame population and the hidden population—so that size estimates from the generalized scale-up method can be compared with estimates from other methods (e.g., see Salganik, Fazito, et al. 2011a). For example, if respondent-driven sampling is used to sample from the hidden population, then researchers could use methods that estimate the size of a hidden population from recruitment patterns in the respondent-driven sampling data (Berchenko, Rosenblatt,

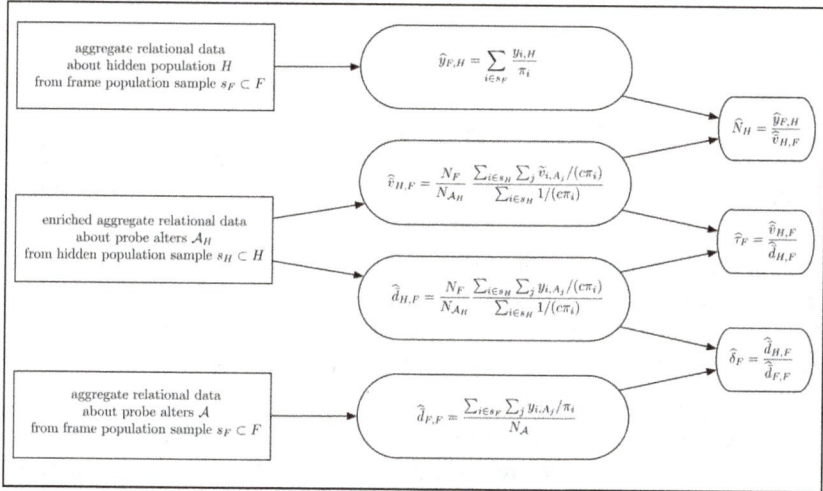

Figure 4. Recommended schematic of inputs and outputs for a study using the generalized scale-up estimator. We recommend that researchers produce size estimates using the generalized scale-up estimator and that researchers produce estimates of the adjustment factors δ_F and τ_F to aid other researchers.

and Frost 2013; Crawford, Wu, and Heimer 2015; Handcock, Gile, and Mar 2014, 2015; Johnston et al. 2015; Wesson et al. 2015).

4.2. *Estimation with Only a Sample from* F

If researchers cannot collect a sample from the hidden population, we have three recommendations. First, we recommend two simple changes to the basic scale-up estimator that remove the need to adjust for the frame ratio, ϕ_F. Recall, that the basic scale-up estimator that has been used in previous studies (see Section 3) is

$$\widehat{N}_H = \frac{\widehat{y}_{F,H}}{\widehat{d}_{F,U}} \times N = \frac{\widehat{y}_{F,H}}{\widehat{d}_{F,U} \big/ N}. \tag{22}$$

Instead of equation 22, we suggest a new estimator, called the modified basic scale-up estimator, that more directly deals with the fact that researchers sample from the frame population F (typically adults), and not from the entire population U (adults and children):

$$\widehat{N}_H = \frac{\widehat{\bar{y}}_{F,H}}{\widehat{d}_{F,F}} \times N_F = \frac{\widehat{\bar{y}}_{F,H}}{\widehat{d}_{F,F} / N_F} . \tag{23}$$

There are two differences between the modified basic scale-up estimator (equation 23) and the basic scale-up estimator (equation 22). First, we recommend that researchers estimate $\widehat{d}_{F,F}$ (i.e., the total number of connections between adults and adults) rather than $\widehat{d}_{F,U}$ (i.e., the total number of connections between adults and everyone). To do so, researchers should design the probe alters for the frame population so that they have similar personal networks to the frame population; in Appendix B.4 in the online journal, we define this requirement formally, and in Section B.4.1 we provide guidance for choosing the probe alters. Second, we recommend that researchers use N_F rather than N.[7] These two simple changes remove the need to adjust for the frame ratio ϕ_F, and thereby eliminate an assumption about an unmeasured quantity. An improved version of the basic scale-up estimator would then be

$$\widehat{N}_H = \underbrace{\frac{\widehat{\bar{y}}_{F,H}}{\left(\widehat{d}_{F,F} / N_F\right)}}_{\text{modified basic scale-up}} \times \underbrace{\frac{1}{\widehat{\delta}_F} \times \frac{1}{\widehat{\tau}_F}}_{\text{adjustment factors}} . \tag{24}$$

Our second recommendation is that researchers using the modified basic scale-up estimator (equation 23) perform a sensitivity analysis using the results in Table 3. In particular, we think that researchers should be explicit about the values that they assume for the adjustment factors δ_F and τ_F. Our third recommendation is that researchers construct confidence intervals using the rescaled bootstrap procedure, while explicitly accounting for the fact that there is uncertainty around the assumed adjustment factors and bearing in mind that this procedure will likely produce intervals that are anticonservative.

5. CONCLUSION AND NEXT STEPS

In this paper, we developed the generalized network scale-up estimator. This new estimator improves upon earlier scale-up estimators in several ways: it enables researchers to use the scale-up method in populations with nonrandom social mixing and imperfect awareness about membership in the hidden population, and it accommodates data collection with complex sample designs and incomplete sampling frames. We also compared the generalized and basic scale-up estimators, leading us to introduce a framework that makes the design-based assumptions of the basic scale-up estimator precise. Finally, researchers who use either the basic or generalized scale-up estimator can use our results to assess the sensitivity of their size estimates to assumptions.

The approach we followed to derive the generalized scale-up estimator has three elements, and these elements may prove useful in other problems related to sampling in networks. First, we distinguished between the network of reports and the network of relationships. Second, using the network of reports, we derived a simple identity that permitted us to develop a design-based estimator free of any assumptions about the structure of the network of relationships. Third, we combined data from different types of samples. Together, these three elements may help other researchers in other situations derive relatively simple, design-based estimators that are an important complement to complex, model-based techniques.

Although the generalized scale-up estimator has many attractive features, it also requires that researchers obtain two different samples, one from the frame population and one from the hidden population. In cases in which studies of the hidden population are already planned (e.g., the behavioral surveillance studies of the groups most at risk for HIV/ AIDS), the necessary information for the generalized scale-up estimator could be collected at little additional cost by appending a modest number of questions to existing questionnaires. In cases in which these studies are not already planned, researchers can either collect their own data from the hidden population, or they can use the modified basic scale-up estimator and borrow estimated adjustment factors from other published studies.

The generalized scale-up estimator, like all estimators, depends on a number of assumptions, and we think three of them will be most problematic in practice. First, the estimator depends on the assumption that there are no false positive reports, which is unlikely to be true in all situations. Although we have derived an estimator that works even in the presence of false positive reports (Appendix A in the online journal), we were not able to design a practical data collection procedure that would allow us to estimate one of the terms it requires. Second, the generalized scale-up estimator depends on the assumption that hidden population members have accurate aggregate awareness about visibility (equation 9). That is, researchers have to assume that hidden population respondents can accurately report whether or not their alters would report them, and we expect this assumption will be difficult to check in most situations. Third, the generalized scale-up estimator depends on having a relative probability sample from the hidden population. Unfortunately, we cannot eliminate any of these assumptions, but we have stated them clearly and we have derived the sensitivity of the estimates to violations of these assumptions, individually and jointly.

Our results and their limitations highlight several directions for further work, in terms of both of improved modeling and improved data collection. We think the most important direction for future modeling is developing estimators in a Bayesian framework, and a recent paper by Maltiel et al. (2015) offers some promising steps in this direction. We see two main advantages of the Bayesian approach in this setting. First, a Bayesian approach would allow researchers to propagate the uncertainty they have about the many assumptions involved in scale-up estimates, whereas our current approach captures only uncertainty introduced by sampling. Furthermore, as more empirical studies produce estimates of the adjustment factors (τ_F and δ_F), a Bayesian framework would permit researchers to borrow values from other studies in a principled way. In terms of future directions for data collection, researchers need practical techniques for estimating the rate of false positive reporting. These estimates, combined with the estimator in Appendix A in the online journal, would permit the relaxation of one of the most important remaining assumptions made by all scale-up studies to date. We hope that the framework introduced in this paper will provide a basis for these and other developments.

Acknowledgments

We thank Alexandre Abdo, Francisco Bastos, Russ Bernard, Neilane Bertoni, Dimitri Fazito, Sharad Goel, Wolfgang Hladik, Jake Hofman, Mike Hout, Karen Levy, Rob Lyerla, Mary Mahy, Chris McCarty, Maeve Mello, Tyler McCormick, Damon Phillips, Justin Rao, Adam Slez, and Tian Zheng for helpful discussions. Many of the methods described in this paper can be implemented using our accompanying open-source R package, called networkreporting, which is available on the Comprehensive R Archive Network. Replication materials for our analysis can be freely downloaded from the Harvard Dataverse (http://dx.doi.org/10.7910/DVN/HHAUDF).

Funding

The author(s) disclosed receipt of the following financial support for the research, authorship, and/or publication of this article: This research was supported by the Joint

United Nations Programme on HIV/AIDS, the National Science Foundation (CNS-0905086), and the Eunice Kennedy Shriver National Institute of Child Health and Human Development (R01-HD062366, R01-HD075666, and R24-HD047879). Some of this research was conducted while Dr. Salganik was an employee at Microsoft Research. The opinions expressed here represent the views of the authors and not the funding agencies.

Notes

1. Throughout the paper, we consider only the case ion which i never reports j more than once.

2. We use the term *essentially unbiased* because equation (7) is not, strictly speaking, unbiased; the ratio of two unbiased estimators is not itself unbiased. However, a large literature confirms that the biases caused by the nonlinear form of ratio estimators are typically insignificant relative to other sources of error in estimate (e.g., Sarndal et al. 1992, chap. 5). Unfortunately, many of the estimators we propose are actually ratios of ratios, sometimes called "compound ratio estimators" or "double ratio estimators." In Appendix E in the online journal, we demonstrate that the bias caused the nonlinear form of our estimators is not a practical cause for concern.

3. Note that because $\bar{d}_{U,F} = (N_F/N)\bar{d}_{F,U}$, an equivalent expression for the frame ratio is $\phi_F = \frac{\bar{d}_{F,F}}{\bar{d}_{F,U}(N_F/N)} = \frac{\bar{d}_{F,F}}{\bar{d}_{F,U}}\frac{N}{N_F}$.

4. Note that the fact that in-reports must equal out-reports means that τ_F can also be defined as $\tau_F = \frac{\text{\# reported edges from } F \text{ actually connected to } H}{\text{\# edges connecting } F \text{ and } H} = \frac{y^+_{F,H}}{d_{F,H}}$. Here we have written $y^+_{F,H}$ to mean the true positive reports among the $y_{F,H}$; see Appendix A in the online journal for a detailed explanation.

5. Computer code to perform the simulations was written in R (R Core Team 2014) and used the following packages: devtools (Wickham and Chang 2013), functional (Danenberg 2013), ggplot2 (Wickham 2009), igraph (Csardi and Nepusz 2006), networkreporting (Feehan and Salganik 2014), plyr (Wickham 2011), sampling (Tillé and Matei 2015), and stringr (Wickham 2012).

6. In addition to the settings in which the basic scale-up model holds, the basic scale-up estimator can also be unbiased when its different biases cancel (e.g., when the product of the adjustment factors is 1).

7. In some cases this difference between N_F and N can be substantial. For example, if F is adults, then in many developing countries, $N \approx 2N_F$.

References

Aramrattan, A., and M. Kanato. 2012. "Network Scale-up Method: Application in Thailand." Presented at Consultation on Estimating Population Sizes through Household Surveys: Successes and Challenges, March 28–30, New York.

Bengtsson, L., and A. Thorson. 2010. "Global HIV Surveillance among MSM: Is Risk Behavior Seriously Underestimated?" *AIDS* 24(15):2301–303.

Berchenko, Y., J. Rosenblatt, and S.D.W. Frost. 2013. "Modeling and Analysing Respondent Driven Sampling as a Counting Process." arXiv:1304.3505 [stat].

Bernard, H. R., T. Hallett, A. Iovita, E. C. Johnsen, R. Lyerla, C. McCarty, M. Mahy, M. J. Salganik, T. Saliuk, O. Scutelniciuc, G. A. Shelley, P. Sirinirund, S. Weir, and D. F. Stroup. 2010. "Counting Hard-to-count Populations: The Network Scale-up Method for Public Health." *Sexually Transmitted Infections* 86(Suppl. 2):ii11–15.

Bernard, H. R., E. C. Johnsen, P. D. Killworth, and S. Robinson. 1989. "Estimating the Size of an Average Personal Network and of an Event Subpopulation." Pp. 159–75 in *The Small World*, edited by M. Kochen. Norwood, NJ: Ablex.

Bernard, H. R., E. C. Johnsen, P. Killworth, and S. Robinson. 1991. "Estimating the Size of an Average Personal Network and of an Event Subpopulation: Some Empirical Results." *Social Science Research* 20(2):109–21.

Crawford, F. W., J. Wu, and R. Heimer. 2015. "Hidden Population Size Estimation from Respondent-driven Sampling: A Network Approach." arXiv:1504.08349 [stat].

Csardi, G., and T. Nepusz. 2006. "The Igraph Software Package for Complex Network Research." *InterJournal, Complex Systems*:1695.

Danenberg, P. 2013. "Functional: Curry, Compose, and Other Higher-order Functions." R package version 0.4.

Ezoe, S., T. Morooka, T. Noda, M. L. Sabin, and S. Koike. 2012. "Population Size Estimation of Men Who Have Sex with Men through the Network Scale-up Method in Japan." *PLoS ONE* 7(1):e31184.

Feehan, D. M., and M. J. Salganik. 2014. "The Network Reporting Package." R package version 0.1.1.

Gile, K. J., and M. S. Handcock. 2010. "Respondent-driven Sampling: An Assessment of Current Methodology." Pp. 285–327 in *Sociological Methodology*, vol. 40, edited by Tim Futing Liao. Hoboken, NJ: Wiley-Blackwell.

Gile, K. J., and M. S. Handcock.2015. "Network Model-assisted Inference from Respondent-driven Sampling Data." *Journal of the Royal Statistical Society, Series A: Statistics in Society* 178(3):619–39.

Gile, K. J., L. G. Johnston, and M. J. Salganik. 2015. "Diagnostics for Respondent-driven Sampling." *Journal of the Royal Statistical Society, Series A: Statistics in Society* 178(1):241–69.

Gilovich, T., K. Savitsky, and V. H. Medvec. 1998. "The Illusion of Transparency: Biased Assessments of Others' Ability to Read One's Emotional States." *Journal of Personality and Social Psychology* 75(2):332–46.

Goel, S., and M. J. Salganik. 2010. "Assessing Respondent-driven Sampling." *Proceedings of the National Academy of Science* 107(15):6743–47.

Guo, W., S. Bao, W. Lin, G. Wu, W. Zhang, W. Hladik, A. Abdul-Quader, M. Bulterys, S. Fuller, and L. Wang. 2013. "Estimating the Size of HIV Key Affected Populations in Chongqing, China, Using the Network Scale-up Method." *PLoS ONE* 8(8): e71796.

Handcock, M. S., K. J. Gile, and C. M. Mar. 2014. "Estimating Hidden Population Size Using Respondent-driven Sampling Data." *Electronic Journal of Statistics* 8(1): 1491–1521.

Handcock, M. S., K. J. Gile, and C. M. Mar. 2015. "Estimating the Size of Populations at High Risk for HIV Using Respondent-driven Sampling Data." *Biometrics* 71(1): 258–66.

Heckathorn, D. D. 1997. "Respondent-driven Sampling: A New Approach to the Study of Hidden Populations." *Social Problems* 44(2):174–99.

Heimer, R. 2005. "Critical Issues and Further Questions about Respondent-driven Sampling: Comment on Ramirez-Valles, et al." *AIDS and Behavior* 9(4):403–8.

Jing, L., C. Qu, H. Yu, T. Wang, and Y. Cui. 2014. "Estimating the Sizes of Populations at High Risk for HIV: A Comparison Study." *PLoS ONE* 9(4):e95601.

Johnston, L. G., K. R. McLaughlin, H. El Rhinila, A. Latifi, A. Toufik, A. Bennani, K. Alami, B. Elomari, and M. S. Handcock. 2015. "Estimating the Size of Hidden Populations Using Respondent-driven Sampling Data: Case Examples from Morocco." *Epidemiology* 26(6):846–52.

Joint United Nations Programme on HIV/AIDS. 2010. "Guidelines on Estimating the Size of Populations Most at Risk to HIV." Geneva, Switzerland: UNAIDS/WHO Working Group on Global HIV/AIDS and STI Surveillance.

Kadushin, C., P. D. Killworth, H. R. Bernard, and A. A. Beveridge. 2006. "Scale-up Methods as Applied to Estimates of Heroin Use." *Journal of Drug Issues* 36(2): 417–40.

Karon, J., and C. Wejnert. 2012. "Statistical Methods for the Analysis of Time-location Sampling Data." *Journal of Urban Health* 89(3):565–86.

Khounigh, A. J., A. A. Haghdoost, S. SalariLak, A. H. Zeinalzadeh, R. Yousefi-Farkhad, M. Mohammadzadeh, and K. Holakouie-Naieni. 2014. "Size Estimation of Most-at-risk Groups of HIV/AIDS Using Network Scale-up in Tabriz, Iran." *Journal of Clinical Research and Governance* 3(1):21–26.

Killworth, P. D., E. C. Johnsen, C. McCarty, G. A. Shelley, and H. Bernard. 1998. "A Social Network Approach to Estimating Seroprevalence in the United States." *Social Networks* 20(1):23–50.

Killworth, P. D., C. McCarty, H. R. Bernard, G. A. Shelley, and E. C. Johnsen. 1998. "Estimation of Seroprevalence, Rape, and Homelessness in the United States Using a Social Network Approach." *Evaluation Review* 22(2):289–308.

Killworth, P. D., C. McCarty, H. R. Bernard, E. C. Johnsen, J. Domini, and G. A. Shelly. 2003. "Two Interpretations of Reports of Knowledge of Subpopulation Sizes." *Social Networks* 25(2):141–60.

Killworth, P. D., C. McCarty, E. C. Johnsen, H. R. Bernard, and G. A. Shelley. 2006. "Investigating the Variation of Personal Network Size under Unknown Error Conditions." *Sociological Methods and Research* 35(1):84–112.

Kurant, M., A. Markopoulou, and P. Thiran. 2011. "Towards Unbiased BFS Sampling." *IEEE Journal on Selected Areas in Communications* 29(9):1799–1809.

Lavallée, P. 2007. *Indirect Sampling*. New York: Springer.

Li, X., and K. Rohe. 2015. "Central Limit Theorems for Network Driven Sampling." arXiv:1509.04704 [math, stat].

Maghsoudi, A., M. R. Baneshi, M. Neydavoodi, and A. Haghdoost. 2014. "Network Scale-up Correction Factors for Population Size Estimation of People Who Inject Drugs and Female Sex Workers in Iran." *PLoS ONE* 9(11):e110917.

Maltiel, R., A. E. Raftery, T. H. McCormick, and A. Baraff. 2015. "Estimating Population Size Using the Network Scale Up Method." *Annals of Applied Statistics* 9(3):1247–77.

McCarty, C., P. D. Killworth, H. R. Bernard, E. Johnsen, and G. A. Shelley. 2001. "Comparing Two Methods for Estimating Network Size." *Human Organization* 60(1):28–39.

McCormick, T., R. He, E. Kolaczyk, and T. Zheng. 2012. "Surveying Hard-to-reach Groups through Sampled Respondents in a Social Network." *Statistics in Biosciences* 4(1):177–95.

McCormick, T. H., and T. Zheng. 2007. "Adjusting for Recall Bias in 'How Many X's Do You Know?' Surveys." *Proceedings of the Joint Statistical Meetings*, Salt Lake City, UT.

McCreesh, N., S. Frost, J. Seeley, J. Katongole, M. N. Tarsh, R. Ndunguse, F. Jichi, N. L. Lunel, D. Maher, L. G. Johnston, et al. 2012. "Evaluation of Respondent-driven Sampling." *Epidemiology* 23(1):138.

Mills, H. L., C. Colijn, P. Vickerman, D. Leslie, V. Hope, and M. Hickman. 2012. "Respondent Driven Sampling and Community Structure in a Population of Injecting Drug Users, Bristol, UK." *Drug and Alcohol Dependence* 126(3):324–32.

Mouw, T., and A. M. Verdery. 2012. "Network Sampling with Memory: A Proposal for More Efficient Sampling from Social Networks." Pp. 206–56 in *Sociological Methodology*, vol. 42, edited by Tim Futing Liao. Thousand Oaks, CA: Sage.

Nikfarjam, A., M. Shokoohi, A. Shahesmaeili, A. A. Haghdoost, M. R. Baneshi, S. Haji-Maghsoudi, A. Rastegari, A. A. Nasehi, N. Memaryan, and T. Tarjoman. 2016. "National Population Size Estimation of Illicit Drug Users through the Network Scale-up Method in 2013 in Iran." *International Journal of Drug Policy* 31:147–52.

Paniotto, V., T. Petrenko, V. Kupriyanov, and O. Pakhok. 2009. "Estimating the Size of Populations with High Risk for HIV Using the Network Scale-up Method." Technical Report. Kiev, Ukraine: Kiev Internation Institute of Sociology.

R Core Team. 2014. "R: A Language and Environment for Statistical Computing." Vienna, Austria: R Foundation for Statistical Computing.

Rao, J., C. Wu, and K. Yue. 1992. "Some Recent Work on Resampling Methods for Complex Surveys." *Survey Methodology* 18(2):209–17.

Rao, J. N., and C.F.J. Wu. 1988. "Resampling Inference with Complex Survey Data." *Journal of the American Statistical Association* 83(401):231–41.

Rohe, K. 2015. "Network Driven Sampling: A Critical Threshold for Design Effects." arXiv:1505.05461 [math, stat].

Rudolph, A. E., C. M. Fuller, and C. Latkin. 2013. "The Importance of Measuring and Accounting for Potential Biases in Respondent-driven Samples." *AIDS and Behavior* 17(6):2244–52.

Rust, K., and J. Rao. 1996. "Variance Estimation for Complex Surveys Using Replication Techniques." *Statistical Methods in Medical Research* 5(3):283–310.

Rwanda Biomedical Center. 2012. "Estimating the Size of Key Populations at Higher Risk of HIV through a Household Survey (ESPHS) Rwanda 2011." Technical Report. Calverton, MD: RBC/IHDPC, SPF, UNAIDS, and ICF International.

Salganik, M. J. 2012. "Commentary: Respondent-driven Sampling in the Real World." *Epidemiology* 23(1):148–50.

Salganik, M. J., D. Fazito, N. Bertoni, A. H. Abdo, M. B. Mello, and F. I. Bastos. 2011. "Assessing Network Scale-up Estimates for Groups Most at Risk of HIV/AIDS:

Evidence from a Multiple-method Study of Heavy Drug Users in Curitiba, Brazil." *American Journal of Epidemiology* 174(10):1190–96.

Salganik, M. J., M. B. Mello, A. H. Abdo, N. Bertoni, D. Fazito, and F. I. Bastos. 2011. "The Game of Contacts: Estimating the Social Visibility of Groups." *Social Networks* 33(1):70–78.

Sarndal, C. E., B. Swensson, and J. Wretman. 1992. *Model Assisted Survey Sampling.* New York: Springer.

Scott, G. 2008. "'They Got Their Program, and I Got Mine': A Cautionary Tale Concerning the Ethical Implications of Using Respondent-driven Sampling to Study Injection Drug Users." *International Journal of Drug Policy* 19(1):42–51.

Scutelniciuc, O. 2012a. "Network Scale-up Method Experiences: Republic of Kazakhstan." Presented at Consultation on Estimating Population Sizes through Household Surveys: Successes and Challenges, March 28–30, New York.

Scutelniciuc, O. 2012b. "Network Scale-up Method Experiences: Republic of Moldova." Presented at Consultation on Estimating Population Sizes through Household Surveys: Successes and Challenges, March 28–30, New York.

Sheikhzadeh, K., M. R. Baneshi, M. Afshari, and A. A. Haghdoost. 2014. "Comparing Direct, Network Scale-up, and Proxy Respondent Methods in Estimating Risky Behaviors among Collegians." *Journal of Substance Use* 21(1):9–13.

Shelley, G. A., H. R. Bernard, P. Killworth, E. Johnsen, and C. McCarty. 1995. "Who Knows Your HIV Status? What HIV + Patients and Their Network Members Know about Each Other." *Social Networks* 17(3–4):189–217.

Shelley, G. A., P. D. Killworth, H. R. Bernard, C. McCarty, E. C. Johnsen, and R. E. Rice. 2006. "Who Knows Your HIV Status II? Information Propagation within Social Networks of Seropositive People." *Human Organization* 65(4):430–44.

Shokoohi, M., M. R. Baneshi, and A. A. Haghdoost. 2012. "Size Estimation of Groups at High Risk of HIV/AIDS Using Network Scale Up in Kerman, Iran." *International Journal of Preventive Medicine* 3(7):471–76.

Sirken, M. G. 1970. "Household Surveys with Multiplicity." *Journal of the American Statistical Association* 65(329):257–66.

Snidero, S., B. Morra, R. Corradetti, and D. Gregori 2007. "Use of the Scale-up Methods in Injury Prevention Research: An Empirical Assessment to the Case of Choking in Children." *Social Networks* 29(4):527–38.

Snidero, S., N. Soriani, I. Baldi, F. Zobec, P. Berchialla, and D. Gregori. 2012. "Scale-up Approach in CATI Surveys for Estimating the Number of Foreign Body Injuries in the Aero-digestive Tract in Children." *International Journal of Environmental Research and Public Health* 9(11):4056–67.

Snidero, S., F. Zobec, P. Berchialla, R. Corradetti, and D. Gregori. 2009. "Question Order and Interviewer Effects in CATI Scale-up Surveys." *Sociological Methods and Research* 38(2):287–305.

Sulaberidze, L., A. Mirzazadeh, I. Chikovani, N. Shengelia, N. Tsereteli, and G. Gotsadze. 2016. "Population Size Estimation of Men Who Have Sex with Men in Tbilisi, Georgia: Multiple Methods and Triangulation of Findings." *PLoS ONE* 11(2):e0147413.

Tillé, Y., and A. Matei. 2015. "Sampling: Survey Sampling." R package version 2.7.

Volz, E., and D. D. Heckathorn. 2008. "Probability-based Estimation Theory for Respondent-driven Sampling." *Journal of Official Statistics* 24(1):79–97.

Wang, J., Y. Yang, W. Zhao, H. Su, Y. Zhao, Y. Chen, T. Zhang, and T. Zhang. 2015. "Application of Network Scale Up Method in the Estimation of Population Size for Men Who Have Sex with Men in Shanghai, China." *PLoS ONE* 10(11):e0143118.

Wasserman, S., and K. Faust. 1994. *Social Network Analysis*. New York: Cambridge University Press.

Wesson, P., M. S. Handcock, W. McFarland, and H. F. Raymond. 2015. "If You Are Not Counted, You Don't Count: Estimating the Number of African-American Men Who Have Sex with Men in San Francisco Using a Novel Bayesian Approach." *Journal of Urban Health* 92(6):1052–64.

White, H. C., S. A. Boorman, and R. L. Breiger. 1976. "Social Structure from Multiple Networks. I. Blockmodels of Roles and Positions." *American Journal of Sociology* 81(4):730–80.

White, R. G., A. J. Hakim, M. J. Salganik, M. W. Spiller, L. G. Johnston, L. R. Kerr, C. Kendall, A. Drake, D. Wilson, K. Orroth, et al. 2015. "Strengthening the Reporting of Observational Studies in Epidemiology for Respondent-driven Sampling Studies: 'STROBE-RDS' Statement." *Journal of Clinical Epidemiology* 68(12):1463–71.

Wickham, H. 2009. "ggplot2: Elegant Graphics for Data Analysis." New York: Springer.

Wickham, H. 2011. "The Split-apply-combine Strategy for Data Analysis." *Journal of Statistical Software* 40(1):1–29.

Wickham, H. 2012. "stringr: Make It Easier to Work with Strings." R package version 0.6.2.

Wickham, H., and W. Chang. 2013. "devtools: Tools to Make Developing R Code Easier." R Package Version 1.4.1.

Yamanis, T. J., M. G. Merli, W. W. Neely, F. F. Tian, J. Moody, X. Tu, and E. Gao. 2013. "An Empirical Analysis of the Impact of Recruitment Patterns on RDS Estimates among a Socially Ordered Population of Female Sex Workers in China." *Sociological Methods and Research* 42(3):392–425.

Zheng, T., M. J. Salganik, and A. Gelman. 2006. "How Many People Do You Know in Prison? Using Overdispersion in Count Data to Estimate Social Structure in Networks." *Journal of the American Statistical Association* 101(474):409–23.

Author Biographies

Dennis M. Feehan is an assistant professor of demography at the University of California, Berkeley. His research lies at the intersection of demography, social networks, and quantitative methodology.

Matthew J. Salganik is a professor of sociology at Princeton University, and he is affiliated with several of Princeton's interdisciplinary research centers: the Office for Population Research, the Center for Information Technology Policy, the Center for Health and Wellbeing, and the Center for Statistics and Machine Learning. His research interests include social networks and computational social science.

Sociological Methodology
2016, Vol. 46(1) 187–211
© American Sociological Association 2016
DOI: 10.1177/0081175016641713
http://sm.sagepub.com

ॐ 6 ॐ

THE GRAPHICAL STRUCTURE OF RESPONDENT-DRIVEN SAMPLING

Forrest W. Crawford*

Abstract

Respondent-driven sampling (RDS) is a chain-referral method for sampling members of hidden or hard-to-reach populations, such as sex workers, homeless people, or drug users, via their social networks. Most methodological work on RDS has focused on inference of population means under the assumption that subjects' network degree determines their probability of being sampled. Criticism of existing estimators is usually focused on missing data: the underlying network is only partially observed, so it is difficult to determine correct sampling probabilities. In this article, the author shows that data collected in ordinary RDS studies contain information about the structure of the respondents' social network. The author constructs a continuous-time model of RDS recruitment that incorporates the time series of recruitment events, the pattern of coupon use, and the network degrees of sampled subjects. Together, the observed data and the recruitment model place a well-defined probability distribution on the recruitment-induced subgraph of respondents. The author shows that this distribution can be interpreted as an exponential random graph model and develops a computationally efficient method for estimating the hidden graph. The author validates the method using simulated data and applies the technique to an RDS study of injection drug users in St. Petersburg, Russia.

*Yale School of Public Health, New Haven, CT, USA

Corresponding Author:
Forrest W. Crawford, Yale School of Public Health, Department of Biostatistics, 60 College Street, New Haven, CT 06510, USA
Email: forrest.crawford@yale.edu

Keywords

hidden population, link tracing, missing data, network inference, respondent-driven sampling, social network

1. INTRODUCTION

Hidden populations, such as drug users, men who have sex with men, sex workers, or homeless people, are often subjected to social stigma or criminalization. Learning about these populations can be challenging for sociologists, epidemiologists, and public health researchers, because potential subjects may fear exposure or prosecution. Several survey techniques have been developed for sampling from hidden populations, including social link tracing and snowball designs (Goodman 1961; Thompson and Frank 2000). Respondent-driven sampling (RDS) is a common survey method for hidden or hard-to-reach populations for which no convenient sampling frame exists (Broadhead et al. 1998; Heckathorn 1997). In RDS, study participants recruit members of their social network who are also members of the hidden population. Starting with a set of "seeds," participants are given a fixed number of coupons tagged with a unique code. Participants then recruit members of their social network by giving them coupons. The recipient of the coupon redeems it at the study site (or over the phone, online, etc.), is interviewed, and receives coupons to recruit others. A dual incentive encourages recruitment: subjects receive a small reward for participating in the study and for each new subject they recruit. Subjects cannot be recruited more than once, and only a small number of coupons are given to new participants, to prevent the local network from being saturated with coupons or the emergence of a secondary market for coupons. To safeguard the privacy of subjects not participating in the study, subjects do not reveal the identities of their social contacts to researchers. The only network information typically reported by subjects is their *network degree*, the number of social contacts who are also members of the study population.

Although RDS is an effective procedure for recruiting members of a hidden population, estimation of population characteristics from data obtained by RDS is controversial. Most methodological work on RDS assumes that the recruitment process takes place in a hidden social network connecting members of the study population. With the understanding that the structure of this hidden network likely affects individual subjects' likelihood of being recruited, many researchers have sought to determine sampling probabilities for design-based estimation of population means (e.g., human immunodeficiency virus [HIV] infection prevalence). Salganik and Heckathorn (2004) constructed a model of the recruitment process in which subjects receive only one coupon and can be recruited infinitely many times. They modeled the recruitment as a random walk *with replacement* on the hidden population social network. When this walk is at "equilibrium," they argued that the probability that a given subject is sampled is proportional to his or her network degree. Salganik and Heckathorn and Volz and Heckathorn (2008) proposed a Horvitz-Thompson type estimator for the population mean, in which observations are weighted by the inverse of the subject's degree. Aronow and Crawford (2015) clarified the conditions under which this estimator has good statistical properties, and Gile (2011) derived a related estimator whereby sampling is without replacement.

Unfortunately, the characterization of the RDS recruitment process as a sampling design, whereby sampling probability is a function of network degree alone, suffers from some fundamental flaws. First, RDS recruitment is always *without replacement*, because subjects cannot be recruited more than once; second, a without-replacement random walk on a network is never at equilibrium with respect to its probability of sampling particular subjects—once a subject is recruited, he or she can never be visited by the recruitment process again; and third, if the recruitment process operates on the social network connecting the sampled individuals and seeds are not chosen at random, the network structure itself determines the probability that a given person will be reached by the recruitment chain. Indeed, for a given sample size *n* on a fixed population network, any potential subject whose minimum path length to a seed is greater than *n* has sampling probability 0, *regardless of his or her network degree*. The random walk characterization of RDS also neglects the fundamental role of coupon depletion in the dynamics of recruitment. Depletion of certain recruiters' coupons can block paths to isolated parts of the network, providing no way for the recruitment chain to reach some members of the population. Researchers have raised serious concerns about the empirical properties of population estimates from data obtained by RDS and the Volz and Heckathorn (2008) estimator in particular (Gile and Handcock 2010; Goel and Salganik 2010; Johnston et al. 2010; Mills et al. 2014; Salganik 2012; White et al. 2012). Studies comparing RDS with traditional sampling or census of the same population have highlighted serious bias in estimates (McCreesh et al. 2012) or problems with variance estimation (Wejnert 2009).

It is difficult to determine the correct sampling probabilities for recruited subjects under RDS because the underlying social network is only partially observed (Gile and Handcock 2010, 2015). The unobserved links between recruited subjects, and between recruited and unrecruited population members, constitute *missing data* in RDS studies. Characterization of the network on which the sampling process takes place is therefore a major methodological frontier in research on estimation from RDS (Handcock and Gile 2010). Remarkably, a typical RDS study reveals a great deal of information about the network of respondents: the observed degrees, recruitment chain, and patterns of coupon allocation and depletion are all readily available and provide valuable information about the local structure of the population network. Insight into the information content of data from RDS studies would clarify exactly which network and population properties researchers can hope to estimate, and which they cannot, in real-world studies. In particular, a better understanding of the network on which RDS recruitment operates could facilitate computation of marginal sampling probabilities similar to those calculated by Gile and Handcock (2015) for use in Horvitz-Thompson-type estimators for population means (e.g., Volz and Heckathorn 2008). Alternatively, specification of a probability model for dependence between trait values of vertices that share an edge in G may allow regression approaches to population estimation and adjustment for dependence in outcomes induced by the network structure (e.g., Bastos et al. 2012). An estimate of the subnetwork of respondents in an RDS study could also be used to estimate the size of the target population in a manner analogous to the "network scale-up" population size estimator (Bernard et al. 2010; Feehan and Salganik 2014; Killworth et al. 1998).

In addition to its statistical uses for population-level inference, the subnetwork of respondents is of inherent sociological and epidemiological interest. The network connecting sampled subjects reveals social links between participants and possible avenues for transmission of ideas, behaviors, practices, or infectious agents. Comprehensive sociometric mapping can be difficult and costly in hidden populations, and many researchers have attempted to estimate epidemiological properties of recruited individuals' networks from recruitment data obtained by

RDS (e.g., Cepeda et al. 2011; Li et al. 2011; Liu et al. 2009; Stein, Steenbergen, Buskens, et al. 2014; Stein, Steenbergen, Chanyasanha, et al. 2014). The ability to estimate features of the subnetwork of respondents in an RDS study would place sociological and epidemiological inquiries about the local network onto firmer theoretical and methodological ground.

In other areas of network theory, researchers have made progress in reconstructing networks from partial observation. When links are missing, some techniques assume that subjects with similar traits are likely to be connected (Atchade 2011; Leskovec, Huttenlocher, and Kleinberg 2010; Lü and Zhou 2011; Koskinen et al. 2013). When vertices, edges, or egocentric networks are sampled, several authors have proposed ways of estimating global network properties (Bliss, Danforth, and Dodds 2014; Goyal, Blitzstein, and de Gruttola 2014; Smith 2012) or when vertices can be observed more than once (Frank and Snijders 1994; Yan and Gregory 2013). Sometimes dynamic or random processes can reveal structural information about networks (Kramer et al. 2009; Shandilya and Timme 2011; Linderman and Adams 2014). Gile (2011) and Gile and Handcock (2015) presented methods for random graph model-assisted inference of the degree distribution from RDS, but they still assume that sampling probability is a function of network degree alone.

In this article, we show how to use data from RDS studies to probabilistically reconstruct the social network of respondents. We first define the observed data under RDS and construct a realistic continuous-time model of the RDS recruitment process on a graph. The model is a simple and natural formalization of the RDS recruitment procedure initially defined by Heckathorn (1997). Interrecruitment waiting times carry information about the network edges linking recruiters to unsampled individuals at each moment in time. We combine this timing information, knowledge of who recruited whom, who had coupons at which times, and the network degrees of recruited subjects to place a well-defined probability distribution on the structure of the recruitment-induced subgraph. A fundamental result of this article is that under simple and realistic assumptions, the likelihood of the recruitment process on a hidden graph can be interpreted as an exponential random graph model (ERGM). We describe a technique for jointly estimating the recruitment-induced subgraph and recruitment rate. An important feature of the algorithm is a computationally efficient method to calculate the likelihood of the recruitment-induced subgraph. We validate the proposed technique using simulated and real networks and apply it to an RDS study of injection drug users in St. Petersburg, Russia. We conclude with a new perspective on the information content of data from RDS studies.

2. DEFINITIONS AND ASSUMPTIONS

We begin by stating definitions and assumptions to ensure that the graph inference problem is well posed. (We use the terms *graph* and *network* interchangeably.) The first is implicit in the foundational work on RDS and guarantees that the objects under study exist (Heckathorn 1997; Salganik and Heckathorn 2004; Volz and Heckathorn 2008).

> *Assumption (A-1):* The hidden population exists and has finite size N. The social network connecting members of the hidden population is an undirected graph $G = (V, E)$ with $|V| = N$ and no parallel edges or self-loops.

Members of the hidden population are *vertices* in V. A vertex is *recruited* if it is known to the study. A vertex is a *recruiter* if it has at least one coupon and at least one unrecruited neighbor; a *susceptible vertex* is unrecruited and has at least one neighbor who is a recruiter. A *susceptible edge* connects a recruiter and a susceptible vertex, and recruitments can take place only across susceptible edges. A recruited vertex cannot be recruited again. At the moment it is recruited, a vertex is endowed with a non-negative number of coupons it may use to recruit its susceptible neighbors. Every recruitment reduces the number of coupons held by the recruiter by one. When all the coupons belonging to a recruiter vertex are depleted, the vertex is no longer a recruiter, and any edges incident to it are no longer susceptible. *Seeds* are recruited vertices chosen from the entire population of vertices by some mechanism, not necessarily random, usually at the beginning of the study. Seeds are not considered to have been recruited by any other vertex.

> *Definition 1 (Recruitment-induced Subgraph):* The recruitment-induced subgraph is $G_S = (V_S, E_S)$, where $V_S \subseteq V$ consists of $n = |V_S|$ sampled vertices (including seeds), and $\{i,j\} \in E_S$ if and only if $i \in V_S$, $j \in V_S$, and $\{i,j\} \in E$.

Definition 2 (Recruitment Graph): The directed recruitment graph is $G_R = (V_R, E_R)$, where $V_R = V_S$ is the set of n sampled vertices and $(i,j) \in E_R$ means i recruited j.

Because subjects cannot be recruited more than once, G_R is acyclic. Assumption (A-1) does not require that G be connected, nor that the RDS sample take place in the largest connected component, or even a single component. Therefore the recruitment-induced subgraph G_S may not be connected. Let $\mathbf{d}$ be the $n \times 1$ vector of recruited subjects' degrees (in the order of their recruitment into the study) and let $\mathbf{t} = (t_1, \ldots, t_n)$ be the $n \times 1$ vector of recruitment times, where $t_1 < \cdots < t_n$.

Definition 3 (Coupon Matrix): Let $\mathbf{C}$ be the $n \times n$ coupon matrix whose element $\mathbf{C}_{ij}$ is 1 if subject i has at least one coupon just before the jth recruitment event, and zero otherwise. The rows and columns of $\mathbf{C}$ are ordered by subjects' recruitment time.

The RDS recruitment process reveals only some of this information to researchers.

Assumption (A-2): The observed data consist of $\mathbf{Y} = (G_R, \mathbf{d}, \mathbf{t}, \mathbf{C})$.

In particular, researchers do not observe the recruitment-induced subgraph G_S of the sampled vertices. Figure 1 shows an example graph G and a realization of the RDS recruitment process on G. The recruitment graph G_R, recruitment-induced subgraph G_S, degree vector $\mathbf{d}$, recruitment times $\mathbf{t}$, and coupon matrix $\mathbf{C}$ are also shown.

We now state three assumptions about the behavior of recruiters and their knowledge of the recruitment status of their neighbors.

Assumption (A-3): Vertices become recruiters immediately upon entering the study and receiving one or more coupons. They remain recruiters until their coupons or susceptible neighbors are depleted, whichever happens first.

Assumption (A-4): When a susceptible neighbor j of a recruiter i is recruited by any recruiter, the edge connecting i and j is immediately no longer susceptible.

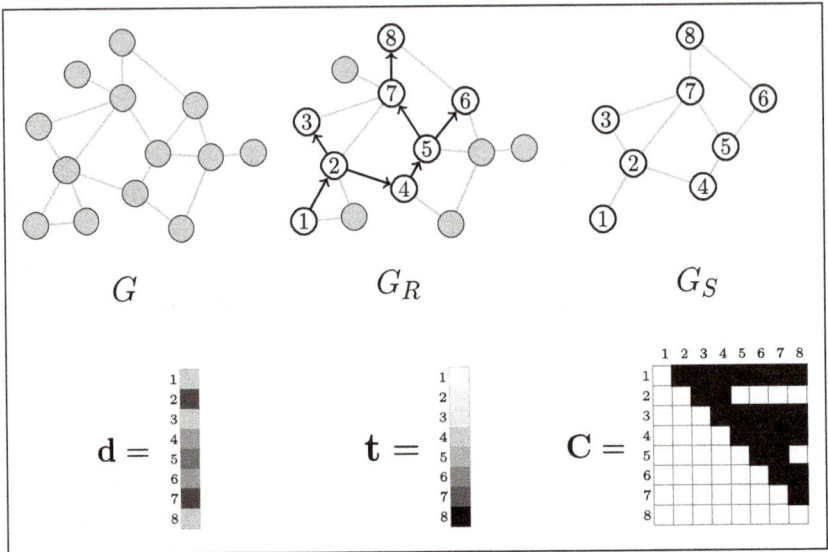

Figure 1. Example of unobserved and observed data in respondent-driven sampling (RDS). The true hidden population network is G. One seed is chosen (the vertex marked 1), and the RDS recruitment proceeds with each recruited vertex receiving two coupons. The directed recruitment graph G_R is shown superimposed on G. The recruitment-induced subgraph G_S is the subgraph of the recruited vertices. The degrees $\mathbf{d} = (d_1, \ldots, d_8)$ of each recruited vertex are observed, along with the recruitment times $\mathbf{t} = (t_1, \ldots, t_8)$. The coupon matrix $\mathbf{C}$ shows which recruiters had at least one coupon just before each recruitment event. In RDS, researchers observe neither G nor G_S; the observed data consist of $\mathbf{Y} = (G_R, \mathbf{d}, \mathbf{t}, \mathbf{C})$.

By assumption (A-4), recruitment is competitive: the first recruiter to recruit a given susceptible vertex immediately removes it from the pool of susceptibles. Finally, we specify a parametric waiting time distribution for the time it takes for a recruiter to recruit a susceptible neighbor.

Assumption (A-5): The time to recruitment along an edge connecting a recruiter to a susceptible neighbor has exponential distribution with rate λ, independent of the identity of the recruiter, neighbor, and all other waiting times.

By assumption (A-5), waiting times to recruitment along susceptible edges are independent and elapse concurrently in continuous time, so recruitment is *simultaneous*. Together, assumptions (A-3), (A-4), and

(A-5) place a well-defined probability distribution on the recruitment-induced subgraph of respondents.

2.1. *Consequences of the Waiting Time Assumption*

The results below follow directly from assumption (A-5). Let R be the set of recruiters with coupons and let S be the set of susceptible vertices at a certain moment in the recruitment process. Let S_u be the set of susceptible vertices that are neighbors of the recruiter $u \in R$. Likewise, let R_v be the set of possible recruiters of a susceptible vertex $v \in S$. Clearly, $v \in S_u$ if and only if $u \in R_v$.

> *Proposition 1:* Given that the recruiter u recruits one of its susceptible neighbors $v \in S_u$ before any other recruiter, the waiting time to this recruitment event is distributed as *Exponential*$(\lambda|S_u|)$. The probability that the susceptible vertex $v \in S_u$ is the next recruit is uniform $1/|S_u|$ and independent of the waiting time to the recruitment event.
>
> *Proposition 2:* The waiting time to the next recruitment of any susceptible vertex is distributed as *Exponential*$\left(\lambda \sum_{u \in R} |S_u|\right)$. The probability that the susceptible vertex $v \in S$ is the next recruit is $|R_v| / \sum_{k \in S} |R_k|$, independent of the waiting time.

Proofs of propositions 1 and 2 are given in the online Appendix. Intuitively, proposition 2 means that the new recruited vertex is chosen with probability proportional to the number of edges along which it can be recruited. These results formalize the consequences of simultaneous and competitive recruitment in continuous time.

Interestingly, assumptions (A-3) to (A-5) and the resulting recruitment probability differ starkly from the recruitment dynamics used in simulations by other researchers to test the performance of estimators for RDS. Gile and Handcock (2010) simulated the RDS recruitment process by first choosing seeds, after which "subsequent sample waves were selected without-replacement by sampling up to two nodes at random from among the unsampled alters of each sampled node" (p. 303). This leads to a brief corollary establishing the difference between these approaches.

> *Corollary 1:* Assumptions (A-3) to (A-5) (simultaneous and competitive recruitment) result in different conditional recruitment probabilities than the RDS recruitment implementation of Gile and Handcock (2010).

A proof is given in the online Appendix. The process defined by Gile and Handcock (2010) requires that recruiters "take turns." This approach implicitly requires that recruiters have knowledge about the behavior of other recruiters—even those to whom they are not connected in the network. This process induces a different distribution on the susceptible degree, and hence on the overall degree, of the new recruit than the model described in assumptions (A-3) to (A-5) of this article. Most existing methods for population inference from RDS data depend intimately on the degree distribution of recruited vertices (e.g., Gile 2011; Salganik and Heckathorn 2004; Volz and Heckathorn 2008), so it is important to highlight scenarios when methods for simulation of recruitment dynamics differ.

3. LIKELIHOOD OF THE RECRUITMENT TIME SERIES

Proposition 2 shows that under assumptions (A-3) to (A-5), the rate of recruitment is proportional to the number of susceptible edges. Given a realization of the recruitment-induced subgraph G_S, it is not immediately obvious how to determine quickly the number of susceptible edges just before each recruitment. When a given susceptible vertex is recruited, all susceptible edges incident to it disappear from the set of susceptible edges (assumption A-4). Furthermore, the newly recruited vertex now has coupons, so there may be new susceptible edges connected to it. Finally, if the new vertex is not a seed, its recruiter has used one coupon; if its coupons are now depleted, any other susceptible edges incident to the recruiter are no longer susceptible. Clearly, the number of susceptible edges can increase, decrease, or stay the same from one recruitment to the next. In this section, we derive a computationally efficient representation of the likelihood of the recruitment time series using matrix algebra. This approach obviates costly enumeration of all $|E_S|$ edges to determine whether they are susceptible at each step in the recruitment process. A preliminary definition will assist in this task. Let $1\{X\}$ be the indicator of an event X, which takes value 1 when X is true and zero otherwise.

Definition 4 (Compatibility): An estimated subgraph $\widehat{G}_S = (\widehat{V}_S, \widehat{E}_S)$ is compatible with the observed data $\mathbf{Y} = (G_R, \mathbf{d}, \mathbf{t}, \mathbf{C})$ if the following conditions are met:

1. The vertices in the estimated subgraph are identical to the set of recruited vertices: $v \in \widehat{V}_S$ if and only if $v \in V_R$.
2. All directed recruitment edges are represented as undirected edges: for each $(i,j) \in E_R$, $\{i,j\} \in \widehat{E}_S$.
3. The number of edges in G_S belonging to each sampled vertex does not exceed the vertex's degree: for all $v \in V_R$, $\sum_{u \in V_R \setminus v} 1\{\{u,v\} \in \widehat{E}_S\} \leq d_v$, where d_v is the degree of vertex v.

Let $\mathcal{C}(G_R, \boldsymbol{d})$ denote the set of all compatible subgraphs. These compatibility conditions provide topological constraints on the structure of G_S. Combining these with the likelihood of the recruitment time series allows probabilistic reconstruction of G_S.

Let $\mathbf{A}$ be the $n \times n$ adjacency matrix (sociomatrix) of a compatible estimate G_S, where the rows and columns of $\mathbf{A}$ correspond to subjects in the order of their recruitment. The product of $\mathbf{A}$ and the coupon matrix $\mathbf{C}$ gives an $n \times n$ matrix whose elements describe the number of recruiters connected to each vertex in G_S over time. Let $\mathbf{w} = (0, t_2 - t_1, \ldots, t_n - t_{n-1})$ be the $n \times 1$ vector of waiting times between recruitments. Let $\mathbf{u}$ be the $n \times 1$ vector of the number of edge ends belonging to each vertex (in the order of recruitment) that are not connected to any other sampled vertex. When $j \leq i$, $\{\mathbf{AC}\}_{ij}$ is the number of recruiters connected to i just before the time t_j of the jth recruitment. Then $\mathrm{lt}(\mathbf{AC})$, the lower triangle of $\mathbf{AC}$, is the number of recruiters connected to each vertex at each time before recruitment of that vertex. Likewise, the jth element of $\mathbf{C}'\mathbf{u}$ is the number of susceptible edges connecting sampled vertices to unsampled vertices at time t_j. Figure 2 shows examples of these matrices. Finally, let M be the set of seeds.

Proposition 3: Under assumptions (A-1) to (A-5), the likelihood of the recruitment time series is

$$L(\mathbf{w}|G_S, \lambda) = \left(\prod_{k \notin M} \lambda \mathbf{s}_k \right) \exp[-\lambda \mathbf{s}'\mathbf{w}], \tag{1}$$

where

$$\mathbf{s} = \mathrm{lt}(\mathbf{AC})'\mathbf{1} + \mathbf{C}'\mathbf{u} \tag{2}$$

is a vector whose elements are the number of susceptible edges just before each recruitment event.

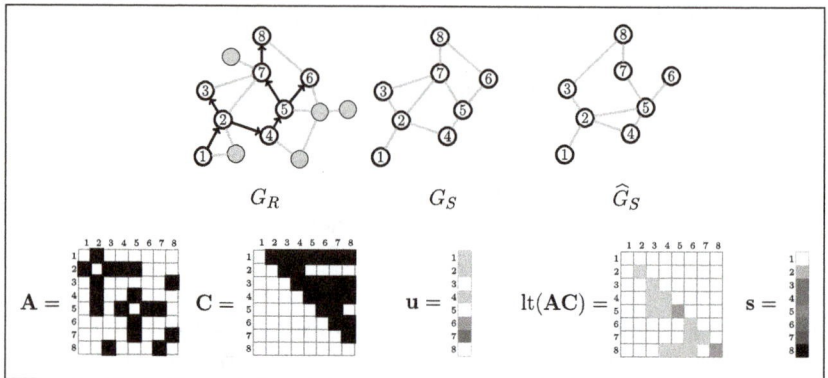

Figure 2. Examples of matrices used to calculate the recruitment time series likelihood. At top left is the recruitment graph G_R overlaid on the population graph G, with recruited vertices numbered and other vertices and edges in gray. The true recruitment-induced subgraph G_S is not directly observed. We estimate G_S by $\widehat{G}_S$ and let $\mathbf{A}$ be the adjacency matrix of $\widehat{G}_S$. The coupon matrix $\mathbf{C}$ and the number of pendant edges attached to each recruited vertex is $\mathbf{u}$. Pendant edges connect recruited vertices to unknown/unsampled vertices. The i,jth element of lt($\mathbf{AC}$) is the number of recruiters connected to i just before the jth recruitment event.

A proof is given in the online Appendix. As before, the rate of recruitment is proportional to the number of susceptible edges, and proposition 3 generalizes proposition 2 by providing an explicit expression for the number of susceptible edges at each step, taking coupons into account and allowing for seeds to be added at any time.

Although equation (1) is the likelihood of the recruitment time series $\mathbf{w}$, we can also view it as a function of the recruitment-induced subgraph adjacency matrix $\mathbf{A}$ with λ and $\mathbf{w}$ held fixed. Consider the statistic $T(\mathbf{A}) = -\lambda\mathbf{s}$, where $\mathbf{s}$ is defined by equation (2), $\theta = \mathbf{w}$, and $B(\mathbf{A}) = \sum_{k \notin M} \log(\lambda\mathbf{s}_k)$. Then we can renormalize the likelihood (equation 1) to form the probability $\Pr(\mathbf{A}|\theta) = \exp[T(\mathbf{A})'\theta + B(\mathbf{A})]/\kappa(\theta)$, where $\kappa(\theta)$ is a normalizing constant that does not depend on $\mathbf{A}$. It is clear that $\Pr(\mathbf{A}|\theta)$ is a member of the exponential family of distributions. In particular, it can be interpreted as an ERGM, also known as a p^* graph (Frank and Strauss 1986; Wasserman and Pattison 1996). Regardless of whether we view equation (1) as the likelihood of the random waiting times $\mathbf{w}$ or as the probability of the random graph G_S, the inference procedure we develop below benefits from Markov chain

Monte Carlo algorithms developed for sampling edges in ERGMs (see Snijders and Van Duijn 2002 for an example).

4. RECONSTRUCTING THE RECRUITMENT-INDUCED SUBGRAPH G_S

Together, the compatibility conditions (definition 4) and proposition 3 make possible simultaneous estimation of the recruitment-induced subgraph G_S and the waiting time parameter λ under the recruitment model. Because proposition 3 implies a probability model for $G_S \in \mathcal{C}(G_R, \mathbf{d})$, we can learn about this distribution by drawing samples from joint posterior

$$p(G_S, \lambda | \mathbf{Y}) \propto L(\mathbf{w} | G_S, \lambda) \, \Pr(G_S) \, \pi(\lambda), \tag{3}$$

where $\mathbf{Y} = (G_R, \mathbf{C}, \mathbf{d}, \mathbf{t})$ is the observed data, and $\Pr(G_S)$ and $\pi(\lambda)$ are prior distributions. We take the uniform prior distribution over the recruitment-induced subgraph: $\Pr(G_S) = 1/|\mathcal{C}(G_R, \mathbf{d})|$ for every $G_S \in \mathcal{C}(G_R, \mathbf{d})$. To draw pairs (G_S, λ) from $p(G_S, \lambda | \mathbf{Y})$, we use a Metropolis-within-Gibbs sampling scheme. To sample G_S conditional on λ, suppose λ is fixed and we have a compatible subgraph G_S. We generate a new compatible subgraph $G_S^* = (V_S, E_S^*)$ using a proposal algorithm given in the online Appendix. To sample λ conditional on G_S, we use a Metropolis-Hastings step based on an approximation of the conditional distribution of λ given in the online Appendix. By alternating these steps, we define a reversible Markov chain whose equilibrium distribution is the given by equation (3).

Computationally efficient Monte Carlo sampling of G_S via the Metropolis-Hastings algorithm depends on rapid evaluation of the likelihood ratio $L(\mathbf{w} | G_S^*, \lambda)/L(\mathbf{w} | G_S, \lambda)$, where G_S^* is a new proposed subgraph. The online Appendix presents simple expressions for these likelihood ratios that depend only on a simple *change statistic* and do not require evaluation of the matrix products required in the likelihood equation (1). More generally, the computational burden of the procedure scales with the sample size n and is not affected by the total size $N = |V|$ of the target population.

When only a single "most likely" subgraph G_S is desired, a faster algorithm is available for maximum likelihood (or maximum *a posteriori* [MAP]) estimate of G_S and λ. This Monte Carlo optimization approach is called "simulated annealing" (e.g., see Robert and Casella

2004 for details) and produces a sequence of estimates (G_S, λ) that tend toward the most likely values under the likelihood or posterior distribution. The simulated annealing procedure is outlined in the online Appendix.

5. VALIDATION BY SIMULATION

In simulation studies, reasonably accurate reconstruction of the recruitment-induced subgraph G_S can be achieved using the proposed recruitment model (equation 1). In the online Appendix, we analyze the performance of reconstruction in simulated networks and a real-world social network. Conditional on the population network, we simulate the RDS recruitment process with n subjects, $|M|$ seeds, and recruitment rate λ, under assumptions (A-3) to (A-5). From the simulated recruitment data, we extract the observed data $\mathbf{Y} = (G_R, t, \mathbf{d}, \mathbf{C})$ in accordance with assumption (A-2). We place a gamma prior distribution on the waiting time parameter, $\pi(\lambda) \propto \lambda^{\eta-1} e^{-\xi\lambda}$, where $\eta > 0$ and $\xi > 0$. We assess the accuracy of reconstruction over 100 repetitions of simulated RDS recruitment over different networks, and for each simulated data set, we find the joint MAP estimate of G_S and λ using the procedure outlined in Section 4. MAP estimates represent the mode of the posterior distribution over (G_S, λ) and provide a convenient point estimate for comparing results over many repetitions of the simulation. We also assess the accuracy of reconstruction under a misspecified waiting time model in which assumption (A-5) is violated; reconstruction remains robust, with corresponding bias in estimates of λ.

6. APPLICATION

The HIV epidemic in St. Petersburg, Russia, is concentrated in people who inject drugs (PWID). At least 12,000 people are registered as drug users, but the number of current PWID is likely much higher (Heimer and White 2010). Injection drug use is highly stigmatized in the Russian Federation, and criminal penalties for drug possession can be severe. PWID suffer from high rates of HIV infection and may lack access to treatment and health-related educational resources (Niccolai et al. 2010, 2011).

As part of a study to assess perceived barriers to use of HIV prevention and treatment services, $n = 813$ PWID were recruited using RDS in

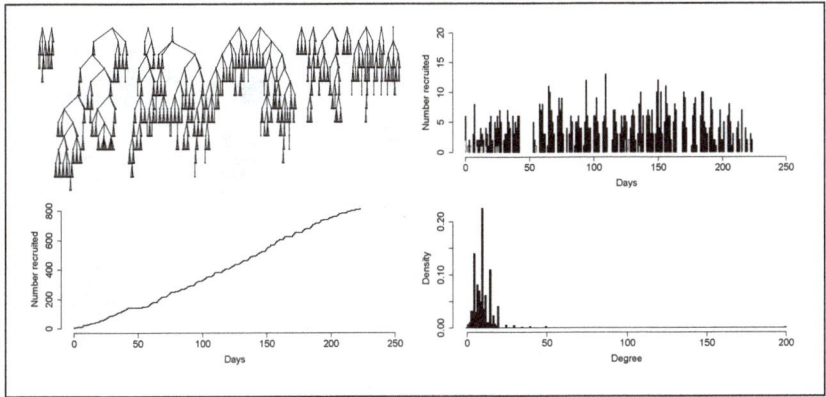

Figure 3. Raw data from a respondent-drivien sampling (RDS) sample of $n = 813$ people who inject drugs in St. Petersburg, Russia. In the top left panel, 14 RDS recruitment chains originating from different seeds are shown. Recruited subjects are organized into "waves" along the vertical axis. The top right panel shows the number of subjects interviewed on each day of the study, with seeds indicated by gray bars. The bottom left panel shows the cumulative number of recruits over the course of the study, and the bottom right panel shows a histogram of the reported degrees of subjects, with bin size 1.

St. Petersburg during 2012 and 2013. Outreach workers identified 17 seed subjects using venue-based sampling in six city districts. Interviews collected demographic information, injection practices, sex practices, mental health measures, and knowledge of HIV/AIDS and tuberculosis resources, but we focus solely on network structure in this analysis. Figure 3 shows the raw RDS data: the recruitment trees, number of new recruits per day, cumulative number of recruits, and reported network degrees.

Participation in the study was limited to current injection drug users over the age of 21 years who had injected within the previous four weeks. Subjects' status as PWID was verified either by inspection of arms for injection marks or explanation of drug preparation. Subjects received a voucher with a value of about US$20 for being interviewed and a secondary reward with value about US$10 for recruiting another eligible subject. Following their interview, each subject received three coupons, and no subject could be recruited more than once. Informed consent was obtained from all participants, and the study was approved by the Yale University and Stellit (St. Petersburg) institutional review boards.

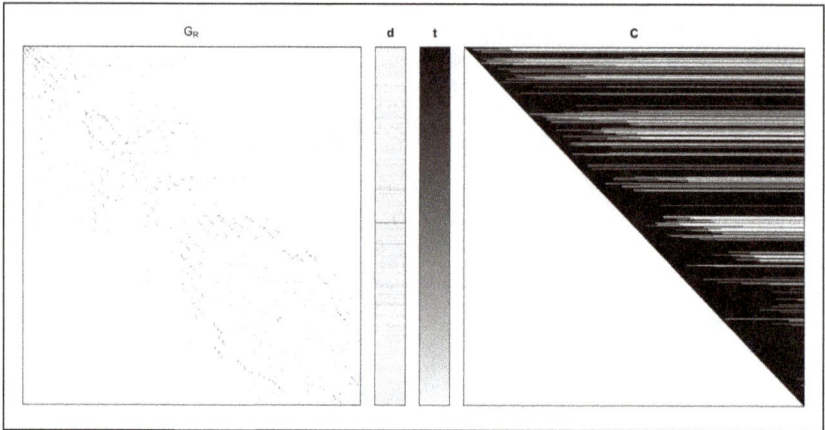

Figure 4. Raw respondent-driven sampling data $\mathbf{Y} = (G_R, \mathbf{d}, \mathbf{t}, \mathbf{C})$ extracted from study recruitment information.

Figure 4 shows the observed data $\mathbf{Y} = (G_R, \mathbf{d}, \mathbf{t}, \mathbf{C})$ from this study. The recruitment graph G_R was constructed by matching participants' coupon ID with the IDs of coupons given to their recruiter. The coupon matrix $\mathbf{C}$ was constructed by calculating the number of coupons held by each subject just before each recruitment event. Interviews assessed network degree by asking, "How many people do you know (you know one another's names) who you have seen within the last 4 weeks who inject drugs?"

A subject's *minimum degree* was defined as the number of undirected edges incident to that subject in the recruitment graph G_R. We assumed a subject's network degree was accurately reported, except when a subject's reported degree was less than his or her minimum degree. In these cases, we replaced the reported degree by the minimum degree. The average reported degree of subjects was 10.3. Interview dates and times were recorded for each subject; the elapsed time between a subject's interview and the next interview (in days) was treated as the interrecruitment waiting time. To estimate the edgewise recruitment rate λ more reliably, we removed weekends and other breaks during which no interviews were scheduled. This slightly changed the units of λ but allowed better estimation of the true waiting time distribution. The online Appendix describes the prior specification for λ. In a few cases, the interview times for a subject and his or her recruit were the same, presumably because both individuals came to the interview site together. In these cases, we resolved the tie by jittering the recruiter's interview time to be slightly earlier than the recruitee's interview time.

Construction of G_R and $\mathbf{C}$ revealed a minor violation of the RDS recruitment specification: we found seven recruits whose coupon IDs matched the IDs of already redeemed coupons. The financial reward for recruiting another eligible subject may provide a strong incentive for participants to fraudulently inflate the number of coupons they hold by creating a facsimile of the original coupon and giving it to another potential subject to redeem. This appears to be what happened: the recruiter photocopied the original coupon, this reproduction was not detected by the interviewer, and both the new recruit and recruiter received their corresponding rewards. Rather than breaking the recruitment chain by omitting data from the seven subjects with duplicated coupon IDs, we instead artificially increased the number of coupons held by the apparent recruiter to be equal to the number of subjects who redeemed coupons bearing the ID of the recruiter.

Overall recruitment of participants in this study was rapid: the mean time between interviews was 0.28 ± 0.74 days. However, the mean time between a particular subject's interview and his or her recruiter's interview was 23.4 ± 18.0 days, indicating that the per edge waiting times for recruited subjects were substantially longer (the maximum waiting time from interview to recruitment was 112 days). Indeed, this calculation is conditional on the subject's actually being recruited within the study time frame, so any longer waiting times are censored by the end of the study. We evaluated the posterior mode with η ranging from 0.1 to 10, and in every case the estimate ranged from 0.0050 to 0.0053. The rate of recruitment across susceptible edges is estimated to be approximately $1/\lambda = 199$ days with posterior quantiles $(186, 215)$, nearly as long as the study duration of 223 days. The apparent discrepancy between the high frequency of interviews and very slow recruitment across susceptible edges is explained by the fact that researchers observe the *minimum* waiting time to recruitment across all susceptible edges at each step in the recruitment process.

Figure 5 shows the MAP estimate of the adjacency matrix for all 813 sampled subjects (left) and inset submatrix (right). Recruitment edges appear in gray. The apparent bands in the adjacency matrix represent high-degree individuals with many nonrecruitment edges. Probabilistic assignment of these edge ends to other recruited individuals depends on the timing of recruitments of other subjects. The blocklike structure

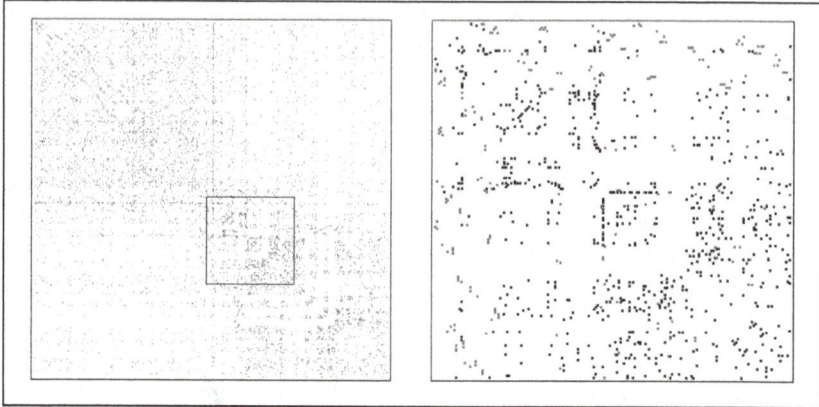

Figure 5. Maximum *a posteriori* (MAP) estimate of G_S for the St. Petersburg respondent-driven sampling study. The left panel shows the MAP estimate of the adjacency matrix of G_S, and the right panel shows the inset submatrix in detail. Edges in the recruitment graph are shown in gray.

evident in this adjacency matrix may indicate subnetworks of highly connected individuals. Subjects recruited nearby in time may be more likely to know one another, even if they are not linked by a recruitment edge.

7. DISCUSSION

Nearly every paper on statistical methods for RDS data states or assumes a version of assumption (A-1): the social network connecting members of the hidden population exists and determines the sampling probabilities. But because this network is only partially observed in real-world RDS studies, assumption (A-1) is usually disregarded in the formulation of statistical estimators. Instead, researchers usually make the simplifying assumption that sampling probability is proportional to degree and does not otherwise depend on subjects' location in the network. This simplification is justified by a thought experiment in which the rules of the game are altered: subjects can be recruited infinitely many times, each subject receives only one coupon, and this process continues for an infinitely long time (Goel and Salganik 2009; Salganik and Heckathorn 2004; Volz and Heckathorn 2008).

In this paper, we have embraced assumption (A-1) and its natural consequence: RDS recruitment happens across edges in the network connecting members of the hidden population. This point of view emphasizes that RDS is more like a stochastic spreading process on a hidden network than a survey sampling method. We define a simple continuous-time model for RDS recruitment on a hidden population graph using the kind of data obtained by every RDS study. The model results in sensible nonuniform conditional recruitment probabilities: the next subject is recruited with probability proportional to the number of edges he or she shares with recruiters (proposition 2), *not their total network degree*. Combining this model with the observed data from an RDS study allows joint estimation of the recruitment-induced subgraph G_S and the waiting time parameter λ. Most important, the model directly connects the observed data to the recruitment process on the underlying network.

This approach yields two computational benefits. First, the time required to evaluate the likelihood via proposition 3 is a function of the sample size *n* alone, and it does not depend on the population size *N*, which is likely to be much larger. In particular, we never simulate unobserved portions of the population network *G*; the ERGM (equation 1) specifies a probability model for the recruitment-induced subgraph G_S only. In contrast, some researchers dealing with partially observed network data marginalize over the entire unsampled portion of the graph, which may be burdensome or impossible for large *N* (Gile and Handcock 2015). Second, the likelihood (outlined in the online Appendix) does not require computation of the matrix products implied by equation (2). Instead, efficient update expressions given in the online Appendix depend only on a *change statistic* that can be efficiently updated.

Our approach is unique because it uses all the available data $\mathbf{Y} = (G_R, \mathbf{d}, \mathbf{t}, \mathbf{C})$ from real-world RDS studies. Several researchers have attempted to estimate the population degree distribution but they use only G_R and $\mathbf{d}$ (and sometimes $\mathbf{d}$ alone), ignoring $\mathbf{t}$ and $\mathbf{C}$ (Gile 2011; Gile and Handcock 2015; Handcock, Gile, and Mar 2015; Salganik and Heckathorn 2004; Volz and Heckathorn 2008). Berchenko, Rosenblatt, and Frost (2013) gave a formulation of recruitment event intensity similar to assumption (A-2) by using a multitype epidemic model in which active recruiters correspond to infective individuals. In their model, the rate of recruitment of a new subject with degree *k* is proportional to the product of the number of active recruiters and the number of susceptible subjects with degree *k*. However, they use only $\mathbf{d}$, $\mathbf{t}$, and $\mathbf{C}$ but do not take advantage of the topological information contained in G_R.

Historically, there have been two major statistical objections to RDS as a survey design for inference of population quantities. First, sampling probabilities cannot be computed directly from the observed data without additional assumptions (Gile 2011; Gile and Handcock 2010). Second, there may be statistical dependence between the traits of a given subject and his or her neighbors (particularly their recruiter) in the network (Fisher and Merli 2014; Heckathorn 1997, 2002; Tomas and Gile 2011). This dependency might be due to homophily—the tendency for people to form social ties with others similar to themselves—or preferential recruitment of certain types of people, conditional on existing social ties. Clearly, the network structure local to the seeds and recruitment chain encodes the sampling probabilities and the statistical dependencies between subjects' attributes. This leads us to the conclusion that a fundamental obstacle to principled statistical inference for RDS is *missing data*: in RDS, not all network neighbors of a vertex i are observed, either because they remain unsampled, or because the recruitment graph G_R does not reveal a tie between i and the sampled vertices to which it is connected. Objections to RDS typically understate the information about this network contained in the recruitment graph G_R and the time series of interviews. Our results—revealing the graphical structure of data obtained by RDS—raise the possibility that researchers can account for both of these sources of missing data without imposing strong prior assumptions about the network.

Although the network may be of interest for sociological reasons, it can also be viewed as a nuisance parameter when population attributes are of primary interest. Marginalizing (integrating) over the recruitment-induced subgraph G_S can be understood as multiple imputation, repeatedly filling in the missing data in accordance with its distribution under the model (Little and Rubin 1986; Huisman 2009; Koskinen, Robins, and Pattison 2010; Koskinen et al. 2013). In the absence of any other information, we could marginalize over compatible graphs in $\mathcal{C}(G_R, \mathbf{d})$ with respect to the uniform distribution. However, the reconstructed graph would be subject to two types of reconstruction inaccuracy. First, for three sampled vertices i, j, and k with at least one pendant edge each, the uniform distribution provides no basis to distinguish an edge $\{i, j\}$ from an edge $\{i, k\}$ unless a recruitment event took place along one of those edges. For any given pendant edge, there are usually many more incorrect ways to connect it to sampled vertices than there are correct ways. Second, marginalization with respect to the uniform distribution usually results in inclusion of too many or too few edges overall in G_S. The waiting time model developed in this paper provides a coherent basis for adding edges to the recruitment subgraph G_R and helps ensure that estimates of G_S have approximately the same number of edges as the true underlying graph.

In conclusion, we offer a mixed message about the prospect for statistically rigorous analysis of data from real-world RDS studies. First, current estimators for population characteristics depend on assumptions that bear little similarity to RDS recruitment processes on social networks, and they do not use all the available data. This may account for their poor performance in empirical studies. Second, and more optimistically, data from RDS studies contain far more information about the social network connecting respondents than has been acknowledged. Estimation of population-level characteristics should therefore proceed from knowledge about the network of sampled subjects: The subgraph G_S is the maximal network object that can be estimated directly from the observed data without further assumptions. Extrapolation to the population network requires stronger assumptions than those given in this article. By introducing a simple technique for probabilistic reconstruction of the recruitment-induced subgraph, we hope to offer researchers a new tool for sociological inquiry: a social network sampling method that delivers the network.

Acknowledgments

I am especially indebted to Edward H. Kaplan, Robert Heimer, Peter M. Aronow, and Leonid Chindelevitch for providing detailed comments on the manuscript. I also thank Yakir Berchenko, Russel Barbour, Alexander Bazazi, Lin Chen, Krista Gile, Mark Handcock, Olga Levina, Aleksandr Sirotkin, Edward White, Jiacheng Wu, Alexei Zelenev, and Li Zeng for valuable suggestions and discussion.

Funding

This work was supported by National Institutes of Health (NIH) grant KL2 TR000140, National Institute of Mental Health grant P30 MH062294, the Yale Center for Clinical Investigation, and the Yale Center for Interdisciplinary Research on AIDS. The Project 90 data were obtained from the Office of Population Research at Princeton University (http://opr.princeton.edu/archive/P90). The RDS data presented in the application are from the Influences on HIV Prevalence and Service Access among IDUs in Russia and Estonia study, funded by NIH/National Institute on Drug Abuse grant 1R01DA029888 to Robert Heimer and Anneli Uusküla (co–principal investigators). I made use of the Yale University Biomedical High Performance Computing Center, funded by NIH grant RR029676-01.

References

Aronow, Peter M., and Forrest W. Crawford. 2015. "Nonparametric Identification for Respondent-driven Sampling." *Statistics and Probability Letters* 106:100–102.

Atchade, Yves F. 2011. "Estimation of Network Structures from Partially Observed Markov Random Fields." arXiv preprint arXiv:1108.2835.

Bastos, Leonardo S., Adriana A. Pinho, Claudia Codeco, and Francisco I. Bastos. 2012. "Binary Regression Analysis with Network Structure of Respondent-driven Sampling Data." arXiv preprint arXiv:1206.5681.

Berchenko, Yakir, Jonathan Rosenblatt, and Simon D. W. Frost. 2013. "Modeling and Analysing Respondent Driven Sampling as a Counting Process." arXiv preprint arXiv:1304.3505.

Bernard, H. Russell, Tim Hallett, Alexandrina Iovita, Eugene C. Johnsen, Rob Lyerla, Christopher McCarty, Mary Mahy, Matthew J. Salganik, Tetiana Saliuk, Otilia Scutelniciuc, Gene A. Shelley, Petchsri Sirinirund, Sharon Weir, and Donna F. Stroup. 2010. "Counting Hard-to-count Populations: The Network Scale-up Method for Public Health." *Sexually Transmitted Infections* 86(Suppl. 2):ii11–15.

Bliss, Catherine A., Christopher M. Danforth, and Peter Sheridan Dodds. 2014. "Estimation of Global Network Statistics from Incomplete Data." *PLoS ONE* 9: e108471.

Broadhead, Robert S., Douglas D. Heckathorn, David L. Weakliem, Denise L. Anthony, Heather Madray, Robert J. Mills, and James Hughes. 1998. "Harnessing Peer Networks as an Instrument for AIDS Prevention: Results from a Peer-driven Intervention." *Public Health Reports* 113(1):42.

Cepeda, Javier A., Veronika A. Odinokova, Robert Heimer, Lauretta E. Grau, Alexandra Lyubimova, Liliya Safiullina, Olga S. Levina, and Linda M. Niccolai. 2011. "Drug Network Characteristics and HIV Risk among Injection Drug Users in Russia: The Roles of Trust, Size, and Stability." *AIDS and Behavior* 15:1003–10.

Feehan, Dennis M., and Matthew J. Salganik. 2014. "Generalizing the Network Scale-up Method: A New Estimator for the Size of Hidden Populations." arXiv preprint arXiv:1404.4009.

Fisher, Jacob C., and M. Giovanna Merli. 2014. "Stickiness of Respondent-driven Sampling Recruitment Chains." *Network Science* 2(2):298–301.

Frank, Ove, and Tom Snijders. 1994. "Estimating the Size of Hidden Populations Using Snowball Sampling." *Journal of Official Statistics* 10(1):53–67.

Frank, Ove, and David Strauss. 1986. "Markov Graphs." *Journal of the American Statistical Association* 81(395):832–42.

Gile, Krista J. 2011. "Improved Inference for Respondent-driven Sampling Data with Application to HIV Prevalence Estimation." *Journal of the American Statistical Association* 106(493):135–46.

Gile, Krista J., and Mark S. Handcock. 2010. "Respondent-driven Sampling: An Assessment of Current Methodology." Pp. 285–327 in *Sociological Methodology*, Vol. 40, edited by Tim Futing Liao. Hoboken, NJ: Wiley-Blackwell.

Gile, Krista J., and Mark S. Handcock. 2015. "Network Model-assisted Inference from Respondent-driven Sampling Data." *Journal of the Royal Statistical Society, Series A (Statistics in Society)* 178(3):619–39.

Goel, Sharad, and Matthew J. Salganik. 2009. "Respondent-driven Sampling as Markov Chain Monte Carlo." *Statistics in Medicine* 28(17):2202–29.

Goel, Sharad, and Matthew J. Salganik. 2010. "Assessing Respondent-driven Sampling." *Proceedings of the National Academy of Sciences* 107(15):6743–47.

Goodman, Leo A. 1961. "Snowball Sampling." *Annals of Mathematical Statistics* 32(1): 148–70.

Goyal, Ravi, Joseph Blitzstein, and Victor de Gruttola. 2014. "Sampling Networks from Their Posterior Predictive Distribution." *Network Science* 2(1):107–31.

Handcock, Mark S., and Krista J. Gile. 2010. "Modeling Social Networks from Sampled Data." *Annals of Applied Statistics* 4(1):5–25.

Handcock, Mark S., Krista J. Gile, and Corinne M. Mar. 2015. "Estimating the Size of Populations at High Risk for HIV Using Respondent-driven Sampling Data." *Biometrics* 71(1):258–66.

Heckathorn, Douglas D. 1997. "Respondent-driven Sampling: A New Approach to the Study of Hidden Populations." *Social Problems* 44(2):174–99.

Heckathorn, Douglas D. 2002. "Respondent-driven Sampling II: Deriving Valid Population Estimates from Chain-referral Samples of Hidden Populations." *Social Problems* 49(1):11–34.

Heimer, Robert, and Edward White. 2010. "Estimation of the Number of Injection Drug Users in St. Petersburg, Russia." *Drug and Alcohol Dependence* 109(1):79–83.

Huisman, Mark. 2009. "Imputation of Missing Network Data: Some Simple Procedures." *Journal of Social Structure* 10(1):1–29.

Johnston, Lisa Grazina, Sara Whitehead, Milena Simic-Lawson, and Carl Kendall. 2010. "Formative Research to Optimize Respondent-driven Sampling Surveys among Hard-to-reach Populations in HIV Behavioral and Biological Surveillance: Lessons Learned from Four Case Studies." *AIDS Care* 22(6):784–92.

Killworth, Peter D., Christopher McCarty, H. Russell Bernard, Gene Ann Shelley, and Eugene C. Johnsen. 1998. "Estimation of Seroprevalence, Rape, and Homelessness in the United States Using a Social Network Approach." *Evaluation Review* 22: 289–308.

Koskinen, Johan H., Garry L. Robins, and Philippa E. Pattison. 2010. "Analysing Exponential Random Graph (p-star) Models with Missing Data Using Bayesian Data Augmentation." *Statistical Methodology* 7(3):366–84.

Koskinen, Johan H., Garry L. Robins, Peng Wang, and Philippa E. Pattison. 2013. "Bayesian Analysis for Partially Observed Network Data, Missing Ties, Attributes, and Actors." *Social Networks* 35(4):514–27.

Kramer, Mark A., Uri T. Eden, Sydney S. Cash, and Eric D. Kolaczyk. 2009. "Network Inference with Confidence from Multivariate Time Series." *Physical Review E* 79: 061916.

Leskovec, Jure, Daniel Huttenlocher, and Jon Kleinberg. 2010. "Predicting Positive and Negative Links in Online Social Networks." Pp. 641–50 in *Proceedings of the 19th International Conference on World Wide Web*. New York: Association for Computing Machinery.

Li, Jian, Hongjie Liu, Jianhua Li, Jian Luo, Nana Koram, and Roger Detels. 2011. "Sexual Transmissibility of HIV among Opiate Users with Concurrent Sexual

Partnerships: An Egocentric Network Study in Yunnan, China." *Addiction* 106:1780–87.

Linderman, Scott W., and Ryan P. Adams. 2014. "Discovering Latent Network Structure in Point Process Data." arXiv preprint arXiv:1402.0914.

Little, Roderick J. A., and Donald B. Rubin. 1986. *Statistical Analysis with Missing Data*. New York: John Wiley.

Liu, Hongjie, Tiejian Feng, Hui Liu, Hucang Feng, Yumao Cai, Anne G. Rhodes, and Oscar Grusky. 2009. "Egocentric Networks of Chinese Men Who Have Sex with Men: Network Components, Condom Use Norms, and Safer Sex." *AIDS Patient Care and STDs* 23:885–93.

Lü, Linyuan, and Tao Zhou. 2011. "Link Prediction in Complex Networks: A Survey." *Physica A: Statistical Mechanics and Its Applications* 390(6):1150–70.

McCreesh, Nicky, Simon Frost, Janet Seeley, Joseph Katongole, M. N. Tarsh, R. Ndunguse, F. Jichi, N. L. Lunel, D. Maher, L. G. Johnston, P. Sonnenberg, A. J. Copas, R. J. Hayes, and R. G. White. 2012. "Evaluation of Respondent-driven Sampling." *Epidemiology* 23(1):138–47.

Mills, Harriet L., Samuel Johnson, Matthew Hickman, Nick S. Jones, and Caroline Colijn. 2014. "Errors in Reported Degrees and Respondent-driven Sampling: Implications for Bias." *Drug and Alcohol Dependence* 142:120–26.

Niccolai, Linda M., Olga V. Toussova, Sergei V. Verevochkin, Russell Barbour, Robert Heimer, and A. P. Kozlov. 2010. "High HIV Prevalence, Suboptimal HIV Testing, and Low Knowledge of HIV-positive Serostatus among Injection Drug Users in St. Petersburg, Russia." *AIDS and Behavior* 14(4):932–41.

Niccolai, Linda M., Sergei V. Verevochkin, Olga V. Toussova, E. White, Russell Barbour, A. P. Kozlov, and Robert Heimer. 2011. "Estimates of HIV Incidence among Drug Users in St. Petersburg, Russia: Continued Growth of a Rapidly Expanding Epidemic." *European Journal of Public Health* 21(5):613–19.

Robert, Christian, and George Casella. 2004. *Monte Carlo Statistical Methods*. 2nd ed. New York: Springer Science & Business Media.

Salganik, Matthew J. 2012. "Commentary: Respondent-driven Sampling in the Real World." *Epidemiology* 23(1):148–50.

Salganik, Matthew J., and Douglas D. Heckathorn. 2004. "Sampling and Estimation in Hidden Populations Using Respondent-driven Sampling." Pp. 193–240 in *Sociological Methodology*, Vol. 34, edited by Ross M. Stolzenberg. Hoboken, NJ: Wiley-Blackwell.

Shandilya, Srinivas Gorur, and Marc Timme. 2011. "Inferring Network Topology from Complex Dynamics." *New Journal of Physics* 13:013004.

Smith, Jeffrey A. 2012. "Macrostructure from Microstructure: Generating Whole Systems from Ego Networks." *Sociological Methodology* 42(1):155–205.

Snijders, Tom A. B., and Marijtje A. J. Van Duijn. 2002. "Conditional Maximum Likelihood Estimation under Various Specifications of Exponential Random Graph Models." Pp. 117–34 in *Contributions to Social Network Analysis, Information Theory, and Other Topics in Statistics: A Festschrift in Honour of Ove Frank on the Occasion of His 65th Birthday*, edited by Jan Hagberg. Stockholm, Sweden: University of Stockholm, Department of Statistics.

Stein, Mart L., Jim E. van Steenbergen, Vincent Buskens, Peter G. M. van der Heijden, Charnchudhi Chanyasanha, Mathuros Tipayamongkholgul, Anna E. Thorson, Linus Bengtsson, Xin Lu, and Mirjam E. E. Kretzschmar. 2014. "Comparison of Contact Patterns Relevant for Transmission of Respiratory Pathogens in Thailand and the Netherlands Using Respondent-driven Sampling." *PLoS ONE* 9:e113711. doi:10.1371/journal.pone.0113711

Stein, Mart L., Jim E. van Steenbergen, Charnchudhi Chanyasanha, Mathuros Tipayamongkholgul, Vincent Buskens, Peter G. M. van der Heijden, Wasamon Sabaiwan, Linus Bengtsson, Xin Lu, Anna E. Thorson, and Mirjam E. E. Kretzschmar. 2014. "Online Respondent-driven Sampling for Studying Contact Patterns Relevant for the Spread of Close-contact Pathogens: A Pilot Study in Thailand." *PLoS ONE* 9:e85256. doi:10.1371/journal.pone.0085256

Thompson, Steven K., and Ove Frank. 2000. "Model-based Estimation with Link-tracing Sampling Designs." *Survey Methodology* 26(1):87–98.

Tomas, Amber, and Krista J. Gile. 2011. "The Effect of Differential Recruitment, Non-response and Non-recruitment on Estimators for Respondent-driven Sampling." *Electronic Journal of Statistics* 5:899–934.

Volz, Erik, and Douglas D. Heckathorn. 2008. "Probability-based Estimation Theory for Respondent-driven Sampling." *Journal of Official Statistics* 24(1):79–97.

Wasserman, Stanley, and Philippa Pattison. 1996. "Logit Models and Logistic Regressions for Social Networks: I. An Introduction to Markov graphs and p^*." *Psychometrika* 61(3):401–25.

Wejnert, Cyprian. 2009. "An Empirical Test of Respondent-driven Sampling: Point Estimates, Variance, Degree Measures, and Out-of-equilibrium Data." *Sociological Methodology* 39(1):73–116.

White, Richard G., Amy Lansky, Sharad Goel, David Wilson, Wolfgang Hladik, Avi Hakim, and Simon D. W. Frost. 2012. "Respondent-driven Sampling—Where We Are and Where Should We Be Going?" *Sexually Transmitted Infections* 88(6): 397–99.

Yan, Bowen, and Steve Gregory. 2013. "Identifying Communities and Key Vertices by Reconstructing Networks from Samples." *PLoS ONE* 8:e61006.

Author Biography

Forrest W. Crawford is an assistant professor in the Department of Biostatistics at the Yale School of Public Health. He is affiliated with the Department of Ecology and Evolutionary Biology, the Center for Interdisciplinary Research on AIDS, the Institute for Network Science, the Computational Biology and Bioinformatics program, and the Operations doctoral program at the Yale School of Management. He received his PhD in biomathematics from the University of California, Los Angeles, in 2012. His interests include network analysis, stochastic processes, and missing data.

Sociological Methodology
2016, Vol. 46(1) 212–251
© American Sociological Association 2015
DOI: 10.1177/0081175015599807
http://sm.sagepub.com
$SAGE

∂ 7 ∂

ROBUST ESTIMATION OF INEQUALITY FROM BINNED INCOMES

*Paul T. von Hippel**
Samuel V. Scarpino[†]
*Igor Holas**

Abstract

Researchers often estimate income inequality by using data that give only the number of cases (e.g., families or households) whose incomes fall in "bins" such as $ 0 to $9,999, $10,000 to $14,999, . . . , ≥ $200,000. We find that popular methods for estimating inequality from binned incomes are not robust in small samples, where popular methods can produce infinite, undefined, or arbitrarily large estimates. To solve these and other problems, we develop two improved estimators: a robust Pareto midpoint estimator (RPME) and a multimodel generalized beta estimator (MGBE). In a broad evaluation using U.S. national, state, and county data from 1970 to 2009, we find that both estimators produce very good estimates of the mean and Gini coefficient but less accurate estimates of the Theil index and mean log deviation. Neither estimator is uniformly more accurate, but the RPME is much faster, which may be a consideration when many estimates must be obtained from many data sets. We have made the methods available as the rpme *and* mgbe *commands for Stata and the* binequality *package for R.*

*University of Texas, Austin, TX, USA
[†]Santa Fe Institute, Santa Fe, NM, USA

Corresponding Author:
Paul T. von Hippel, LBJ School of Public Affairs, University of Texas, 2315 Red River, Box Y, Austin, TX 78712, USA
Email: paulvonhippel.utaustin@gmail.com

Keywords

grouped data, income brackets, interval censored, top-coded, top-coding

1. INTRODUCTION

Social scientists often estimate income inequality, not just for large areas such as nations and states but within small areas such as neighborhoods, school districts, and counties. There are numerous examples in sociology and economics:

- Jargowsky (1996) estimated 20-year trends in the segregation of metropolitan neighborhoods by income. To estimate his neighborhood segregation index, he had to estimate the mean and variance of income within each neighborhood.
- Corcoran and Evans (2010) asked whether school districts with greater income inequality were more likely to vote for redistribution in the form of high local school taxes. To estimate inequality within districts, the authors had to estimate the mean and median income within each school district.
- Galbraith and Hale (2004, 2009) estimated 40-year trends in the component of the Theil inequality statistic that lies between U.S. counties. It would be informative to also estimate trends in the Theil component that lie within counties.

All these studies of small-area inequality encountered the same practical challenge: Within small areas, researchers rarely have data on the incomes of individual cases (e.g., households or families). Instead, researchers must estimate the distribution of incomes from income *bins* (also known as brackets, groups, or intervals). As an example of binned income data, Table 1 summarizes the distribution of household income within the richest county in the United States—Nantucket County, Massachusetts—and the poorest county—Maricao County, Puerto Rico. For each county, Table 1 estimates how many households have incomes in each of 16 bins. The top bin ($200,000, ∞) is one sided because it has no upper bound. Technically, the bottom bin is also one sided because it has no lower bound, but it is customary and innocuous to treat the bottom bin as though its lower bound is zero (i.e., $ 0, $10,000).[1]

It is clear that Nantucket is much richer than Maricao. Yet it is challenging to be more specific. How much inequality lies between the two

Table 1. Distribution of Household Income in Two U.S. Counties

			Households in County	
Bin	Minimum	Maximum	Maricao	Nantucket
1		$10,000	781	165
2	$10,000	$15,000	245	109
3	$15,000	$20,000	140	67
4	$20,000	$25,000	156	147
5	$25,000	$30,000	85	114
6	$30,000	$35,000	60	91
7	$35,000	$40,000	37	148
8	$40,000	$45,000	61	44
9	$45,000	$50,000	9	121
10	$50,000	$60,000	57	159
11	$60,000	$75,000	19	358
12	$75,000	$100,000	0	625
13	$100,000	$125,000	0	338
14	$125,000	$150,000	0	416
15	$150,000	$200,000	0	200
16	$200,000		0	521
Total			1,650	3,623

Note: The number of households in each bin is estimated from a one-in-eight sample of households that took the American Community Survey from 2006 through 2010. Incomes reported in the survey have been inflated to 2010 dollars.

counties; for example, how much larger is the mean or median household income in Nantucket than in Maricao? And how do the two counties differ with respect to a measure of within-county inequality, such as the Gini coefficient, the Theil index, the coefficient of variation (CV), or the mean log deviation (MLD)? Questions such as these are fundamental to research on small-area inequality, and their answers must often be estimated from binned data.

A variety of methods have been proposed for estimation with binned incomes. In fact, much more effort has been spent on developing estimators than on evaluating their performance, especially in small samples. Most methodological articles simply describe a binned-data estimator and then demonstrate its use in a handful of data sets, typically with large samples (for exceptions, see Bandourian, McDonald, and Turley 2002; Evans, Hout, and Mayer 2004; von Hippel, Holas, and Scarpino 2012; Minoiu and Reddy 2009, 2012). The limited nature of past evaluations was due in part to the limited availability of published inequality statistics, especially for small areas. Until recently, no inequality statistics were available below the level of U.S. states, and even at the state level, no statistic was available except for the Gini coefficient.

It is now possible to evaluate binned-data estimators more thoroughly in small samples, as the U.S. Census Bureau has published the household Gini coefficients of all 3,221 U.S. counties and the household Gini coefficient, Theil index, and MLD of every U.S. state (Bee 2012; Hisnanick and Rogers 2006). In this paper, we use these and other statistics to carry out the largest evaluation of binned-data estimators that has been conducted to date.

Our evaluation focuses on two of the most popular estimators. One is the simple Pareto midpoint estimator (PME) (Henson 1967). The PME assigns cases to the midpoints of their bins, except for the top bin, for which there is no midpoint and the PME assigns cases to the arithmetic mean of a Pareto distribution. The other method involves fitting distributions from the generalized beta (GB) family, which includes the Pareto, log normal, log logistic, Dagum, and Singh-Maddala distributions, as well as different forms of the beta and gamma distributions (McDonald and Xu 1995).

Our evaluation uncovers problems that have escaped notice in more limited evaluations. In particular, we find that neither estimator, as commonly implemented, is robust in small samples. In small samples, the PME sometimes yields estimates that are arbitrarily large or even infinite. Some GB distributions can also yield infinite or undefined estimates, at least on occasion, and some can fail to yield an estimate at all because of nonidentification or nonconvergence. In addition, past literature gives incomplete guidance regarding which distribution from the GB family should be fit to a particular data set.

We modify both estimators to make them more robust in small samples. To improve the PME, we develop a *robust PME* (RPME), which is just like the PME except that the mean of the top bin is guaranteed to be defined and not arbitrarily large. We offer several flavors of the RPME, which achieve robustness by constraining the Pareto shape parameter and/or by replacing the arithmetic mean with a robust alternative such as the harmonic mean. These changes make little difference in large samples but dramatically improve the estimates in small samples.

To improve the utility of the GB family, we develop a *multimodel GB estimator* (MGBE). The MGBE fits 10 distributions from the GB family, discards any fitted distributions with undefined estimates, and selects or averages among the remaining distributions according to some criterion of fit. By default our criterion is the Akaike information criterion (AIC), but we get similar results using the Bayesian information criterion (BIC).

Between the RPME and the MGBE, neither is uniformly more accurate than the other. Theoretically, the RPME is a nearly nonparametric estimator that, with enough bins, can fit the nooks and crannies of any income distribution, regardless of its shape. RPME estimates can be very accurate if there are many bins but much less accurate if there are few. The MGBE, by contrast, is a parametric estimator which assumes that incomes follow one of the distributions in the GB family. The GB family is quite flexible, but all of its distributions are unimodal and positively skewed. If the true income distribution approximates one of the GB distributions, then the resulting estimates can be accurate, even if the bins are coarse. However, if the true income distribution is a poor fit to the GB family (e.g., if the true income distribution is bimodal) then estimates from the GB family will be biased (Minoiu and Reddy 2009).

Empirically, the two estimators are about equally accurate in typical data with 8 to 20 bins. In artificial data with 4 bins, the MGBE has an advantage, though both estimators are inaccurate. Both estimators are very good at estimating the mean, fairly good at estimating the Gini coefficient, and potentially poor at estimating the Theil index and MLD. The MGBE is better than the RPME at estimating the median, though neither estimator is bad. The MGBE seems to be slightly more accurate in data from 1970 to 1990, while the RPME is slightly more accurate in data from 2000 and later. This could mean that since 1990 the income distribution has changed in ways that worsen its fit to distributions in the GB family.

A further consideration is that the RPME is about 1,000 times faster than the MGBE. The difference in speed does not matter if you are fitting a single binned data set, but if you are fitting thousands of binned data sets (e.g., from every county or school district in the United States), using the RPME can save hours.

Because either estimator can be better under different circumstances, when practical it would be wise for applied researchers to try both approaches and check the results against known reference statistics. To encourage adoption, we have made the methods available as the *rpme* and *mgbe* commands for Stata, which can be installed using Stata's *ssc* command (Duan and von Hippel 2016; von Hippel and Powers 2014). The methods are also available as the *binequality* package for R, which can be downloaded from the Comprehensive R Archive Network at http://cran.r-project.org/web/packages/binequality/index.html (Scarpino, von Hippel, and Holas 2014).

In the rest of this article, we define the RPME and MGBE estimators in more detail and then present our evaluation.

2. METHODS

In this section, we motivate and define the RPME and MGBE and related estimators.

2.1. *Notation and Terminology*

Throughout this section, we use the following notation and terminology. There are n cases (e.g., families or households) with different values of a continuous variable X (e.g., income or wealth). The distribution of X is summarized by a frequency table like Table 1, which gives the number of cases n_b whose X values lie in each of several bins ($[l_b, u_b]$, $b = 1$, 2, . . . , where l_b and u_b are the lower and upper bounds of bin b).

In small samples or very poor populations, it is not unusual for some of the bins to be empty, with $n_b = 0$, as the upper bins are in Maricao (Table 1). The empty bins contribute nothing to the estimates and can cause trouble for some calculations, so we will find it helpful to restrict the calculations to the populated bins with $n_b > 0$. The number of populated bins is called B. The total number of bins, both populated and unpopulated, will be called B_{all}.

2.2. *The Midpoint Estimators*

The most straightforward approach to binned data involves assigning cases to their bin midpoints. In this section, we start with the most basic type of midpoint estimator (ME), then review how the Pareto distribution can be used to estimate the mean of the top bin, and finally discuss how to make estimating from the top bin more robust.

2.2.1. *The Basic ME.* Within each bin, the basic ME assigns the n_b cases to the bin midpoint $m_b = (l_b + u_b)/2$ and then calculates statistics using the midpoints m_b. For example, the mean μ and variance σ^2 of X are estimated as

$$\hat{\mu}_{ME} = \frac{1}{n} \sum_{b=1}^{B} n_b m_b,$$

$$\hat{\sigma}^2_{ME} = \frac{1}{n-1} \sum_{b=1}^{B} n_b (m_b - \hat{\mu}_{ME})^2. \tag{1}$$

Naturally, the ME is defined only if every populated bin has a defined upper and lower bound. In Table 1, for example, the ME is defined for Maricao but not for Nantucket.

Despite its simplicity, the ME has a very nice statistical property that we call *bin consistency*. Bin consistency means that as the bins get narrower and more numerous, the ME estimate converges to its estimand. The ME is bin consistent because it is nonparametric and makes no assumptions about the shape of the underlying distribution.

The properties of the ME get even nicer if certain assumptions are met. One assumption is that the distribution of X is "smooth" near the tails, which is a reasonable assumption but not one that can be verified from binned data. Another assumption is that the bin widths $w_b = u_b - l_b$ are roughly equal and not too much larger than the standard deviation σ. (Heitjan [1989] suggested $w_b < 1.6\sigma$.) This assumption holds for some data but not for others. Consider Table 1. In Nantucket, the populated bins are far from equal, ranging in width from \$5,000 to \$50,000, but in Maricao, the widest bins are empty and the populated bins are roughly equal in width. So in Maricao, the assumptions about bin width approximately hold, but in Nantucket they do not.

If the assumptions hold, $\hat{\mu}_{ME}$ will be nearly unbiased, and $\hat{\sigma}^2_{ME}$ will have a bias that is correctable and typically small. The relative bias of $\hat{\sigma}^2_{ME}$ is approximately $(w_b/\sigma)^2/12$ (Heitjan 1989). For example, in Maricao (Table 1), the populated bins are approximately $w_b \approx \sigma/2$ wide,[2] so the relative bias of $\hat{\sigma}^2_{ME}$ is about 2 percent. That is, the estimate $\hat{\sigma}^2_{ME}$ is expected to be just 2 percent larger than the estimand σ^2.

In addition to having little bias, under the stated assumptions, the estimates $\hat{\mu}_{ME}$ and $\hat{\sigma}^2_{ME}$ will be almost fully efficient. That is, they will be almost as efficient as the maximum likelihood estimates that would be obtained by fitting the binned data to the true distribution of X. This is a very nice property. Because we rarely know the true distribution of

X, it is reassuring that knowing it would not greatly improve on the ME estimates.

The literature on the ME focuses on the estimands μ and σ, but if μ and σ are well estimated, then the inequality statistic CV = σ/μ will be well estimated, too. The Gini coefficient should also be well estimated, because the Gini coefficient is ½MD/μ, where MD is the mean difference, which is similar to σ (Hosking 1990). It is not clear how good the ME estimates will be for other inequality statistics, such as the Theil index and MLD.

2.2.2. The PME. If the top bin is populated and has no upper bound, as in Nantucket (Table 1), then its midpoint is undefined, and some statistic other than the midpoint must be plugged in for the top bin. Intuitively the statistic should be some multiple of the top bin's lower bound l_B, and the multiple should be larger if there is evidence that the tail is longer. To estimate the length and shape of the tail, we must slightly compromise the nonparametric nature of the ME and assume that the incomes in the top two bins follow some parametric distribution.

An arbitrary but convenient and popular approach is to assume that the top two bins follow a Pareto distribution[3] with shape parameter $\alpha >$ 0. Then the mean μ_B of the top bin is a simple function of α:

$$\mu_B = \begin{cases} l_B \frac{\alpha}{\alpha-1} & \text{if } \alpha > 1 \\ \infty & \text{if } \alpha \leq 1 \end{cases}, \tag{2}$$

and an estimate of μ_B can be used in place of the midpoint. An estimate of μ_B can be obtained by plugging an estimate of α into the formula for μ_B. The most popular estimator for α is

$$\hat{\alpha} = \frac{\ln\left((n_{B-1}+n_B)/n_B\right)}{\ln(l_B/l_{B-1})} = \frac{\ln(n_{B-1}+n_B) - \ln(n_B)}{\ln(l_B) - \ln(l_{B-1})} \tag{3}$$

(Henson 1967; Quandt 1966). If the top two bins really follow a Pareto distribution, then $\hat{\alpha}$ is the maximum likelihood estimator. Past attempts to improve on the maximum likelihood estimator $\hat{\alpha}$ have not been successful (West 1986; West, Kratzke, and Butani 1992).[4]

2.2.3. The RPME. Our results will show that the PME can perform well in large samples, but it is not robust or even usable in some small samples. The problem is the sensitivity of μ_B to the value of α. As equation (2) shows, if $\alpha \leq 1$, then the mean is infinite, and as α approaches

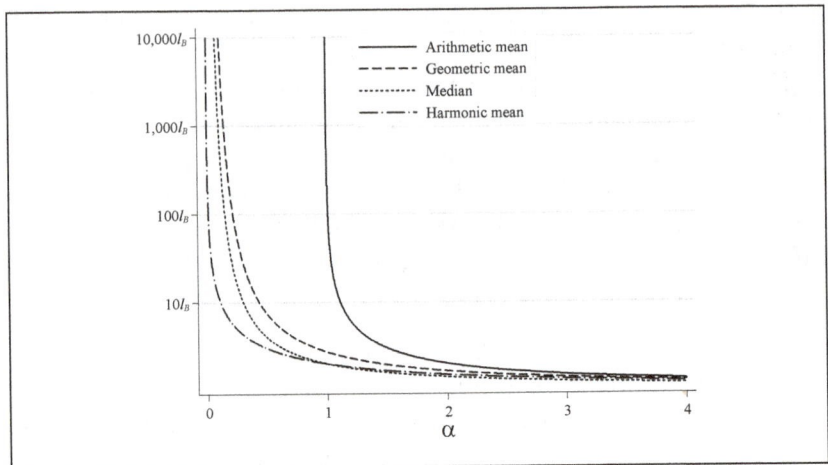

Figure 1. The median and three types of mean graphed as functions of the Pareto shape parameter α.
Note: The vertical axis is on a log scale.

1 from above, the mean grows arbitrarily large very quickly. In large samples, the estimate $\hat{\alpha}$ is rarely problematic, but in small samples, where $\hat{\alpha}$ has more error, $\hat{\alpha}$ can easily be close to or less than 1. For example, $\hat{\alpha}$ gets close to 0 as n_{B-1} gets small relative to n_B.

Instead of the mean μ_B of the top bin, some authors have suggested using the median M_B (Parker and Fenwick 1983). One could also use the geometric mean g_B or the harmonic mean h_B. Under a Pareto distribution, these statistics are all simple functions of α (Johnson, Kotz, and Balakrishnan 1994):

$$M_B = l_B 2^{1/\alpha}$$

$$g_B = l_B e^{1/\alpha}$$

$$h_B = l_B(1 + 1/\alpha). \tag{4}$$

As shown in Figure 1, M_B, g_B, and h_B are far less sensitive to α than μ_B is. In particular, M_B, g_B, and h_B are defined for all positive values of α, and M_B, g_B, and h_B do not get arbitrarily large until α gets close to 0. h_B is the least sensitive to α, and that makes h_B the *a priori* best candidate for robust estimation. h_B is also the smallest statistic when $\alpha < 1$, and that could introduce some negative bias, but the bias will be compensated by some reduction in variance, a trade-off that we will evaluate empirically. For larger values of α, the differences between the statistics are relatively small so the choice of statistic will matter less.

Because all the statistics can get arbitrarily large as α gets small, further robustness can be added by setting a lower bound α_{min} on the estimate of α. That is, we can replace the estimate $\hat{\alpha}$ with

$$\tilde{\alpha} = \max(\alpha_{min}, \hat{\alpha}) \tag{5}$$

and plug $\tilde{\alpha}$ into the formulas for μ_B, M_B, g_B, and h_B. If we are estimating the arithmetic mean, then α_{min} must be greater than 1, but if we are estimating one of the other statistics, then α_{min} need only be positive.

The best value for α_{min} must be estimated empirically, and that is a weakness to this approach. However, the h_B statistic will be relatively insensitive to the value of α_{min} and will tolerate an α_{min} that is close to 0. That is another point in favor of using h_B instead of one of the other statistics.

Our *rpme* command for Stata implements the RPME with a default of h_B and $\alpha_{min} = 1$. There are options to change the value of α_{min} and an option to choose among μ_B, M_B, g_B, and h_B.

2.3. *Continuous Densities*

A discomfiting feature of the RPME is that it models income as a discrete variable. We might hope to model income better by fitting a continuous density. Many continuous densities have been used to model income, although relatively few have been implemented in software that can fit binned incomes.

2.3.1. *The GB Family.*
Some of the most popular densities for modeling income are members of the GB family (McDonald 1984; McDonald and Xu 1995). Part of the GB family tree is shown in Figure 2.[5] The tree starts with the flexible GB2 density, which has four positive parameters μ, σ, ν, τ (Stasinopoulos, Rigby, and Akantzilioutou 2008).[6] From the GB2, the tree branches out into three-parameter and two-parameter distributions, including the log normal, log logistic, Pareto (type 2), beta (type 2), Dagum, Singh-Maddala, gamma, and generalized gamma distributions. All of these distributions are either *nested* in the GB2 (as special cases that fix some parameters to a value of 1) or *embedded* in the

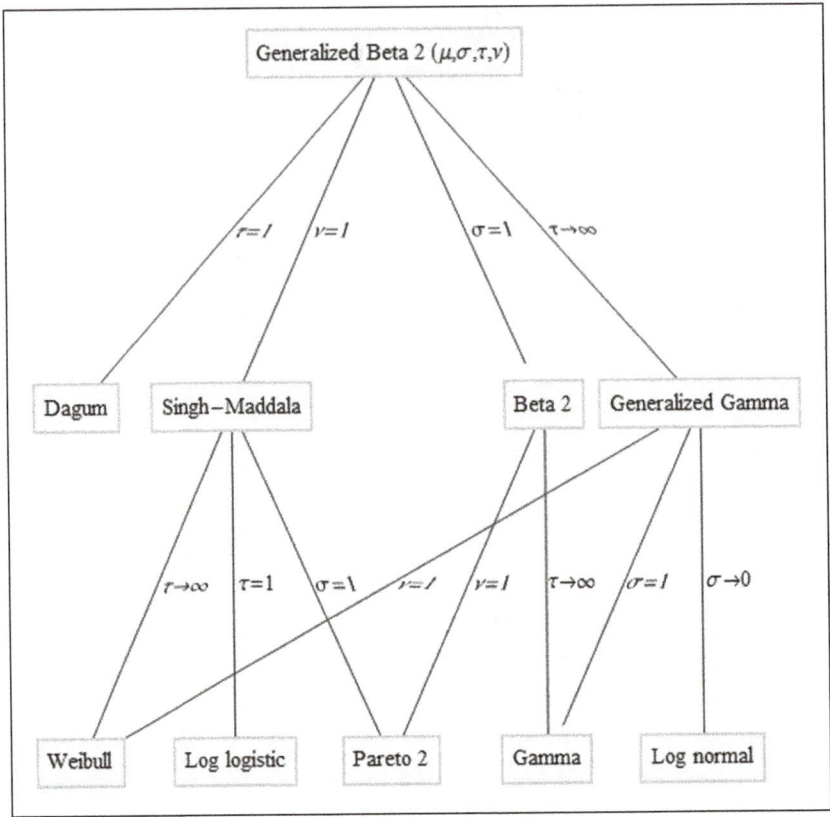

Figure 2. Part of the generalized beta family tree.

GB2 (as limiting cases that are approached as some parameter goes to 0 or ∞).

Continuous densities can be fit to binned data by maximum likelihood (McDonald and Ransom 2008). A binned-data maximum likelihood estimation of the GB family densities is available in R's *gamlss* package (Stasinopoulos et al. 2008), which we have used as the basis for our own *binequality* package (Scarpino et al. 2014). Formulas have been derived to transform maximum likelihood estimates of the GB parameters into estimates of the mean, variance, Gini coefficient, and Theil index (McDonald and Ransom 2008), but some of these formulas require functions that are difficult to implement, and some statistics, such as the MLD, cannot readily be calculated from the GB parameters.

A relatively easy, general, and accurate alternative is to approximate the statistics using numerical methods (McDonald and Ransom 2008). Our software implements a numerical recipe that draws 1,000 evenly spaced quantiles (i.e., the 0.05th, 0.15th, . . . , 99.95th percentiles) from the fitted distribution and plugs them into sample formulas for statistics including the mean, median, variance, CV, Gini coefficient, Theil index, and MLD.

2.3.2. *Practical Problems in Fitting the GB Family.*

A liability of GB family distributions is that they occasionally produce an undefined estimate of the mean or variance.[7] The mean of a GB family distribution is undefined whenever $-\hat{\nu}<1/\hat{\sigma}<\hat{\tau}$, and the variance is undefined whenever $-\hat{\nu}<2/\hat{\sigma}<\hat{\tau}$ (McDonald and Xu 1995; Stasinopoulos et al. 2008). If the mean is undefined, all inequality statistics will be undefined as well, because all inequality statistics are functions of the mean. If the mean is defined but the variance is undefined, some inequality statistics (such as the CV) will be undefined, and other inequality statistics may be very imprecise. Undefined estimates are more likely to occur in small samples, where parameter estimates are more variable and more likely to wander into parts of the parameter space where the mean or variance is undefined.

Another issue that arises in fitting the GB family is *model uncertainty. A priori* it is not clear which distribution from the GB family will provide the best fit to the data. Past research has approached the problem of model uncertainty in several ways. One approach is to fit only the most general distribution, which is the four-parameter GB2 (e.g., Chotikapanich, Rao, and Tang 2007). This can produce good results in large samples, but it has several problems in small ones. First, the fitted GB2 distribution can have undefined moments, especially in small samples. Second, in small samples, the four parameters of the GB2 distribution may be estimated imprecisely, so that the resulting estimates are worse than if a simpler two-parameter or three-parameter distribution had been fit. Third, even in large samples, the GB2 distribution can only fit the distributions that it nests (the Dagum, Singh-Maddala, beta 2, Pareto 2, and log logistic); it cannot fit the distributions that it embeds (the Weibull, log normal, gamma, and generalized gamma), because fitting those distributions would require a GB2 parameter to go to ∞ or 0.

Another approach to model uncertainty is to test the null hypothesis that the fitted density is in fact the true distribution of income.

Bandourian et al. (2002) did this using a Pearson X^2 statistic; we do it using a likelihood-ratio (LR) statistic (which is asymptotically equivalent to X^2):

$$G^2 = -2\ln(LR) = -2\left(\ell - \sum_{b=1}^{B} n_b \left(\ln\left(\frac{n_b}{n}\right)\right)\right), \qquad (6)$$

Here ℓ is the log likelihood of the fitted distribution. Under the null hypothesis that the fitted distribution is the true distribution, G^2 (or X^2) will follow an asymptotic chi-square distribution with

$$df = \min(B, B_{all} - 1) - k \qquad (7)$$

degrees of freedom, where k is the number of parameters, B is the number of populated bins, and B_{all} is all the bins, including the empty ones (cf. Stirling 1986). If all the bins are populated, this expression for degrees of freedom simplifies to $df = B - k - 1$.[8]

The limitation of the null hypothesis test is that it rarely highlights a single distribution as fitting the data. In small samples, the power of the test is low, and there may be several distributions that it fails to reject. In large samples, the power of the test is high, and it may reject every model that is fit. This is in fact what happened when Bandourian et al. (2002) fit GB family distributions to 23 countries in various years from 1967 to 1997. In every country and every year, they rejected every distribution in the GB family by a X^2 test.[9] In some countries, these rejections might have meant that the distributions fit poorly, but in other countries, it might have been that several distributions provided good though imperfect approximations and might have produced good estimates of statistics such as the Gini coefficient. A null hypothesis test cannot distinguish between these situations.

Another approach to model uncertainty is to decide between two models using a classical LR test (McDonald 1984). However, classical LR tests apply only to nested models (White 1982) such as the GB2 and Dagum; they cannot choose between embedded models such as the GB2 and generalized gamma, or between models with the same number of parameters, such as the Dagum and Singh-Maddala. In addition, classical LR tests assume that one of the distributions being compared is in fact the true distribution of income (White 1982). This is not a plausible assumption; as mentioned earlier, Bandourian et al. (2002) rejected the hypothesis that any distribution in the GB family was the true distribution of income in any of 23 countries.

2.3.3. *The MGBE.* The MGBE is a principled approach to handling uncertainty about the true distribution of income. The MGBE fits every GB family distribution in Figure 2, discards any fitted distributions with an undefined variance (i.e., with $-\hat{v}<2/\hat{\sigma}<\hat{\tau}$), and then uses fit statistics to derive estimates from the remaining distributions.

Two fit statistics can be used, the AIC and the BIC. We can use the AIC or BIC to *select* a single best model or to construct weights that we use to *average* estimates across models. Theoretically, model averaging should produce better estimates than model selection (Burnham and Anderson 2004). Whether it is better to use the AIC or BIC is a subtler issue that depends on how well the different GB family distributions actually fit the data (Burnham and Anderson 2004). This is something we can assess only empirically.

3. DATA AND RESULTS

In this section, we evaluate the RPME and MGBE in data on family and household incomes. We tune the estimators to data from U.S. counties, then validate the tuned estimators in data from U.S. states and from the United States as a whole.

To evaluate the quality of an estimator, we would like to compare its estimates with the true value of the estimand in the population. For example, we would like to compare the true population Gini coefficient with the Gini coefficient estimated from binned data. Unfortunately, the population Gini coefficient is unknown because the Census Bureau does not record income for every U.S. family or household. So instead of population statistics, we use sample estimates calculated by the Census Bureau from unbinned incomes. Although these estimates are not population values, they are more accurate than binned-data estimates, and they can therefore serve as a gold standard. For brevity's sake, we refer to them as "true" values or estimands in the evaluation.

We evaluate the accuracy of the estimates as follows. Let $\hat{\theta}_j$ and θ_j be the estimate and estimand for state or county j. Then $e_j = 100(\hat{\theta}_j - \theta_j)/\theta_j$ is the percentage relative error of the estimate— that is, the error expressed as a percentage of the estimand. Across all the counties or states in the United States, the *percentage relative bias* is the mean of e_j, the *percent root mean squared error* (RMSE) is the square root of the mean of e_j^2, and the *reliability* of the estimate is the squared correlation between θ_j and $\hat{\theta}_j$. We also plot $\hat{\theta}_j$ against θ_j to visualize the accuracy of the estimates at different values of the estimand.

3.1. *Counties in 2006–2010*

Our evaluation begins with household income data from each of the 3,221 counties in the United States and Puerto Rico. We estimate the mean, median, and Gini coefficient from the county bins and compare these estimates with the "true" values of the county mean, median, and Gini coefficient published by the Census Bureau (Bee 2012; U.S. Census Bureau N.d.). The Census Bureau has never published county Gini coefficients before.

The county data come from responses to the American Community Survey (ACS). To increase the number of households sampled per county, the ACS pooled data across the five-year period from 2006 to 2010 to accumulate a one-in-eight sample of U.S. households, with incomes inflated to 2010 dollars. For each county, the Census Bureau summarized county incomes using 16 bins.

The binned data for two counties, Nantucket and Maricao, are given in Table 1. The counts in Table 1 are not sample counts but estimated population counts. Because the sampling fraction was one in eight, we approximated the sample counts by dividing the estimated population counts by 8. This makes no difference to the RPME but does slightly affect the MGBE, because the AIC, BIC, and G^2 statistics are functions of sample size. The ACS is a complex sample survey, but unfortunately the binned data provide no information about the sample design, and we thus analyze the data as if they came from a simple random sample. This means that p values, likelihoods, and the AIC and BIC are only approximate (Rao and Scott 1981).

County data present a challenging estimation task because some counties have small populations, small sample sizes, empty bins, and lumpy income distributions. Maricao is a fairly small county, but some counties are much smaller; Loving County, Texas, is the smallest, with only 31 households in 2010. County data will highlight the properties of our estimators when there are low bin counts, empty bins, and idiosyncratic income distributions.

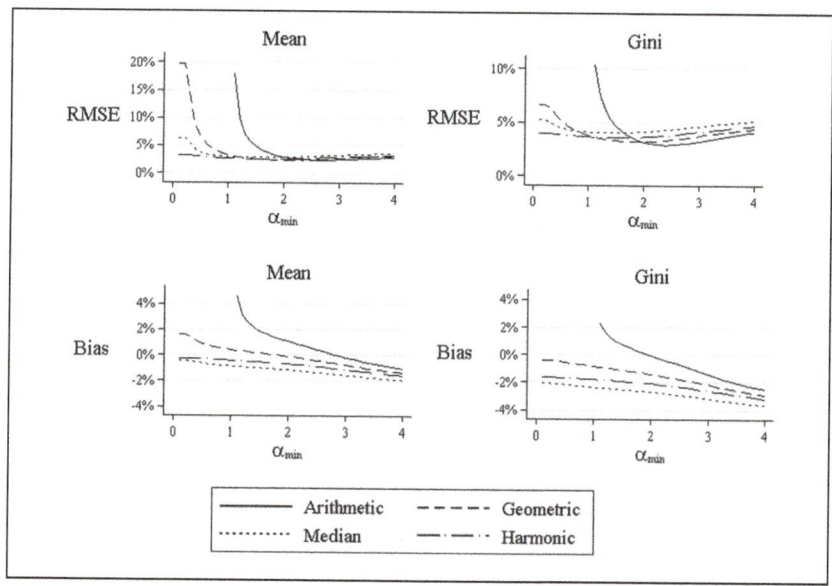

Figure 3. Root mean squared error (RMSE) and bias for variants of the robust Pareto midpoint estimator in estimating the county means and Gini coefficients.

3.1.1. *RPME Estimates.*

We begin by tuning the RPME estimator. Figure 3 shows the RMSE and bias for estimates of the Gini coefficient and mean produced by different flavors of the RPME. There are flavors that use the arithmetic mean, the harmonic mean, the geometric mean, and the median, and for each flavor α_{min} can take values from 0 to 4 or beyond. Note that the choice $\alpha_{min} = 0$ means that the estimate of α is not constrained at all.

Although nearly all past research used the arithmetic RPME with $\alpha_{min} = 0$, that choice produces the worst estimates. It has an infinite RMSE when $\alpha_{min} < 1$ and a very large RMSE when α_{min} is greater than but close to 1. The arithmetic RPME can achieve a decent RMSE when $\alpha_{min} > 2$, but overall it is distressingly sensitive to the value chosen for α_{min}.

The harmonic RPME is the most robust choice. It is the least sensitive to the specific value chosen for α_{min} and yields the smallest RMSE when $\alpha_{min} < 1$. It also yields a negative bias, but the bias is small when α_{min} is small, and the bias-variance trade-off is favorable. The median

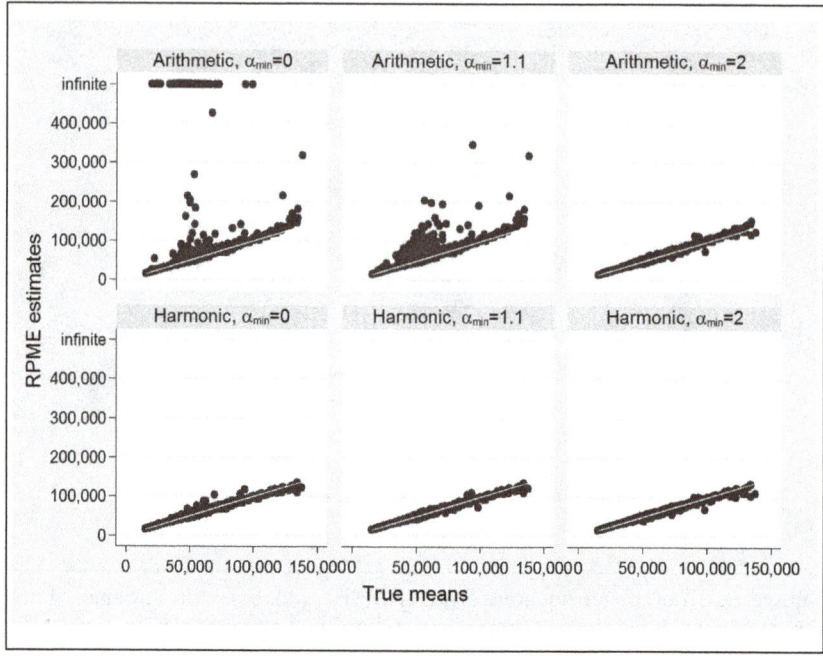

Figure 4. True versus estimated county means provided by different variants of the robust Pareto midpoint estimator (RPME). An $X = Y$ reference line shows what perfect estimation would look like.

and geometric RPME are also more robust than the arithmetic RPME but not as robust as the harmonic RPME.

When the harmonic RPME is used, the most accurate estimates are obtained near $\alpha_{min} = 1$, although the precise value of α_{min} does not matter much.

The value $\alpha_{min} = 1$ is also a reasonable choice with the geometric RPME or the median RPME, but $\alpha_{min} = 2$ is a better choice with the arithmetic RPME. In addition to being empirically supported (at least in these data), all these choices have a common and intuitive interpretation. They imply that the statistic used for the top bin cannot be more than twice that bin's lower bound.

Figure 4 illustrates the robustness that is achieved by using the harmonic RPME instead of the arithmetic RPME. The figure is a graphics grid that contains scatterplots of RPME-estimated county means against true county means. At the upper left, we see the RPME estimates obtained using the arithmetic RPME with $\alpha_{min} = 0$. Many of the estimates are much too large, and some are infinite. Increasing α_{min} to 1.1 gets rid of the infinite estimates but still leaves many estimates that are much too large. Increasing α_{min} to 2 gives good results, but it is troubling that the results are so sensitive to the choice of α_{min}.

By contrast, the lower row of Figure 4 shows the estimates produced by the harmonic RPME. The results are far more robust and much less sensitive to α_{min}. The best results are achieved at $\alpha_{min} = 1.1$, but the results at $\alpha_{min} = 0$ or $\alpha_{min} = 2$ are only a little different.

Table 2 summarizes the accuracy of the arithmetic RPME with $\alpha_{min} = 2$ and the geometric, median, and harmonic RPMEs with $\alpha_{min} = 1$. In estimating the Gini coefficient, the arithmetic RPME is the most accurate, with the smallest bias, the smallest RMSE, and the highest reliability. However, the performance of the arithmetic RPME is highly dependent on the choice of α_{min}, as we have seen. The harmonic RPME with $\alpha_{min} = 1$ provides estimates that are almost as good and less sensitive to the choice of α_{min}. The geometric and median RPMEs are a little less accurate than the harmonic RPME and more sensitive to α_{min}. In estimating the mean, all four flavors of RPME provide very similar estimates, although again, the harmonic RPME is the least sensitive to α_{min}. In estimating the median, all RPME variants are identical because the median is never in the top bin.

Overall, any flavor of RPME can provide excellent estimates of the mean (98 percent to 99 percent reliability and 0 percent to 1 percent bias), very good estimates of the median (97 percent reliability and 1 percent bias), and good estimates of the Gini coefficient (83 percent to 87 percent reliability and 0 percent to 2 percent bias). However, the arithmetic RPME requires careful tuning of α_{min} to achieve good results, while the harmonic RPME is much more robust.

3.1.2. MGBE Estimates. Table 2 gives county estimates obtained by fitting various GB family distributions. Some distributions yield very poor estimates, which drives home the point that continuous distributions are useful only if they approximate the true distribution of income. Estimates from the Pareto distribution are the worst; Pareto-estimated Gini coefficients have a bias of 20 percent and are only 23 percent reliable. Fortunately the fit statistics would never lead us to choose the Pareto distribution; in 98 percent of counties, a G^2 test rejects the Pareto as the true distribution of income, and in 100 percent of counties, there is some other distribution that fits better according to the AIC and BIC.

Table 2. County Estimates in 2006–2010

Estimator	Type	Median % Bias	Median % RMSE	Median % Reliable	Mean % Bias	Mean % RMSE	Mean % Reliable	Gini Coefficient % Bias	Gini Coefficient % RMSE	Gini Coefficient % Reliable	% Selected by AIC	% Selected by BIC	Fit Rejected
RPME	Arithmetic	1	5	97	1	3	99	0	3	87			
	Geometric	1	5	97	0	3	98	−1	4	83			
	Median	1	5	97	−1	3	99	−2	4	84			
	Harmonic	1	5	97	0	3	99	−2	4	85			
GB family distributions	Weibull	2	4	98	−2	4	98	−4	5	83	6	7	85
	Log logistic	−5	6	99	12	13	97	13	14	47	1	1	90
	Pareto 2	−16	17	91	−2	6	93	20	22	23	0	0	98
	Gamma	0	4	98	−2	4	98	−4	5	81	21	24	81
	Log normal	−9	9	98	3	5	98	6	8	73	3	3	91
	Dagum	1	3	99	1	3	99	0	4	83	28	29	77
	Singh-Maddala	−2	4	97	−1	3	97	−1	6	77	7	8	80
	Beta 2	−5	6	98	0	4	98	1	5	73	1	1	86
	Generalized gamma	−1	3	99	−2	3	99	−4	5	86	19	16	77
	GB2	0	3	98	−1	3	97	−1	3	87	15	11	76
MGBE, 10 models	AIC selected	0	3	99	−1	3	99	−2	4	87			
	BIC selected	0	3	99	−1	3	99	−2	4	87			
	AIC averaged	0	3	99	−1	3	99	−2	4	88			
	BIC averaged	0	3	99	−1	3	99	−2	4	88			
MGBE, 3 models	AIC selected	0	3	99	−1	3	99	−2	4	86			
	BIC selected	0	3	99	−1	3	99	−3	4	86			
	AIC averaged	0	3	99	−1	3	99	−2	4	87			
	BIC averaged	0	3	99	−1	3	99	−3	4	87			

Note: The arithmetic robust Pareto midpoint estimator (RPME) was used with $\alpha_{min} = 2$; the harmonic, geometric, and median RPMEs were used with $\alpha_{min} = 1$. Rejections of fit were based on a likelihood-ratio chi-square test with a significance level of .05. AIC = Akaike information criterion; BIC = Bayesian information criterion; GB = generalized beta; MGBE = multimodel generalized beta estimator; RMSE = root mean squared error.

Table 3. Estimation Problems When Fitting GB Family Distributions

Parameters	Model	% Undefined Estimates	% Nonconvergence	% Total Runtime
2	Weibull	.0	0	.5
	Log logistic	53.1	0	1.3
	Pareto 2	1.3	1	34.0
	Gamma	.0	0	.4
	Log normal	.0	0	.5
3	Dagum	4.2	0	8.8
	Singh-Maddala	.4	2	16.2
	Beta 2	.0	11	20.1
	Generalized gamma	.6	0	1.7
4	GB2	4.4	1	16.5

Note: GB = generalized beta.

The best-fitting distributions, according to the AIC and BIC, are the GB2, Dagum, gamma, and generalized gamma. The Weibull also fits well, although it rarely fits as well as the others. Even the best-fitting distributions are rejected by the G^2 test in at least three quarters of counties. So they fit the data only approximately, although the approximation may be useful.

The bottom two sections of Table 2 summarize the MGBE estimates we get when we use the AIC or BIC to select the best-fitting of the 10 distributions. It also shows what happens when we average estimates across distributions using weights that are functions of the AIC and BIC. The average weight assigned each a model is not shown (to save space), but it is very similar to the percentage of counties in which that model is chosen as best. The model-weighted averages are slightly more accurate than the model-selected estimates, a result that accords with statistical theory (Burnham and Anderson 2004). The AIC and BIC produce almost identical estimates because they nearly always select the same distributions.

Looking again at Table 2, one might get the impression that the 10-distribution MGBE is hardly worth the trouble, because its estimates are barely better than those obtained from fitting the GB2 distribution alone. But this is not quite right. As Table 3 shows, fitting just the GB2 distribution would fail to yield defined estimates in 4 percent of counties, whereas the MGBE is guaranteed to always produce a defined estimate. This implies that the accuracy statistics in Table 2 do not mean quite the same thing for the MGBE as they do for the GB2 distribution. In estimating the Gini coefficient, for example, the GB2's RMSE of 3 percent applies only to the 96 percent of counties where the GB2 produces a defined estimate. The MGBE achieves just as good an RMSE in those counties but also provides estimates for the remaining 4 percent. It is only when those 4 percent of counties are included that the MGBE has a RMSE of 4 percent.

Table 3 shows that several distributions in addition to the GB2 occasionally yield undefined estimates. The problem is worst for the log logistic distribution, which yields undefined estimates in more than half of counties. Table 3 also shows that several distributions occasionally have trouble with convergence. This is partly a software issue, but it is not unique to our software package in R. We also encountered similar convergence issues when we tried fitting GB2-family distributions to the same data in Stata and SAS. A possible explanation is that convergence is tricky to define with binned data because the likelihood can be relatively flat near the maximum.

Table 3 also highlights another problem that arises when using the MGBE: the problem of runtime. The MGBE is much slower than the RPME, because the MGBE requires iteratively maximizing the likelihood of several distributions, some of which may have trouble with convergence. On our computer,[10] it takes 32 hours to fit all 3,221 counties to all 10 GB distributions, but the RPME provides estimates for the same counties in a couple of seconds.

Table 3 also highlights several opportunities to substantially reduce runtime. The table shows that just three distributions—the Pareto 2, the beta 2, and the Singh-Maddala—account for 70 percent of the runtime in R. It is not really necessary to run these three distributions, because they get very little weight in the 10-model estimates. Some other distributions, notably the log logistic, are also unnecessary, because they so often produce inaccurate or undefined estimates.

We can get faster results by fitting just two to four models. For example, the Dagum, generalized gamma, and gamma account for just 11 percent of the total runtime in R, but they get almost 70 percent of the weight in the county multimodel inferences. Running just these three distributions, we fit all 3,221 U.S. counties in three hours using R. The bottom rows of Table 2 show that the resulting three-model estimates are only slightly less accurate than the 10-model estimates. An even more radical solution would be to fit just the Weibull distribution, which

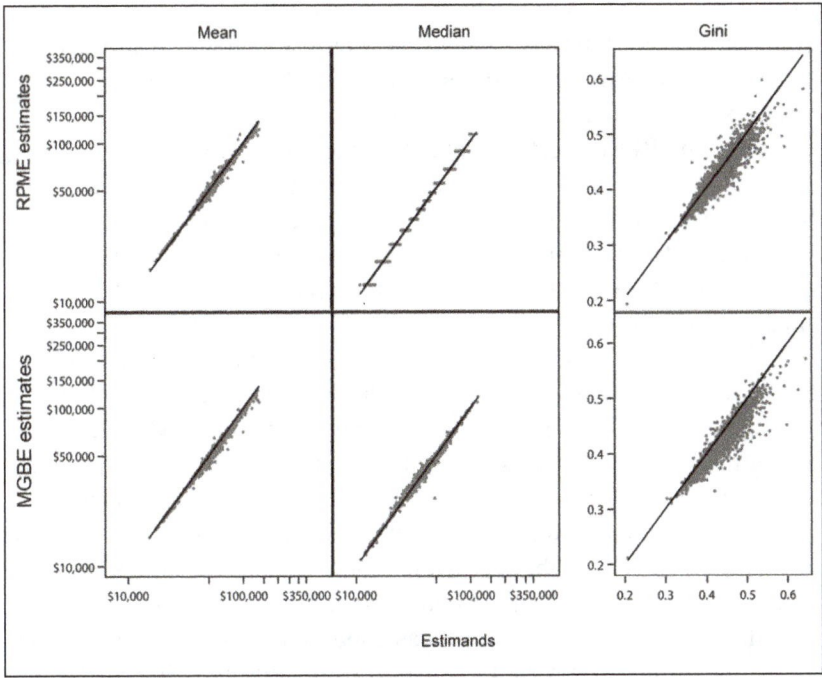

Figure 5. True versus estimated county means, medians, and Gini coefficient means for the harmonic robust Pareto midpoint estimator (RPME) (with α_{min} = 1) and the multimodel generalized beta estimator (MGBE) (using the Akaike information criterion to select the best of 10 models). An $X = Y$ reference line shows what perfect estimation would look like.

consumes only 0.5 percent of the total runtime, always converges, and always provides a defined estimate. If only the Weibull distribution is fit, estimates for all 3,221 counties could be obtained in about 10 minutes. Even at 10 minutes, though, the Weibull estimates would be much slower than the RPME, and no more accurate.

The scatterplots in Figure 5 help visualize relative performance of the harmonic RPME and the MGBE. It is evident from the scatterplots that the MGBE mean and Gini coefficient estimates have a slight negative bias; that is, the points on the scatterplot tend to fall below the line. No such bias is evident in RPME estimates of the mean and Gini coefficient. On the other hand, RPME median estimates are slightly unreliable, because they can take values only at the bin midpoint; this is evident from the stripes in the scatterplot.

3.1.3. Fewer Bins. As we wrote earlier, the RPME is a bin-consistent estimator that is at its best when the bins are numerous and narrow. With, say, 100 narrow bins, the RPME would be impossible for a parametric estimator such as the MGBE to beat. Even with 16 bins, as we have seen, the RPME performs about as well as the MGBE.

How few bins can there be before the relative performance of the RPME starts to break down? Table 4 addresses this question by rebinning the county data. First it merges adjacent bins to reduce the number of bins from 16 to 8. Then it merges adjacent bins again to reduce the number of bins from 8 to 4.

With 8 bins, the performance of both estimators worsens, but the RPME remains about as accurate as the MGBE when estimating the mean and Gini coefficient. The RPME is worse than the MGBE at estimating the median; that was true with 16 bins as well, but with 8 bins, the difference between the median estimates is more pronounced. In general, the RPME produces worse median estimates because it is discrete and assumes that the median can only be at a bin midpoint. This problem becomes more consequential when the bins are coarse.

With 4 bins, the RPME is almost as good as the MGBE when estimating the mean, but it is notably worse in estimating the Gini coefficient and median. These 4-bin results will have limited interest to researchers using U.S. census data, which typically have at least 10 bins. But 4-bin results will be of interest to researchers who use data from the developing world, which sometimes has as few as 5 bins.

Which flavor of the RPME performs best with fewer bins? The answer is not clear cut. With eight bins, the arithmetic RPME is best at estimating the Gini, but the harmonic RPME is best at estimating the mean, and all RPME flavors are identical in estimating the median. With four bins, the ranking of RPME flavors depends on the criterion of accuracy (RMSE or reliability), but all flavors of RPME are inferior to the MGBE.

3.2. States in 1980

Our evaluation continues with state-level summaries of family incomes reported on the long-form census in 1980, which asked about family incomes received during 1979. For each U.S. state and the District of Columbia, the census reported the mean,[11] median, and Gini coefficient of family income (U.S. Census Bureau 1982; U.S. Census Bureau, Data

Table 4. Accuracy of County Estimates, with Fewer Bins

Bins	Estimator	Type	Median			Mean			Gini Coefficient		
			% Bias	% RMSE	% Reliable	% Bias	% RMSE	% Reliable	% Bias	% RMSE	% Reliable
8	RPME	Arithmetic	2	11	89	2	4	98	−1	4	78
		Geometric	2	11	89	1	4	97	−2	5	78
		Harmonic	2	11	89	0	3	98	−4	5	76
		Median	2	11	89	−1	3	98	−5	6	74
	MGBE	10 distributions, best AIC	0	3	99	−4	5	98	−4	6	79
4	RPME	Arithmetic	9	28	64	12	13	95	0	7	38
		Geometric	9	29	62	11	14	94	−2	7	44
		Harmonic	9	28	64	4	6	96	−7	9	53
		Median	9	28	64	3	6	96	−9	10	55
	MGBE	10 distributions, best AIC	0	3	99	−5	6	96	−6	8	65

Note: The arithmetic robust Pareto midpoint estimator (RPME) uses $\alpha_{min} = 2$, while the geometric, harmonic, and median RPMEs use $\alpha_{min} = 1$. AIC = Akaike information criterion; MGBE = multimodel generalized beta estimator; RMSE = root mean squared error.

Integration Division 2010) and also summarized family incomes using 17 bins (National Historical Geographic Information System 2005; U.S. Census Bureau 1982). The data give estimated bin counts for the population, and because the long-form census was given to a one-in-six sample, we divide the population counts by 6 to estimate the sample counts. The long-form census was given to a complex sample of households (Navarro and Griffin 1990), but unfortunately the binned data provide no information about the sample design. We again analyze the data as if they came from a simple random sample, and this means that p values, likelihoods, and the AIC and BIC are only approximate (Rao and Scott 1981).

Evans et al. (2004) previously analyzed these same data by fitting the Dagum distribution. We extend the analysis by fitting additional distributions from the GB family, combining the resulting estimates to obtain the MGBE, and comparing the MGBE with the RPME.

Table 5 and Figure 6 summarize estimates of the median, mean, and Gini coefficient of family income. Replicating the results from Evans et al. (2004), we find that excellent estimates can be obtained by fitting the Dagum distribution. The Dagum distribution had an RMSE of 1 percent and a reliability of 98 percent to 99 percent in estimating the mean, median, and Gini coefficient.

Although the Dagum results leave little room for improvement, we find that the estimates are just as good if we fit the Weibull, generalized gamma, or GB2 distributions. Other distributions are worse. Despite the good results obtained from some of the GB family distributions, the results show that all the GB distributions are only approximate fits to the binned data. In every state, a chi-square test rejected every fitted distribution as the true distribution of income. Estimates obtained from the MGBE are practically the same as estimates obtained from either the Dagum or the GB2 alone. In fact, the MGBE gives 12 percent of its weight to the Dagum estimates and 88 percent to the GB2 estimates, with other distributions getting no weight.

The RPME is almost but not quite as accurate as the MGBE. In estimating the mean, the RPME is every bit as good as the MGBE, and in estimating the Gini coefficient, the RPME is just as reliable as the MGBE but has a small positive bias. In estimating the median, the RPME is unbiased but less reliable than the MGBE, because the RPME median estimates can take only discrete values at the bin midpoints.

Table 5. Accuracy of State Estimates in 1980

| Estimator | Flavor | Estimand | | | | | | | | | % Selected | | |
| | | Median | | | Mean | | | Gini Coefficient | | | | | |
		% Bias	% RMSE	% Reliable	% Bias	% RMSE	% Reliable	% Bias	% RMSE	% Reliable	by AIC	by BIC	% Fit Rejected
RPME	Arithmetic	0	4	94	1	1	99.7	3	3	97.2			
	Geometric	0	4	94	1	1	99.7	2	2	97.2			
	Harmonic	0	4	94	0	1	99.7	2	2	97.2			
	Median	0	4	94	0	1	99.7	1	1	97.2			
GB family distributions	Weibull	−1	1	98	−1	2	99.6	0	1	99.6	0	0	100
	Log logistic	−6	6	99	11	12	97.0	16	16	84.3	0	0	100
	Pareto 2	−23	23	91	−2	6	86.1	39	40	29.6	0	0	100
	Gamma	−3	4	98	−1	1	99.6	1	2	95.0	0	0	100
	Log normal	−11	11	98	4	4	98.9	14	14	83.8	0	0	100
	Dagum	−1	1	99	−1	1	99.7	0	1	98.1	12	12	100
	Singh-Maddala	−3	4	97	−1	2	99.4	1	4	64.2	0	0	100
	Beta 2	−6	6	98	−1	1	99.7	4	4	95.4	0	0	100
	Generalized gamma	−3	3	99	−1	1	99.7	0	0	99.8	0	0	100
	GB2	0	0	98	0	1	99.9	0	1	97.7	88	88	100
MGBE (10 models)	AIC selected	0	0	99	0	1	99.9	0	1	97.8			
	BIC selected	0	0	99	0	1	99.9	0	1	97.8			
	AIC weighted	0	0	99	0	1	99.9	0	1	97.7			
	BIC weighted	0	0	99	0	1	99.9	0	1	97.7			

Note: AIC = Akaike information criterion; BIC = Bayesian information criterion; GB = generalized beta; MGBE = multimodel generalized beta estimator; RMSE = root mean squared error; RPME = robust Pareto midpoint estimator.

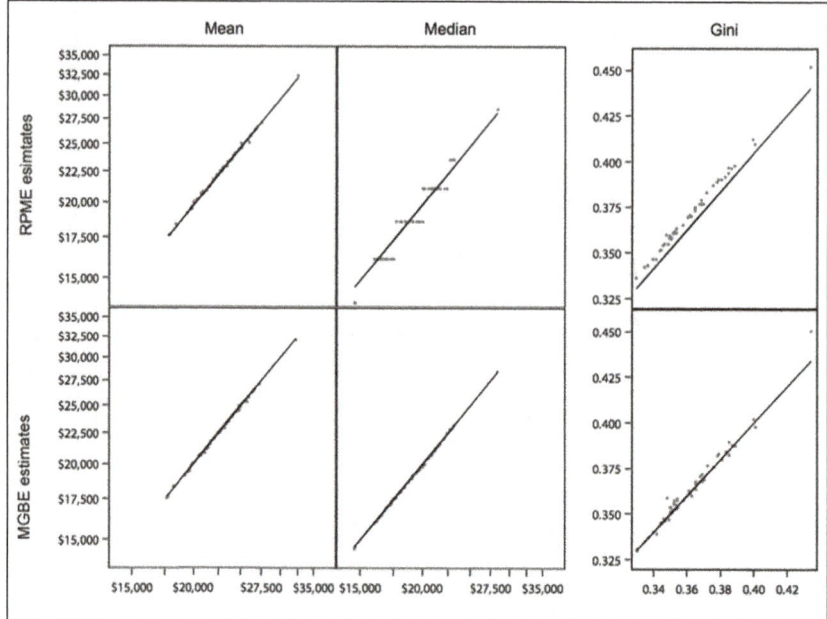

Figure 6. State estimates in 1980. If the estimates were perfectly accurate, all points would fall on the diagonal reference lines. When all the points are above the reference line, there is positive bias.

Note: MGBE = multimodel generalized beta estimator; RPME = robust Pareto midpoint estimator.

The discreteness of the RPME median estimates is evident from the stripes in the scatterplot.

All the flavors of RMPE—arithmetic, geometric, harmonic, and median—produce very similar estimates. As in the county estimates, we used the arithmetic RPME with α_{min} = 2 and the harmonic, geometric, and median RPMEs with α_{min} = 1. These choices had no effect on the estimates, because all the states had $\hat{\alpha} > 2$ anyway. In short, the robust features of the RPME, which are so necessary in fitting small county samples, were not needed for these large state samples.

3.3. *States in 2005*

Our evaluation continues with state-level summaries of household incomes reported to the ACS in 2005. For each state, the District of Columbia, and Puerto Rico, the Census Bureau published income distributions using the same 16 bins as those in Table 1 (U.S. Census Bureau N.d.). The data give estimated bin counts for the population, but because the ACS was given to a 1-in-40 sample, we divide the population counts by 40 to estimate the counts in the sample. Again, although the ACS is a complex sample, the binned data provide no information about the sample design, so we analyze the data as if they came from a simple random sample. This means that p values, likelihoods, and the AIC and BIC are only approximate (Rao and Scott 1981).

What is especially valuable about the 2005 state data is that, for the first time, the Census Bureau published not just state Gini coefficients but also state Theil indices and MLDs (Hisnanick and Rogers 2006),[12] providing an opportunity to evaluate how well our estimators reproduce these less common inequality statistics. Another valuable aspect of the 2005 data is that the distribution of family income changed substantially between 1980 and 2005, becoming much more unequal and stretching the upper tail (Western, Bloome, and Percheski 2008). It is important to check whether the GB family distributions fit as well in 2005 as they did in 2008.

Table 6 and Figure 7 summarize state estimates of the mean, Gini coefficient, Theil index, and MLD in 2005. As in 1980, the best-fitting GB family distributions are the GB2 and, less often, the Dagum, although all the distributions are rejected by the goodness-of-fit test. In 1980, the rejection of the distributions' fit did not prevent them from providing good estimates, but in 2005 the estimates are worse. Perhaps that means that the GB family approximated the true income distribution worse in 2005 than in 1980.

For all estimators, the state estimates are worse in 2005 than they were in 1980 and worse for the Theil index and MLD than for the Gini coefficient. The 2005 RPME estimates are a little better than the MGBE estimates, which is a reversal of the findings for 1980, but both estimators have similar problems. In estimating the mean, both the RPME and the MGBE are excellent. In estimating the Gini coefficient, the MGBE has a small negative bias but is highly reliable, while the RPME is a little less reliable but has less bias and a smaller RMSE.

In estimating the Theil index and MLD, both estimators have large negative biases. Why are the relative biases so much worse for the Theil index and MLD than they are for the Gini coefficient? We can get some intuition for the answer by pointing out that, empirically, the true Theil and MLD can be very well approximated by linear functions of the true

Table 6. Accuracy of State Estimates in 2005

		Mean			Gini Coefficient			Theil Index			MLD			% Selected		
		% Bias	% RMSE	% Reliable	% Bias	% RMSE	% Reliable	% Bias	% RMSE	% Reliable	% Bias	% RMSE	% Reliable	by AIC	by BIC	% Fit Rejected
RPME	Arithmetic	3	3	99	−1	1	87	−6	7	72	−18	18	88			
	Geometric	1	2	99	−2	3	87	−11	11	72	−20	20	88			
	Harmonic	1	1	99	−3	3	87	−13	14	72	−21	21	88			
	Median	0	1	99	−4	4	87	−15	16	72	−22	22	88			
GB family distributions	Weibull	−2	3	99	−6	6	89	−23	23	89	−24	24	94	0	0	100
	Log logistic	10	10	99	8	8	69	11	13	73	−14	15	73	0	0	100
	Pareto 2	−6	8	90	13	14	53	14	18	76	17	23	76	0	0	100
	Gamma	−2	3	99	−6	6	91	−24	24	90	−28	28	96	0	0	100
	Log normal	3	3	99.5	3	4	88	−3	6	90	−23	23	93	0	0	100
	Dagum	−1	1	99.7	−4	4	96	−17	18	94	−24	24	96	12	12	100
	Singh-Maddala	−2	2	99	−5	5	93	−20	21	91	−28	28	96	0	0	100
	Beta 2	−2	2	99	−4	4	95	−18	19	92	−28	28	96	0	0	100
	Generalized gamma	−2	3	99	−5	6	92	−22	23	90	−29	29	96	0	0	100
	GB2	−2	2	99.6	−5	5	96	−19	20	93	−25	26	96	88	88	100
MGBE (10 models)	AIC selected	−2	2	99.6	−5	5	96	−19	19	93	−25	25	96			
	BIC selected	−2	2	99.6	−5	5	96	−19	19	93	−25	25	96			
	AIC weighted	−2	2	99.6	−5	5	96	−19	19	93	−25	25	96			
	BIC weighted	−2	2	99.6	−5	5	95	−19	19	93	−25	25	96			

Note: AIC = Akaike information criterion; BIC = Bayesian information criterion; GB = generalized beta; MGBE = multimodel generalized beta estimator; MLD = mean log deviation; RMSE = root mean squared error; RPME = robust Pareto midpoint estimator.

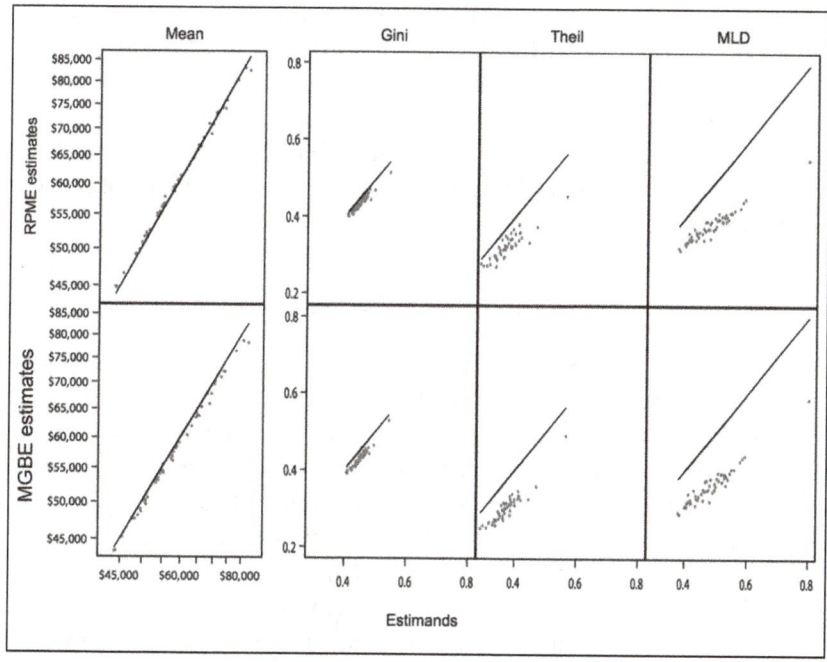

Figure 7. State estimates in 2005. If the estimates were perfectly accurate, all points would fall on the diagonal reference lines. When the points are below the reference line, there is negative bias.

Note: MGBE = multimodel generalized beta estimator; MLD = mean log deviation; RPME = robust Pareto midpoint estimator.

Gini coefficient. The following are the least squares lines for the states in 2005:

$$Theil \approx -.46 + 1.86Gini \ (R^2 = .96),$$

$$MLD \approx -.78 + 2.83Gini \ (R^2 = .92). \tag{8}$$

Consider first the equation for the Theil index. The fact that the slope is 1.86 suggests that the bias of the Theil index estimate will be approximately 1.86 times that of the Gini coefficient estimate. However, the fact that the intercept is negative means that the true Theil index is less than 1.86 times the true Gini coefficient. In relative terms, then, the bias will be a larger percentage of the Theil index than it is of the Gini coefficient. A similar argument holds for the MLD.

The different flavors of RPME performed similarly. As in the county estimates, we used the arithmetic RPME with α_{min} = 2 and the harmonic, geometric, and median RPMEs with α_{min} = 1. These choices had no effect on the estimates because all the states (except for the District of Columbia) had $\hat{\alpha} > 2$ anyway. Again, the robust features of the RPME, which are so necessary in fitting small county samples, were not needed for these large state samples.

3.4. *National Data from 1970 to 2006–2009*

Up to this point we have found that the RPME and MGBE have comparable accuracy in the common situation in which there are at least eight bins. It is tempting to conclude that it does not matter which estimator you use. That conclusion might be valid for a cross-sectional analysis.

In a longitudinal analysis, however, the choice of estimator can matter. Although the errors in Gini coefficient estimates are small compared with the cross-sectional variation in the Gini coefficient, the errors are not trivial compared with the trends in the Gini coefficient over time.

To illustrate this point, we use national data on family incomes in every decade from 1970 to 2006–2009. Binned data are available from the long-form censuses from 1970, 1980, 1990, and 2000 and from the ACS in 2006–2009 (adjusted to 2009 dollars). There are 25 bins in 1990 and 15 to 17 bins in other years. In each year, we estimate the Gini coefficient using the MGBE and RPME and compare the resulting estimates with the "true" Gini coefficient calculated by the Census Bureau using unbinned data from the Current Population Survey.

As in previous analyses, we use the arithmetic RPME with α_{min} = 2 and the harmonic, geometric, and median RPMEs with α_{min} = 1. Our choice of α_{min} has no effect on the estimates, because in every year the estimate $\hat{\alpha}$ is nevertheless greater than 2. Our MGBE estimates select the best of 10 distributions according to the AIC. In 2006–2009, the best distribution was the GB2; in all other years it was the Dagum.

Table 7 gives true and estimated values for the national Gini coefficient in every decade from 1970 to 2006–2009. Clearly the MGBE estimates do a better job than the RPME in capturing the true uptrend in the Gini coefficient. The true Gini coefficient increases in every decade, and so does the MGBE estimate, but the RPME estimates do not. On the contrary, every flavor of RPME estimates that the Gini coefficient decreased between 1970 and 1980 and that the Gini coefficient did not

Table 7. National Estimates for the Gini Coefficient of Family Income

| Decade | True | MGBE | | RPME | | | | | | | |
| | | | | Arithmetic | | Geometric | | Median | | Harmonic | |
		Estimate	Error (%)	Estimate	Error (%)	Estimate	Error (%)	Estimate	Error (%)	Estimate	Error (%)
1970	.349	.353	1	.385	10	.382	9	.378	8	.380	9
1980	.365	.362	−1	.380	4	.376	3	.371	2	.374	2
1990	.401	.392	−2	.419	4	.410	2	.403	0	.406	1
2000	.429	.404	−6	.431	0	.420	−2	.410	−4	.414	−3
2006–2009	.439	.411	−6	.431	−2	.419	−5	.407	−7	.413	−6
Bias			−3		4		2		0		1
RMSE			4		6		6		6		6

Note: MGBE = multimodel generalized beta estimator; RMSE = root mean squared error; RPME = robust Pareto midpoint estimator.

increase between 2000 and 2006–2009. Although every estimator captures the big increases in the Gini coefficient between 1980 and 2000, the smaller increases before and after can be swamped by estimation error. The estimation errors are not huge, 10 percent at the most, but because the errors change from year to year, they can obscure the underlying trends.

Overall, the MGBE has a smaller RMSE, but this is entirely because of its better performance in 1970. Excluding that one year, the MGBE and RPME have very similar RMSEs. The real advantage of the MGBE is that it does a better job of capturing the trend.

The fact that the MGBE captures the trend in these particular data does not mean that it would always capture trends better than the RPME. Before conducting any analysis, it is useful to consult a few known "true" values, as we do in Table 7, to see which estimator best captures the known trends in a particular data set.

4. DISCUSSION

Although the RPME and MGBE often produce very good estimates, they do leave room for improvement, particularly in estimating the Theil index and MLD.

4.1. *Ideas for Improvement*

Although the RPME is bin consistent, its major weakness is that it treats income as a discrete variable. Alternatives are possible that treat income as continuous while preserving the property of bin consistency. The simplest idea is to spread each bin's cases evenly across their bin, so that each bin has its own uniform density, except for the top bin, which is given a Pareto density. This approach is simple but tends to exaggerate inequality because the fitted uniform distributions are too flat in the tails (Cloutier 1988).

The idea of fitting each bin to its own density can be improved. One improvement is to fit each of the right bins to its own downward-sloping density (Jargowsky 1995) and fit each of the left bins to its own upward-sloping density. In addition, alternatives to the Pareto density may be considered for the top bin. Unfortunately, distributions that fit each bin to its own density are not continuous at the bin boundaries. They are also not identified because they have as many parameters as bins. An attempt to solve both problems would be to fit each bin to its own density but to constrain the densities to be continuous at the bin boundaries.

Although the MGBE treats income as continuous, its major weakness is that it uses only unimodal GB family densities. These densities are usually rejected as the true distribution of income, and in some data, they may be poor approximations. The multimodel approach could be improved by adding densities from outside the GB family, especially bimodal densities that the GB family cannot mimic.

These suggested improvements converge on the idea of fitting a density that is as smooth as the GB family densities but as flexible and assumption free as the RPME. The idea is attractive, but it may be difficult to achieve in practice. A very flexible continuous distribution would have many parameters, and it gets hard to achieve identification or convergence as the number of parameters approaches the number of bins. In our experience, it can be difficult getting the four-parameter GB2 distribution to converge; a distribution with more parameters would probably be even more challenging. It is helpful to parameterize distributions so that each parameter captures a distinct aspect of the distribution's shape (Stasinopoulos and Rigby 2013), but this can be hard to achieve when there are many parameters.

There have been two attempts to fit binned incomes to flexible non-GB distributions that can be either unimodal or multimodal. One attempt is the logspline method, which fits binned data to a density whose logarithm is a polynomial spline (Kooperberg and Stone 1992). A second attempt is kernel density estimation, which requires bin means as well as bin counts (e.g., Sala-i-Martin 2006). Although the ideas behind these methods are appealing, unfortunately neither has performed well in practice. Both methods can display substantial biases when fit to real or simulated income distributions, and both methods perform worse than simply fitting the GB2 distribution (von Hippel et al. 2012; Minoiu and Reddy 2012).[13]

4.2. *Using Auxiliary Statistics*

Further improvements are possible if the data offer statistics beyond the bin counts. As mentioned previously, some data give the bin means as well as the bin counts. These can be used to improve the RPME, because we can simply replace the bin midpoints with the bin means, and there is no need for a Pareto assumption in the top bin. Theoretically, the bin means can also be used to improve the estimation of GB family distributions using a generalized method of moments (GMM) estimator. However, the GMM estimator is difficult to optimize and can break down when the second moment does not exist or the estimates get close to the edge of the parameter space (Hajargasht et al. 2012). Bin means are also needed for kernel density estimation; however, the results obtained by applying kernel density estimation to binned data have been disappointing, as previously discussed (Minoiu and Reddy 2012). Bin means are also needed for interpolation methods which fit a simple density within each bin, subject to the constraint that the fitted density must reproduce the known bin mean (Cowell and Mehta 1982).

Although bin means are available only occasionally, it is quite common for the Census Bureau to provide the grand mean. Then binned-data estimates can be improved by constraining the estimates to match the grand mean. The improvement will be very small, however, because even without constraints, RPME and MGBE already match the mean with high accuracy. In addition, the constraints on the GB family distributions will be a nonlinear function of several parameters and will be difficult to implement in software.[14]

It is also common for data to provide the grand median. Having the grand median is equivalent to having an extra bin, because the investigator can split one of the bins at the median and easily figure out the counts for the split bins. Adding an extra bin will improve estimates from either the RPME or MGBE, but the improvement will be small if there are already many bins or if the median is close to a bin boundary.

Methods for improving the estimation of GB parameters with auxiliary statistics will not be fruitful if the true income distribution is not well approximated by a GB family distribution. For example, if the true income distribution is bimodal, then fitting a unimodal GB distribution is a procrustean effort that will result in bias even if auxiliary statistics are available.

5. CONCLUSION

This is the broadest evaluation of binned-data estimators that has ever been carried out. We evaluated two of the most popular estimators and found that neither was robust in small samples. We developed robust versions of the estimators, which we called the RPME and the MGBE. We have implemented these estimators in the *rpme* and *mgbe* commands for Stata and the *binequality* package for R (Duan and von Hippel 2016; von Hippel and Powers 2014).

The MGBE is typically more accurate if there are fewer than eight bins, but with more bins, the two methods typically produce very similar and very good estimates of the mean and Gini coefficient. It is likely that both methods also produce good estimates of the CV because the CV and Gini coefficient are closely related (Hosking 1990; Milanovic 1997). However, both methods can produce poor estimates of the Theil index and MLD.

Although errors in estimating the Gini are typically small, they can be large enough to affect estimated trends in the Gini coefficient over time. When using the methods to estimate trends, it is helpful to use reference statistics to check which method is giving better estimates.

An advantage of the RPME is that it runs much faster than the iterative MGBE. This can be an important consideration when working with hundreds or thousands of binned data sets, each of which represents a different county, school district, or neighborhood.

Our discussion suggests potential improvements to both methods. Some of the improvements would not be easy but if attempted could be tested using the relatively new data in this paper.

Authors' Note

Paul T. von Hippel conceived the study, led the writing, and developed and ran the Stata code. Samuel V. Scarpino developed and ran the R code. Igor Holas summarized the results.

Funding

The authors disclosed receipt of the following financial support for the research, authorship, and/or publication of this article: Paul T. von Hippel and Igor Holas were supported in part by a New Scholar Grant from the Stanford Center on Poverty and Inequality and by a grant from the Policy Research Institute at the University of Texas. Samuel V. Scarpino was supported by a National Science Foundation graduate research fellowship, by the Santa Fe Institute, and by the Omidyar Group.

Notes

1. There are two justifications for treating the bottom of the lower bin as zero. First, the distributions that are commonly fit to incomes only support non-negative values. Second, according to data from the Integrated Public Use Microdata Series, only 1.6 percent of bottom-bin households, or 0.05 percent of all households, have negative incomes, so rounding those incomes up to zero has negligible effects on the estimates. In our results, for example, we will not find that treating the bottom bin as bounded at zero results in an overestimate of the mean or median, even in poor communities.

2. For Maricao, the average width of the populated bins is $6,818, and the estimated S.D. is $\hat{\sigma}_{ME}$ = $14,307. The ratio of these quantities is about 1/2.

3. The Pareto assumption is arbitrary. We worked on an alternative that fit the top two bins to a truncated Weibull distribution, but the resulting formulas were more complicated and did not produce better results.

4. We tried West's alternative estimator in our data and confirmed that it produced results worse than $\hat{\alpha}$.

5. A more complete family tree would include the four-parameter GB1 distribution and a five-parameter GB distribution that nests both the GB1 and the GB2 (McDonald and Xu 1995). We have excluded the GB and GB1 because (1) they rarely fit income distributions better than the GB2 (Bandourian et al. 2002), and (2) they are not implemented in available software. In fact, the GB and GB1 would be difficult to implement because their parameters include a lower bound, which would be difficult to estimate from binned data (Robert Rigby, personal communication).

6. Stasinopoulos et al. (2008) allowed σ to be negative, but we constrained it to be positive by using a log link, that is, by estimating $\ln(\sigma)$.

7. It is possible to sample from a distribution with an undefined mean and calculate an average from the sample. But this sample average is meaningless and, if calculated, can be arbitrarily large and variable.

8. Notice that the fit of a model can be tested only if df is at least 1. For example, if fewer than B = 5 bins are populated, it will not be possible to test a distribution with more than k = 3 parameters.

9. The authors did not remark on these rejections, but you can see them by comparing the reported X^2 values to the critical values from a chi-square distribution.

10. We used an Intel i7 Quad Core MacBook Pro laptop, which has a 2.0-GHz clock speed, 16 GB of random-access memory, and a 500-GB hard drive.

11. The census actually reported total ("aggregate") family income. We calculated mean family income by dividing total income by the number of families.

12. The state Theil indices and MLDs are "unofficial" statistics published in a report by Census Bureau employees (Hisnanick and Rogers 2006), which was not vetted as thoroughly as an official Census Bureau report. Initially, typographical errors in the unofficial report made us concerned about the accuracy of the unofficial estimates. Later, however, we noticed that the national Theil index and MLD in the unofficial report agreed closely with the national Theil index and MLD given in an official Census Bureau report based on the current population survey (DeNavas-Walt, Proctor, and Smith 2013).

13. It is not clear whether the observed bias of logspline estimates is due to the logspline method itself or to the way that it is implemented in legacy software (namely, the *oldlogspline* package for R).

14. The *gamlss* package for R, which we used to implement the MGBE, has no provision for multiparameter constraints (Stasinopoulos et al. 2008).

References

Bandourian, Ripsy, James McDonald, and Robert S. Turley. 2002. "A Comparison of Parametric Models of Income Distribution across Countries and over Time."

Luxembourg Income Study Working Paper 305. Retrieved June 1, 2012 (http://papers.ssrn.com/sol3/papers.cfm?abstract_id=324900).

Bee, Adam. 2012. *Household Income Inequality within U.S. Counties: 2006–2010.* Washington, DC: U.S. Census Bureau.

Burnham, Kenneth P., and David R. Anderson. 2004. "Multimodel Inference Understanding AIC and BIC in Model Selection." *Sociological Methods and Research* 33(2):261–304.

Chotikapanich, Duangkamon, D. S. Prasada Rao, and Kam Ki Tang. 2007. "Estimating Income Inequality in China Using Grouped Data and the Generalized Beta Distribution." *Review of Income and Wealth* 53(1):127–47.

Cloutier, Norman R.1988. "Pareto Extrapolation Using Rounded Income Data." *Journal of Regional Science* 28(3):415.

Corcoran, Sean, and William N. Evans. 2010. "Income Inequality, the Median Voter, and the Support for Public Education." National Bureau of Economic Research. Retrieved October 15, 2012 (http://www.nber.org/papers/w16097).

Cowell, Frank A., and Fatemeh Mehta. 1982. "The Estimation and Interpolation of Inequality Measures." *Review of Economic Studies* 49(2):273–90.

DeNavas-Walt, Carmen, Bernadette D. Proctor, and Jessica C. Smith. 2013. *Income, Poverty, and Health Insurance Coverage in the United States: 2012.* Washington, DC: U.S. Census Bureau.

Duan, Yutong, and Paul T. von Hippel. 2016. "MGBE: Stata Module to Compute Multimodel Generalized Beta Estimator." Retrieved from http://econpapers.repec.org/software/bocbocode/s458189.htm

Evans, William N., Michael Hout, and S. E. Mayer. 2004. "Assessing the Effect of Economic Inequality." Pp. 933–68 in *Social Inequality*, edited by Kathryn M. Neckerman. New York: Russell Sage.

Galbraith, James K., and J. Travis Hale. 2004. "Income Distribution and the Information Technology Bubble." Working Paper 57. Austin: University of Texas Inequality Project.

Galbraith, James K., and J. Travis Hale. 2009. "The Evolution of Economic Inequality in the United States, 1969–2007: Evidence from Data on Inter-industrial Earnings and Inter-regional Incomes." Working Paper 57. Austin: University of Texas Inequality Project.

Hajargasht, Gholamreza, William E. Griffiths, Joseph Brice, D. S. Prasada Rao, and Duangkamon Chotikapanich. 2012. "Inference for Income Distributions Using Grouped Data." *Journal of Business and Economic Statistics* 30(4):563–75.

Heitjan, Daniel F. 1989. "Inference from Grouped Continuous Data: A Review." *Statistical Science* 4:164–79.

Henson, Mary F. 1967. *Trends in the Income of Families and Persons in the United States, 1947–1964.* Washington, DC: U.S. Department of Commerce, Bureau of the Census.

Hisnanick, John J. and Annette L. Rogers. 2006. *Household Income Inequality Measures Based on the ACS Data: 2000–2005.* Washington, DC: U.S. Census Bureau.

Hosking, J.R.M. 1990. "L-moments: Analysis and Estimation of Distributions Using Linear Combinations of Order Statistics." *Journal of the Royal Statistical Society, Series B (Methodological)* 52(1):105–24.

Jargowsky, Paul A. 1995. "Take the Money and Run: Economic Segregation in U.S. Metropolitan Areas." Madison, WI: Institute for Research on Poverty. Retrieved October 17, 2012 (http://www.jstor.org/stable/2096304).

Jargowsky, Paul A. 1996. "Take the Money and Run: Economic Segregation in U.S. Metropolitan Areas." *American Sociological Review* 61(6):984–98.

Johnson, Norman L., Samuel Kotz, and N. Balakrishnan. 1994. *Continuous Univariate Distributions, Vol. 1.* 2nd ed. New York: Wiley-Interscience.

Kooperberg, Charles, and Charles J. Stone. 1992. "Logspline Density Estimation for Censored Data." *Journal of Computational and Graphical Statistics* 1(4):301–28.

McDonald, James B. 1984. "Some Generalized Functions for the Size Distribution of Income." *Econometrica* 52(3):647–63.

McDonald, James B., and Michael Ransom. 2008. "The Generalized Beta Distribution as a Model for the Distribution of Income: Estimation of Related Measures of Inequality." Pp. 147–66 in *Modeling Income Distributions and Lorenz Curves: Economic Studies in Equality, Social Exclusion and Well-being*, edited by Duangkamon Chotikapanich. New York: Springer.

McDonald, James B., and Yexiao J. Xu. 1995. "A Generalization of the Beta Distribution with Applications." *Journal of Econometrics* 66(1–2):133–52.

Milanovic, Branko. 1997. "A Simple Way to Calculate the Gini Coefficient, and Some Implications." *Economics Letters* 56(1):45–49.

Minoiu, Camelia, and Sanjay G. Reddy. 2009. "Estimating Poverty and Inequality from Grouped Data: How Well Do Parametric Methods Perform?" *Journal of Income Distribution* 18(2). Retrieved March 27, 2014 (http://papers.ssrn.com/abstract= 925969).

Minoiu, Camelia, and Sanjay G. Reddy. 2012. "Kernel Density Estimation on Grouped Data: The Case of Poverty Assessment." *Journal of Economic Inequality* 12(2):1–27.

National Historical Geographic Information System. 2005. "1980_STF3." Retrieved June 16, 2014 (https://data2.nhgis.org/main).

Navarro, Alfredo, and Richard Griffin. 1990. "Sample Design for the 1990 Decennial Census." Retrieved June 16, 2014 (http://scholar.google.com/scholar?cites=1295625 8902425784063&as_sdt=5,44&sciodt=0,44&hl=en).

Parker, Robert Nash, and Rudy Fenwick. 1983. "The Pareto Curve and Its Utility for Open-ended Income Distributions in Survey Research." *Social Forces* 61(3):872–85.

Quandt, R. E. 1966. "Old and New Methods of Estimation and the Pareto Distribution." *Metrika* 10(1):55–82.

Rao, J.N.K., and A. J. Scott. 1981. "The Analysis of Categorical Data from Complex Sample Surveys: Chi-squared Tests for Goodness of Fit and Independence in Two-way Tables." *Journal of the American Statistical Association* 76(374):221–30.

Sala-i-Martin, Xavier. 2006. "The World Distribution of Income: Falling Poverty and . . . Convergence, Period." *Quarterly Journal of Economics* 121(2):351–97.

Scarpino, Samuel V., Paul T. von Hippel, and Igor Holas. 2014. "Binequality: Methods for Analyzing Binned Income Data." Retrieved November 4, 2014 (http://cran.r-project.org/web/packages/binequality/index.html).

Stasinopoulos, Mikis, and Robert Rigby. 2013. "Gamlss: Generalised Additive Models for Location Scale and Shape." The Comprehensive R Archive Network (CRAN). Retrieved (http://cran.r-project.org/web/packages/gamlss/index.html).

Stasinopoulos, Mikis, Robert Rigby, and Calliope Akantzilioutou. 2008. *Instructions on How to Use the Gamlss Package in R.* 2nd ed.

Stirling, W. Douglas. 1986. "A Note on Degrees of Freedom in Sparse Contingency Tables." *Computational Statistics and Data Analysis* 4(1):67–70.

U.S. Census Bureau. 1982. "1980_STF3." Retrieved June 16, 2014 (http://www2.census.gov/census_1980/).

U.S. Census Bureau. N.d. "American FactFinder." Retrieved June 16, 2014 (http://factfinder2.census.gov/faces/nav/jsf/pages/index.xhtml).

U.S. Census Bureau, Data Integration Division. 2010. "Income—Table S4. Gini Ratios by State: 1969, 1979, 1989, 1999—U.S Census Bureau." Retrieved June 16, 2014 (https://www.census.gov/hhes/www/income/data/historical/state/state4.html).

von Hippel, Paul T., Igor Holas, and Samuel V. Scarpino. 2012. "Estimation with Binned Data." arXiv:1210.0200. Retrieved October 8, 2012 (http://arxiv.org/abs/1210.0200).

von Hippel, Paul T. and Daniel A. Powers. 2014. "RPME: Stata Module to Compute the Robust Pareto Midpoint Estimator." Boston College Department of Economics. Retrieved January 7, 2015 (https://ideas.repec.org/c/boc/bocode/s457863.html).

West, Sandra. 1986. "Estimation of the Mean from Censored Wage Data." In *Proceedings of the Survey Research Methods Section, Meeting of the American Statistical Association*, Chicago, IL.

West, Sandra, Diem-Tran Kratzke, and Shail Butani. 1992. "Measures of Central Tendency for Censored Wage Data." In *Proceedings of the Survey Research Methods Section, Meeting of the American Statistical Association*, Boston, MA.

Western, Bruce, Deirdre Bloome, and Christine Percheski. 2008. "Inequality among American Families with Children, 1975 to 2005." *American Sociological Review* 73(6):903–20.

White, Halbert. 1982. "Maximum Likelihood Estimation of Misspecified Models." *Econometrica* 50(1):1–25.

Author Biographies

Paul T. von Hippel is an assistant professor at the LBJ School of Public Affairs, University of Texas, Austin. His research interests include family and school influences on academic achievement. He started the project in this article in order to estimate the distribution of family income within and between U.S. school districts.

Samuel V. Scarpino is an Omidyar Fellow at the Santa Fe Institute and an incoming assistant professor in the Department of Mathematics and Statistics and the Complex Systems Center at the University of Vermont. His research primarily focuses on the evolutionary and population dynamics of infectious diseases. He earned a PhD in integrative biology from the University of Texas at Austin.

Igor Holas is a PhD candidate in human development and family science at the University of Texas. He is cofounder and CEO of Mentegram, an organization that helps mental health professionals provide remote monitoring and interventions.

Sociological Methodology
2016, Vol. 46(1) 252–282
© American Sociological Association 2015
DOI: 10.1177/0081175015581379
http://sm.sagepub.com

ॐ 8 ॐ

GOODNESS-OF-FIT OF MULTILEVEL LATENT CLASS MODELS FOR CATEGORICAL DATA

Erwin Nagelkerke*
Daniel L. Oberski*
Jeroen K. Vermunt*

Abstract

In the context of multilevel latent class models, the goodness-of-fit depends on multiple aspects, among which are two local independence assumptions. However, because of the lack of local fit statistics, the model and any issues relating to model fit can only be inspected jointly through global fit statistics. This hinders the search for model improvements, as it cannot be determined where misfit originates and which of the many model adjustments may improve its fit. Also, when relying solely on global fit statistics, assumption violations may become obscured, leading to wrong substantive results. In this paper, two local fit statistics are proposed to improve the understanding of the model, allow individual testing of the local independence assumptions, and inspect the fit of the higher level of the model. Through an application in which the local fit statistics group-variable residual and paired-case residual are used as guidance, it is shown that they pinpoint misfit, enhance the search for model improvements, provide substantive insight, and lead to a model with different substantive conclusions, which would likely not have

*Tilburg University, Tilburg, The Netherlands

Corresponding Author:
Erwin Nagelkerke, Tilburg University, Department of Methodology and Statistics, P.O. Box 90153, 5000 LE, Tilburg, The Netherlands
Email: e.nagelkerke@tilburguniversity.edu

been found when relying on global information criteria. Both residuals can be obtained in the user-friendly Latent GOLD 5.0 software package.

Keywords

latent class analysis, multilevel, goodness-of-fit, local fit, bivariate residual, BVR

1. INTRODUCTION

Latent class (LC) analysis is often used to detect and develop a latent unobserved classification of subjects on the basis of multiple observed categorical characteristics. The usefulness of this application in many scientific fields combined with favorable properties, such as the ability to handle multiple dependent variables and measurement error, have recently caused a growing interest in LC analysis. This in turn has resulted in the development of several extensions to the regular model in an attempt to relax assumptions and make the method more widely applicable. An important extension that has gathered considerable attention is the multilevel LC model (e.g., Muthén and Asparouhov 2009; Vermunt 2003, 2008).

Substantively, the major benefit of this multilevel extension is that it allows simultaneous classification of groups and individuals. The regular LC model may be used either to distinguish typologies of the units under study that are systematically similar (e.g., Harrell et al. 2012) or to find the most common characteristics of predetermined classes (e.g., Finch and Bronk 2011; Laudy et al. 2005). The multilevel extension now makes it possible for nested categorical data in which a natural grouping is observed to also classify the groups on the basis of the similarity of their members.

For example, employees can be classified in terms of job variety, which in turn is associated with job satisfaction and turnover intent (Lambert, Hogan, and Barton 2001). However, the effect is likely to be moderated by the team context whereby correspondence rather than the absolute task variety is of importance. Perceiving far lower task variety compared with the team may cause diminished confidence and boredom, whereas far higher variety may induce stress. A simultaneous classification of both employees and the teams in which they are nested would allow the importance of this team context to be evaluated, providing more insight into outcomes such as frictional unemployment, employee burnout, and declines in overall job satisfaction.

In addition to the substantive application, the multilevel approach solves the statistical problem of dependent observations. Analogous to a multitude of statistical methods, LC analysis assumes that the units under study are independent of one another. However, this assumption does not hold when observing cases nested within a certain grouping, whether they are persons that belong to a particular group or repeated measures that belong to the same unit (Hox 2010; Snijders and Bosker 2012). In the example, the responses of employees from the same team cannot be assumed to be independent. An earlier solution to this dependency problem is the multiple-group approach (Clogg and Goodman 1984), but it requires all parameters to be estimated separately for all groups, causing the method to lose its value when a large number of groups is observed.

Compared with the regular LC model, the multilevel LC model thus has additional substantive applications and offers a solution for categorical data in which there is dependency between observations. However, testing whether or not the model is correctly specified and actually captures all the dependency is currently not possible in its own right, as inspecting model fit is limited to global tests, such as the chi-square (X^2) or log-likelihood-ratio (L^2), and model comparisons through information criteria, such as the Bayesian information criterion (BIC) and Akaike information criterion (AIC). Although these tests and criteria can identify a well-fitting model, or the best fitting out of a series of alternative models, their global nature limits the control they provide. Especially when models become increasingly complex, the information available on the cause of better or worse fit becomes obscured. This in turn not only hinders the search for possible model improvements but also limits substantive understanding of the data.

To gain insight, understand the result of model adjustments, and detect specific misfit or violations of assumptions, these global criteria should ideally be supplemented with local fit statistics that single out and test one particular area of the model. In a regular LC model, such local fit measures exist in the form of the bivariate residual (BVR) (Vermunt and Magidson 2005; see also Mavridis, Moustaki, and Knott 2007) and a score-test approach that leads to modification indices (Glas 1999; Oberski, Van Kollenburg, and Vermunt 2013). Both test the local independence assumption that is central to the LC model and evaluate the degree to which the model captures the association between all pairs of observed variables. As such, these measures indicate why one model fits better or worse, pinpoint violations of the local independence assumption, and facilitate the search for model improvements. For the multilevel LC model, however, there are currently no local fit statistics that give these insights on the group level.

Here we propose two complementary diagnostic measures that enhance exactly these abilities to detect a particular type of model misfit and increase the understanding of the fitted model for multilevel LC analysis. Both take the form of a Pearson residual and relate to the higher level of a multilevel LC model. The first residual, BVR_{group}, relates to the item distributions and is considered a between-group measure. It can be used to evaluate the difference in responses between groups and to detect misfit that originates from the model not fitting particular groups as well as others. The second residual, BVR_{pair}, is a within-group measure in the sense that it can be used to evaluate the degree of similarity among cases within a group, and it is indicative of misfit that originates from any leftover dependency among the units within groups.

The remainder of this paper is structured as follows. In section 2, we introduce the multilevel LC model. In section 3, we discuss the problems with model fit statistics more elaborately, as well as the existing BVR, and introduce the proposed residuals. In section 4, we demonstrate the use of the residuals as local fit measures and the way in which they may affect substantive conclusions by applying them to the job variability data used by Vermunt (2003). In section 5, we use a small simulation study to demonstrate that the proposed measures have adequate power and type I error.

2. THE MULTILEVEL LATENT CLASS MODEL

The multilevel LC model can be expressed using two equations: one for the lower level denoting the conditional probability of all responses given by a unit and one for the higher level marginal probability of all response patterns per group (Lukočiené, Varriale, and Vermunt 2010; Vermunt 2003). The expression for the lower level is essentially that of an LC model, but in the case of a multilevel structure it is made conditional on the LC membership of the group (Vermunt 2003; Vermunt 2008).

Let the response of individual i in group j on item k be denoted as y_{ijk}, with a total of J groups, each having n_j individual members summing to N, and a total of K items, each having R_k categories. All responses to the K items of person i in group j are denoted as the vector $\boldsymbol{y}_{ij}$, with r referring to one particular answer pattern of length K when no values are missing and r_k referring to a particular response to item k. The latent variable η_{ij} that classifies the units within groups has C latent classes and the latent variable ζ_j that classifies the groups has G latent classes, with c and g referring to one of these classes. Assuming conditional independence, the lower level of the multilevel LC model is expressed as[1]

$$\Pr\left(\boldsymbol{y}_{ij}=r|\zeta_j=g\right) = \sum_{c=1}^{C} \Pr\left(\eta_{ij}=c|\zeta_j=g\right) \prod_{k=1}^{K} \Pr\left(y_{ijk}=r_k|\eta_{ij}=c,\zeta_j=g\right). \quad (1)$$

Removing the conditioning on the group-level latent variable (ζ_j) from equation (1) results in the standard LC model, in which the probability of observing a particular response pattern r is a combination of the prevalence of LC c on the latent variable η_{ij} and the probabilities of observing the combination of the responses r_k conditional on the unit's class membership. In the multilevel LC model, all these terms are made conditional on the LC membership of the group a unit belongs to $(\zeta_j=g)$, such that groups can be classified along G LCs and the probability of an individual response pattern is affected by the group-level class membership.

The expression for the higher level of the model then denotes the marginal probability of all response patterns of individuals within group j as $\boldsymbol{y}_j$, with s denoting a particular combination of response patterns of length $n_j * K$. Here an assumption of independence is required as well, but now the full response patterns of individuals rather than the responses to one item should be independent:

$$\Pr\left(\boldsymbol{y}_j=s\right) = \sum_{g=1}^{G} \Pr\left(\zeta_j=g\right) \prod_{i=1}^{n_j} \Pr\left(\boldsymbol{y}_{ij}=r|\zeta_j=g\right). \quad (2)$$

The probability of observing the vector $\boldsymbol{y}_j$ of all individual response patterns s in group j is a combination of the prevalence, or size, of a particular group-level LC g on the latent variable ζ_j and the probabilities of observing the combination of the individual answer patterns r conditional on the LC membership of the group.

It should be noted that these two expressions result in a model in which both the lower level class prevalence and the response probabilities can differ between all higher level classes. Although a multitude of constraints is possible, two are most commonly used in practice, the first of which leads to the most used model that simultaneously classifies higher and lower level units. The first constraint $\Pr(y_{ijk} = r_k | \eta_{ij} = c, \zeta_j = g) = \Pr(y_{ijk} = r_k | \eta_{ij} = c)$ causes the response probabilities on the lower level to be independent of the higher level class membership but the class sizes to be estimated freely (Lukočiené et al. 2010; Vermunt 2003, 2008). The second possibility is to constrain the model by setting $\Pr(\eta_{ij} = c | \zeta_j = g) = \Pr(\eta_{ij} = c)$, causing the response probabilities to be estimated freely but the lower level class membership to be independent of higher level class membership (Lukočiené et al. 2010; Vermunt 2004).

3. GOODNESS-OF-FIT

In this multilevel LC model, there are several key issues relating to model fit. There are the two central assumptions—namely, the local independence of item responses on the lower level and the conditional independence of response patterns of individuals on the higher level— and there are the goals of correctly reproducing the item distributions or observed frequencies for both the individual observations as well as for the groups. These latter goals relate to arriving at a correct classification on both levels and to obtaining the conditional probabilities of interest depending on the substantive goal and specification of the model (Goodman 2002).

Improving the fit of this model can be achieved in a multitude of ways that improve the quality of the prediction, or relax an assumption. An LC or group-level LC can be added, for example. Or, when keeping the same number of classes, a covariance between any combination of observed variables may be modeled, as well as any direct effect from the group-level latent variable to an observed variable. Although it is also possible to add additional categorical or continuous latent variables to the model, for conciseness, these options are not explored in the application.

Unfortunately, despite these different sources of misfit and the many ways to adjust the model, there is little information available as to where model misfit originates and what the effects are of model adjustments. Currently only the local independence assumption on the lower level of the model—the independence of responses conditional on the latent variable—can be inspected through the BVR. The analogous assumption on the higher level—the independence of response patterns conditional on the group-level latent variable—the quality with which the model describes the individual responses, and the degree to which the model correctly describes the groups can only be assessed jointly through global statistics. That is, the fit of the model as a whole is considered, rather than any of the individual aspects of the model.

As a result local misfit may go unnoticed, because even when a model shows adequate global fit, it may still be misspecified. In such cases, a type of local misfit averages out with other, correctly specified, areas of the model. This problem is reinforced when using information criteria, such as the BIC and the AIC, which only compare estimated models. As long as all estimated models in such cases violate one or more assumptions, selecting the best one will still result in using a model that does not fit the data correctly. Ultimately this may lead to a wrong classification and wrong substantive conclusions, especially when the classification is used in subsequent analyses to relate classes to outcomes.

Of course, these problems with global fit measures apply to almost all statistical methods, but they do become more pressing in complex models as the possible sources of misfit are abundant. This is especially clear in multilevel models, for which both levels are considered simultaneously. For multilevel structural equation modeling, several solutions have been offered to evaluate the fit separately for different levels. Yuan and Bentler (2007) did so by estimating the saturated covariance matrices for each level of the full model and subsequently treat these as observed single-level data to test the hypothesized model one level at a time. As such they obtained common fit indices for each level individually. Ryu and West (2009) developed a similar approach whereby the model is initially estimated as hypothesized and subsequently reestimated several times, each time saturating one of the levels.

Although both are elegant solutions to localize model misfit, such methods do not apply to LC analysis, as the higher level cannot be estimated independently from the lower level. As was shown in equation

(2), the vector of group-level patterns is directly related to the estimated answer patterns for respondents. When the lower level is saturated, this also greatly improves the fit on the higher level of the model. Furthermore, even though these methods are able to separate the misfit on different levels, they still are not local fit statistics in the true sense that they are able to pinpoint the assumption violation, misspecification, or variable that causes the misfit. That is, even when the level at which misfit occurs can be determined, the possibilities to improve the model remain plentiful and require more precise measures to be detected.

To address this problem, two local fit statistics for multilevel LC models are proposed in the sections that follow, which aim to test specific areas of the model individually. The first tests the reproduction of univariate item distributions in all the groups and provides a partial test of how well the higher level of the model fits the data. The second is aimed at testing the conditional independence of response patterns and in combination with the BVR allows a test of two central assumptions of the model. Both provide information on the location and extent of misfit.

3.1. *BVR*

To show how the proposed statistics fit the LC framework, and for the sake of completeness, the existing BVR is briefly introduced. Vermunt and Magidson (2013) constructed the BVR to test the assumption of local independence for all pairs of observed variables in a regular LC model, but the test can be applied identically to the lower level of a multilevel model. The BVR assesses the difference between the observed frequencies ($n_{rr'}$) and the model expected frequencies ($m_{rr'}$) in the two-way cross-tabulation of items k and k' by a Pearson statistic divided by its number of degrees of freedom (see also Bartholomew and Leung 2002; Vasdekis, Cagnone, and Moustaki 2012); that is,

$$BVR_{kk'} = \frac{1}{(R_k - 1)(R_{k'} - 1)} \sum_{r=1}^{R_k} \sum_{r'=1}^{R_{k'}} \frac{(n_{rr'} - m_{rr'})^2}{m_{rr'}}. \tag{3}$$

The expected frequencies follow from the LC model, which assumes conditional independence of item responses given LC membership. More specifically, they are obtained by multiplying the class-specific probabilities of the response r on item k and response r' on item k' and summing these over the LCs using the class membership probabilities as weight. For an LC model without a multilevel structure, we obtain

$$m_{rr'} = \sum_{i=1}^{N} \sum_{c=1}^{C} \Pr\,(y_{ik} = r_k | \eta_i = c)\,\Pr\,(y_{ik'} = r_{k'} | \eta_i = c)\,\Pr\,(\eta_i = c | y_i = r). \quad (4)$$

When no values are missing, the same $m_{rr'}$ can be obtained by using $\Pr\,(\eta_i = c)$ as a weight instead of $\Pr\,(\eta_i = c | y_i = r)$ and multiplying the sum over classes by N rather than summing it over N, because $\Pr\,(\eta_i = c)$ equals the average $\Pr\,(\eta_i = c | y_i = r)$ for the complete sample. However, in the case of missing values, the observed frequencies contain only those cases for which both variables are observed. To obtain the corresponding expected frequencies, the class membership probabilities should be based on this subsample. That is, using $\Pr\,(\eta_i = c)$ is not appropriate, and the frequency should be obtained by summing over the cases with both variables observed, using $\Pr\,(\eta_i = c | y_i = r)$ as a weight.

The above formulation for $m_{rr'}$ can be easily generalized to be applicable in a multilevel LC analysis. The sum over LCs must then contain the joint posterior probability of the lower and higher level latent variables, and the sum over individuals must be over both groups and individuals within groups:

$$m_{rr'} = \sum_{j=1}^{J} \sum_{i=1}^{n_j} \sum_{g=1}^{G} \sum_{c=1}^{C} \Pr\,(y_{ijk} = r_k | \eta_{ij} = c, \zeta_j = g)$$

$$\Pr\,(y_{ijk'} = r_{k'} | \eta_{ij} = c, \zeta_j = g)\,\Pr\,(\eta_{ij} = c, \zeta_j = g | y_j = s). \quad (5)$$

Any deviation between the observed and the predicted frequency, which assumes local independence of items given LC membership, is now contained in the residual.

3.2. Group-variable Residual

To further deconstruct global misfit, we here propose a group-variable residual (BVR$_{\text{group}}$). As was shown in equation (2), the response vector y_j containing all individual response patterns is a function of the size of the group-level class and the individual answer patterns. This implies that, among other things, the univariate response frequencies within each group should be modeled correctly for the LC solution to be correct. Because the observed group membership can be understood as a nominal covariate in a multilevel LC model, the BVR can be adapted to assess the response to a nominal dependent variable and group membership:

$$BVR_{group.k} = \frac{1}{(R_k - 1)(J - 1)} \sum_{j=1}^{J} \sum_{r=1}^{R_k} \frac{(n_{jr} - m_{jr})^2}{m_{jr}}. \tag{6}$$

The observed frequency n_{jr} here is simply the number of units in group j with response r_k. The expected frequencies m_{jr} can be obtained from the individual probabilities $\Pr (y_{ijk} = r_k)$:

$$\Pr(y_{ijk} = r_k) = \sum_{g=1}^{G} \Pr(y_{ijk} = r_k | \zeta_j = g) \Pr(\zeta_j = g | y_j = s), \tag{7}$$

where

$$\Pr(y_{ijk} = r_k) = \sum_{c=1}^{C} \Pr(y_{ijk} = r_k | \eta_{ij} = c, \zeta_j = g) \Pr(\eta_{ij} = c | \zeta_j = g). \tag{8}$$

Then

$$m_{jr} = \sum_{i=1}^{n_j} \Pr(y_{ijk} = r_k) = \sum_{i=1}^{n_j} \sum_{g=1}^{G} \Pr(y_{ijk} = r_k | \zeta_j = g) \Pr(\zeta_j = g | y_j = s). \tag{9}$$

Thus, the probability of a particular response is summed over all group members to obtain its frequency within the group, and it is itself a function of the group-class response probabilities and the group-class membership probabilities. It should be noted that for the class membership on the group level, the posterior probability $\Pr(\zeta_j = g | y_j = s)$ is used. Because the interest lies in testing the group by variable relationships and aggregating these over the groups, all available information on the groups should be used, as contained in the posterior.

The statistic itself is computed for all groups separately and summed over the groups to test the assumption of correct model fit in each of the groups. This sum is additionally divided by $(R_k - 1)(J - 1)$. The BVR_{group} now equals the average contribution to the residual per degree of freedom. That is, the dimension of the matrix to which equation (6) is applied is $R_k \times J$, resulting in $(R_k - 1)(J - 1)$ nonredundant parameters. Correcting for both R_k and J standardizes the BVR_{group} such that it is not affected by the number of groups or the number of categories on the variable.

As can be seen in equations (7) through (9), a special case exists when the nested structure of the data is ignored by estimating the multilevel LC model with only one group-level class. The results are identical to omitting the group-level latent variable altogether and ensures that the BVR_{group} is independent from the number of lower level classes to obtain its baseline value, which is substantively indicative of the between-group heterogeneity or the between-group variance. For this model, the residual is then broadly comparable to the empirical Bayes estimates as used in linear multilevel models. Although their common use is to test the normality assumption on the higher level, they can also be used to construct influence diagnostics (Snijders and Berkhof 2008) and as such are indicative of misfit.

3.3. Paired-case Residual

In a multilevel LC model, the higher level has a local independence assumption similar to that of the lower level. Where the assumption in equation (1) is that the responses r_k are independent for all the K items per individual, in equation (2) the response patterns r are assumed independent for all the n_j individuals per group. However, to capture this dependency among units within a group, the responses of the individual members should be related to one another. This cannot be done as straightforwardly as is the case for the dependency between item pairs. Where the response frequencies for the latter can be cross-tabulated directly, the cross-tabulation of dependency among units requires all units within a group to be related. An intuitive approach to do so is to create all pairs of units within every group and obtain the pairwise response frequencies. The expected and observed response frequencies can then be compared again:

$$
BVR_{pair} = \frac{J}{N} \frac{1}{R_k(R_k - 1)/2}
$$
$$
\left[\sum_r^{R_k} \sum_{r'>r}^{R_{k'}} \frac{((n_{krr'} + n_{kk'r}) - (m_{krr'} + m_{kr'r}))^2}{m_{krr'} + m_{kr'r}} + \sum_r \frac{(n_{krr} - m_{krr})^2}{m_{krr}} \right]. \tag{10}
$$

To illustrate, consider a group containing five observations, with responses to one of multiple variables, as in Figure 1. The residual can

Figure 1. Obtaining the Observed Pairwise Response Frequencies

Data		
Obs	Var	Group
A	0	1
B	0	1
C	1	1
D	0	1
E	1	1
F	0	2
G	1	2
H	...	...

B

	0	1
A 0	1	0
1	0	0

C

	0	1
A 0	0	1
1	0	0

D

	0	1
A 0	1	0
1	0	0

E

	0	1
A 0	0	1
1	0	0

C

	0	1
B 0	0	1
1	0	0

D

	0	1
B 0	1	0
1	0	0

E

	0	1
B 0	0	1
1	0	0

D

	0	1
C 0	0	0
1	1	0

E

	0	1
C 0	0	0
1	0	1

E

	0	1
D 0	0	1
1	0	0

be understood as considering the combined responses r and r' of cases i and i' to item k as one element. To obtain the observed frequencies, a square contingency table of which the order is equal to the number of categories on the variable of interest can then be made per pair. The cell that identifies the actual answer pattern of that pair of cases has a frequency of one and all else equals zero.

The corresponding predicted probability of a certain pair of responses follows from the combined probability of person i giving response r and person i' giving response r' conditional on the group-level class:

$$\Pr(y_{ijk}=r_k, y_{i'jk}=r'_k) = \sum_{g=1}^{G} \Pr(y_{ijk}=r_k|\zeta_j=g) \Pr(y_{i'jk}=r'_k) \Pr(\zeta_j=g|y_j=s), \quad (11)$$

where $\Pr(y_{ijk}=r_k|\zeta_j=g)$ can be obtained by equation (8). Because these probabilities are only conditional on the group-level latent variable in a model without covariates, they are identical for identical patterns, and the order of the responses is interchangeable. That is, within a group only the probabilities on the diagonal and either the upper or lower off-diagonal need to be obtained. Aggregating these probabilities to arrive at the expected frequencies can then be done by multiplying the probability of a pair with the number of pairs $n_j(n_j - 1)/2$:

$$m_{krr'} = \sum_{j=1}^{J} (n_j(n_j - 1)/2) \Pr(y_{ijk}=r_k y_{i'jk}=r'_k) \quad (12)$$

Again, as is done for the BVR$_{\text{group}}$, the posterior probability is used in equation (11) to obtain this estimated frequency. In this case the main

Table 1. Obtaining the Pairwise Residual Contribution per Answer Pattern

		Observed				Probability				Expected				Residual Contr.	
		i'				r'				i'				i'	
		0	1			0	1			0	1			0	1
i	0	3	5	r	0	.415	.225	i	0	4.152	2.249	i	0	.32	.056
	1	1	1		1	.225	.135		1	2.249	1.351		1	–	.091

reason is that this weighting is more appropriate in cases in which groups are of different sizes and thus contain different numbers of pairs per group. As can be seen from equations (10) and (12), in comparison with equations (6) and (9), the BVR_{pair} is not obtained for each group separately and is only subsequently summed over the groups, but the aggregation already occurs when computing the expected frequencies. By weighting according to the posterior probability $\Pr(\zeta_j = g | y_j = s)$, the expected frequencies account in the best manner for unequal group sizes. With equal group sizes, using posterior or unconditional class membership probabilities will give the same expected frequencies.

The observed frequency of pairs can now be obtained by summing the pairwise tables from Figure 1, as is done in Table 1. The probability of a pair follows from equation (11) and the expected frequency from equation (12). For the illustration, the probabilities from the first model in the application section are used.

$$BVR_{pair} = \frac{1}{6} \frac{1}{2(2-1)/2}(0.320 + 0.056 + 0.091) = 0.079$$

Here, the structure of equation (10) also becomes clear. Note that because the order of the observations within a group is arbitrary, observing a 0-1 pair is in fact the same as observing a 1-0 pair. This is why the symmetric off-diagonal elements of the table are combined in the first summation in equation (10). The latter part of equation (10) adds the discrepancy between the observed and expected frequencies on the diagonal.

To finally arrive at the BVR_{pair}, the resulting residual is divided in such a way that the statistic equals the contribution to the residual per degree of freedom, in this case $R_k(R_k - 1)/2$ given the symmetry on the off-diagonals. In addition, the raw residual is divided by the average group size to avoid extremely large values, which are likely to occur because the theoretical maximum value of the statistic increases as a triangular sequence with n_j.

Unfortunately, the univariate marginal values for the resulting tables are not reproduced correctly when groups differ in size, in which case $(n_{krr'} + n_{kr'r}) \neq (m_{krr'} + m_{kr'r})$, which is also the case in the illustration. The cause is simply that an observation in a larger group is in more pairs than an observation in a smaller group. Differences between the observed (n) and expected (m) frequencies would then not only reflect the degree to which the model captures dependency between cases, but the residual would also partly reflect the difference in the univariate distribution. This changes the interpretation of the BVR_{pair}, which is unnecessary because the univariate distributions are always correctly reproduced by the model.

Therefore, a number of iterative proportional fitting (IPF) cycles are used to equate the reproduced and observed marginal frequencies and reduce the BVR_{pair} to zero when there is no residual dependency. The pairwise contingency table is made symmetrical first, such that answer patterns that differ only in respect to the order of the responses have the same frequency. As mentioned, the probability and thus the expected frequency of a certain pair of responses are identical regardless of order, but this is not necessarily the way in which they are observed.

In the IPF procedure the cells in the expected frequency table are adjusted so that its marginals match the observed marginals. The subsequent iterations alternate between row and column adjustments where each cell is multiplied by the ratio between the observed and the expected row (column) marginal. This process converges to a table with marginals equal to the observed marginal frequencies while retaining the cross-product ratios within the table (Bishop, Fienberg, and Holland 1975). An example can be found in Table 2.

The resulting BVR_{pair} statistic reduces to zero when the model captures all the dependency among cases within a group. Identical to the BVR_{group}, its baseline value can be obtained by estimating the model where the nesting of the data is ignored by modeling only one group-level class. The statistic is broadly comparable with the residual intraclass correlation in mixed models, which is the degree of dependency that is not captured by the model when controlling for the independent

Table 2. Iterative Proportional Fitting Cycles

		Observed				Expected				IPF Cycle 1—Row[a]	
		i'				i'				i'	
		0	1			0	1			0	1
i	0	3	(5+1)/2	6	i 0	4.152	2.249	6.401	i 0	3.892	2.108	6
	1	(1+5)/2	1	4	1	2.249	1.351	3.599	1	2.499	1.501	4
		6	4	10		6.401	3.599	10		6.391	3.609	—

		IPF Cycle 1—Column[b]				IPF Cycle 2—Row				IPF Cycle 2—Column	
		i'				i'				i'	
		0	1			0	1			0	1
i	0	3.654	2.336	5.99	i 0	3.66	2.34	6	i 0	3.66	2.34	6
	1	2.346	1.664	4.01	1	2.34	1.66	4	1	2.34	1.66	4
		6	*4*	—		*6*	*3.999*	—		*6*	*4*	*10*

Note. IPF = iterative proportional fitting.

[a]Row operation: Multiply cell by the ratio between the observed and expected row marginals: cell(observed row/expected row).

[b]Column operation: Multiply cell by the ratio between the observed and expected column marginals; cell(observed column/expected column).

variables (Snijders and Bosker 2012). The BVR_{pair} is similarly related to the uncaptured dependency and indicative of the homogeneity within groups that is ignored when the nested structure of the data is not or is only partially reproduced.

3.4. Bootstrap

The BVR, BVR_{group}, and BVR_{pair} residuals are all obtained identically to Pearson residuals. However, for the BVR, it is known that it does not follow the chi-square distribution, and the same is expected to be true for the two proposed measures. To still obtain p values for the residuals, a parametric bootstrap can be used (Langeheine, Pannekoek, and Van de Pol 1996), which is known to work for the BVR (Oberski et al. 2013). On the basis of the maximum likelihood estimate, the bootstrap in this instance samples group-class membership, class membership conditional on group-class membership, and the responses conditional on the membership of both. This results in alternative data sets with the same structure as the original to which the model is fitted. For each of these refitted models, the BVR values are obtained. The estimated p value then is the proportion of replicated models in which the BVR residuals are larger than in the original model (Vermunt and Magidson 2013). As such the BVR_{group} and BVR_{pair} are compared not with an asymptotic distribution but rather with an empirical distribution constructed by simulation. The bootstrap p values can be used for hypothesis testing, that is, for determining whether potential assumption violations are statistically significant.

4. APPLICATION: IMPROVING THE JOB VARIETY CLASSIFICATION

To illustrate the usefulness of the BVR_{pair} and BVR_{group}, we apply them here to a data example in which both employees and the teams in which they are nested are classified on the basis of task variety. This is one of the examples Vermunt (2003) used when introducing multilevel LC analysis, which provides the opportunity to see whether the original solution can be improved on the basis of the two residuals.

The variety in the tasks of employees, as well as the degree to which they feel that their capacities are put to good use, has been found to affect job satisfaction and turnover intent (Lambert et al. 2001; Fila et al. 2014). Although these outcomes are inherently individual, the broader context of the team, department, or organization plays an important role in shaping these effects. Gunter and Furnham (1996), for example, found that job variety has an opposite effect on job satisfaction in two different organizations, and van Mierlo et al. (2005) gave a broad overview of studies in which individual and team tasks affected several outcomes.

One of the ways in which context may affect job satisfaction and turnover intent may be through peer perceptions (e.g., Liu et al. 2012). When direct coworkers perceive their jobs as highly varied when individuals do not, this may adversely affect job satisfaction. In contrast, teams with larger differences in task variety may be better able to distribute the work, improving individual job satisfaction and reducing turnover intent. By obtaining a classification of teams through multilevel LC analysis on the basis of the perceived job variety classification of the employees, it becomes possible to detect such differences in team composition and investigate these questions.

Table 3. BIC Values for 29 Models Assuming Local Independence of Items and Indirect Effects of the Group-level Latent Variable[a]

	Lower Level Classes				
Group-level Classes	2	3	4	5	6
1	4,820	4,818	4,837	4,861	—[b]
2	4,786	*4,785*	4,799	4,819	4,844
3	4,794	4,795	4,794	4,814	4,836
4	4,802	4,806	4,809	4,826	4,850
5	4,811	4,818	4,822	4,839	4,865
6	4,820	4,831	4,838	4,857	4,880

Note. BIC = Bayesian information criterion; N = number of groups. Best-fitting model in bold.
[a]Constraint: $\Pr(y_{ijk} = r_k \mid \eta_{ij} = c, \zeta_j = g) = \Pr(y_{ijk} = r_k \mid \eta_{ij} = c)$.
[b]Unidentified.

Although relating the classification to an outcome variable is beyond the scope of this example, the use of simultaneous classification can be easily extended. For example, when job design is aggregated, it may explain frictional unemployment caused by a mismatch between companies and the workforce in a region, a classification of countries on the basis of the degree of religiosity of their populations may form an explanation for policy differences, or a classification of pupils and their groups may be used to enhance school class composition (see also Bennink et al. 2014).

However, when the LC model is incorrectly specified or violates assumptions, there is a possibility not only that teams and employees may be wrongly classified but also that the relationship between an outcome and the classification may be similarly unsound. This first step of classification is clearly an important one, because a wrong classification may result in wrong substantive conclusions on the actual goal of the study. Here the classification is reexamined using the proposed BVR_{group} and BVR_{pair} statistics to demonstrate their use. After excluding all cases with missing values and two teams with only one member, the data contain 848 cases in 86 teams and are similar to the data used by Vermunt (2003, 2005), as collected by van Mierlo (2003). For all employees, the perception of task variety in their jobs was measured with five categorical items, of which the four categories are collapsed to make them dichotomous. The variable measuring task repetitiveness is coded inversely with the other variables, such that a higher score reflects lower repetition and all scores are substantively in accordance. All

Table 4. Latent Class Profile for the Three-class, Two-group-class Model

	Group Class 1 Diverse	Group Class 2 Uniform	Class 1 Diverse	Class 2 Structure	Class 3 Creative	Overall
Nonrepetitive	.428	.279	.515	.125	.225	.385
Creative	.631	.382	.707	.065	.914	.558
Diverse	.792	.480	.961	.146	.483	.700
Capacity	.730	.578	.837	.439	.350	.685
Variation	.754	.461	.964	.192	.000[a]	.668
Class 1	.752	.371				
Class 2	.150	.537				
Class 3	.098	.092				
Prevalence	.707	.293	.640	.263	.097	

Note. Bayesian information criterion = 4,785.3
[a]Boundary solution.

models are estimated in Latent GOLD 5.0. The Latent GOLD syntax and survey wording are provided in Appendix A and Appendix B, respectively, available in the online journal. The data set itself is included in Latent GOLD as example data.

Because the BIC is currently the main criterion for model selection, selecting the best fitting from a series of alternative models, Table 3 depicts the BIC values for 29 models with differing numbers of classes. All of these models assume conditional independence between the five items, contain one latent variable on both levels (η and ζ), and allow only an indirect effect of the group-level latent variable ζ on the items through the lower level latent variable η (see also Vermunt 2003). It should be noted that these BIC values are computed using the number of groups as the sample size, rather than the number of cases, as this is found to be the more appropriate sample size to determine the number of classes in multilevel LC models (Lukočiené et al. 2010; Lukočiené and Vermunt 2010).

On the basis of these values, the model with two group-level and three low-level classes would be the best fitting, resulting in the profile depicted in Table 4. On the lower level, the largest of the three classes is one in which people report high levels of task variation and creativity. The second class is one in which people report having repetitive, uncreative, and unvaried tasks. The third is a class with highly creative tasks, yet quite unvaried and repetitive. On the group level, the classes are less distinguished in their overall profile. Members of teams in the first group-level class are most likely to belong to the first individual-level

Table 5. BVR, BVR_{group}, and BVR_{pair} Residuals for the Three-class, Two-group-class Model

	Nonrepetitive	Creative	Diverse	Capacities	Variation
Creative	.763 (.242)				
Diverse	.248 (.282)	.028 (.442)			
Capacities	.183 (.570)	.359 (.308)	.504 (.106)		
Variation	.010 (.706)	.036 (.272)	.153 (.016)	.011 (.790)	
BVR_{group}	1.586 (.000)	1.051 (.058)	.788 (.164)	1.072 (.132)	.816 (.316)
BVR_{pair}	1.740 (.000)	.570 (.028)	.123 (.296)	.366 (.098)	.000 (.974)

Note. Bayesian information criterion = 4,785.3.

class and those of the second higher level class to the second lower level class. Overall then the team profile of the first group-level class is mostly that of diverse, varied, and challenging tasks, whereas the second class has more repetitive tasks that allow less creativity.

However, the two problems laid out in section 3 would arise when this model would be accepted solely on the basis of the BIC value. First, the BIC identifies the best alternative out of the models presented, but it does not guarantee that no assumptions are violated, that is, that the model picks up all relevant aspects in the data. If this is not the case, the classification described in Table 4 could be faulty, and any further analysis to relate this classification to outcomes may also be affected negatively. Second, many alternative models can be specified, other than those with differing numbers of classes.

In all the estimated models, conditional independence of the observed items is assumed, which can be relaxed by allowing one or more covariances between the observed variables. Furthermore, the effect of the group-level LC on the observed variables is assumed to be fully mediated by lower level class membership. This too can be relaxed by allowing direct effects from the higher level latent variable on any of the items. The prohibitive difficulty of improving the model through trial and error, or even considering the option of estimating all possible models, now quickly becomes clear. When keeping the number of classes constant, there are 1,024 different combinations of allowable covariances and, for each of these combinations, another 32 possible combinations of direct effects. If the possibility of equating certain parameters to one another is also considered, this model can be adjusted in 17 factorial different ways.

Table 6. Residuals for the Three-class, Two-group-class Model, Covariance between Variation and Diverse

	Nonrepetitive	Creative	Diverse	Capacities	Variation
Creative	.101 (.642)				
Diverse	.602 (.104)	.022 (.514)			
Capacities	.871 (.184)	.001 (.938)	.178 (.264)		
Variation	.062 (.400)	.042 (.316)	.000 (.999)	.028 (.670)	
BVR_{group}	1.576 (.000)	.973 (.140)	.776 (.264)	1.037 (.194)	.842 (.312)
BVR_{pair}	1.523 (.000)	.294 (.130)	.128 (.296)	.256 (.138)	.011 (.780)

Note. Bayesian information criterion = 4,783.2.

To illustrate how the local fit measures may largely resolve the problem of identifying misfit without the need to estimate many additional models, the residual measures for the model with the lowest BIC are presented in Table 5 with bootstrapped p values for all BVR measures in parentheses. The regular BVR indicates that the variable measuring the diversity of a person's job shows some residual dependency with the variable measuring job variation, which substantively should come as no surprise. On the higher level, the BVR_{group} and BVR_{pair} also show assumption violations, whereby the repetitive and creative variables both show dependency between cases that is not captured by the model, as well as an incorrectly reproduced item distribution between the groups. So, even though it is the best alternative out of 30 models, the three-individual-level, two-group-level class model violates the three tested assumptions to some extent.

From Table 3, it can be concluded that improving this model is not achieved by increasing the number of classes. Inspecting the BVR measures for these models leads to the same conclusion, as a combination of problems on both levels of the model persists when increasing either the number of classes on the lower level, the higher level, or both.

Thus, to improve this model, a solution other than increasing the number of classes is required. Starting model improvements on the lower level of the model is often the most fruitful, as it is more likely that group-level dependency is introduced by having a wrong specification on the lower level than the reverse (Lukočienė et al. 2010). This is due to the higher level classification being partly determined by the classes on the lower level, as can be seen in equation (2).

Table 7. Residuals for the Three-class, Two-group-class Model, Covariance between Variation and Diverse and Direct Effect from Group-level Latent Variable on Repetitive

	Nonrepetitive	Creative	Diverse	Capacities	Variation
Creative	.004 (.922)				
Diverse	.737 (.082)	.068 (.204)			
Capacities	.962 (.180)	.026 (.732)	.046 (.670)		
Variation	.019 (.664)	.034 (.212)	.000 (.999)	.090 (.432)	
BVR_{group}	1.544 (.000)	1.405 (.000)	1.356 (.000)	1.194 (.040)	1.125 (.048)
BVR_{pair}	1.657 (.000)	.930 (.006)	1.325 (.002)	.458 (.048)	.280 (.070)

Note. Bayesian information criterion = 4,777.1.

Substantively, the significant dependency between the self-reported variation and diversity of work is sensible, and including a covariance between these two variables seems justified. As shown in Table 6, adding this covariance removes any problematic bivariate dependency on the lower level of the model.

Considering the BVR_{group} and BVR_{pair} statistics, the logical next step is to add a direct effect from the group-level latent variable on the repetitive variable. Such a direct effect is the most parsimonious solution in an attempt to capture more dependency and improve within-group model fit regarding the repetitive variable, adding only one parameter. Substantively too, there is evidence that the differences in repetitive work between teams reflect on that of the individual tasks (van Mierlo 2003).

After adding this effect, problems arise in all five variables, as depicted in Table 7, causing the model to no longer describe the within-team item distributions correctly; nor does it adequately capture the dependency between cases. Yet despite the large shift on the group level of the model, the lower level does not show any problems. The interpretations of the individual-level classes (not reported) also do not change, indicating that the problems are largely the result of a failure to capture team differences correctly. Given that there are problems with all five variables on the group level of the model, adding an additional group-level class is the best option here.

Adding a third group-level class indeed solves most problems on the higher level of the model, as can be seen from Table 8. In this model, the covariance between the variation and diverse variable, as well as the

Table 8. Residuals for the Three-class, Three-group-class Model, with Covariance between Variation and Diverse and Direct Effect from Group-level Latent Variable on Repetitive

	Nonrepetitive	Creative	Diverse	Capacities	Variation
Creative	.073 (.720)				
Diverse	.315 (.214)	.054 (.362)			
Capacities	.620 (.274)	.170 (.536)	.003 (.880)		
Variation	.046 (.378)	.114 (.154)	.000 (.999)	.053 (.546)	
BVR$_{group}$	1.041 (.046)	1.185 (.012)	.843 (.316)	1.150 (.054)	.931 (.290)
BVR$_{pair}$	.138 (.214)	.589 (.020)	.051 (.496)	.326 (.118)	.092 (.454)

Note. Bayesian information criterion = 4,768.9.

Table 9. Residuals for the Three-class, Three-group-class Model, with Covariance between Variation and Diverse and Direct Effects from Group-level Latent Variable on Repetitive and Creative

	Nonrepetitive	Creative	Diverse	Capacities	Variation
Creative	.001 (.950)				
Diverse	.530 (.108)	.085 (.192)			
Capacities	.837 (.186)	.005 (.858)	.090 (.454)		
Variation	.003 (.890)	.048 (.238)	.000 (.999)	.023 (.716)	
BVR$_{group}$	.771 (.260)	.739 (.452)	.927 (.112)	1.083 (.136)	.914 (.216)
BVR$_{pair}$	.016 (.628)	.011 (.696)	.202 (.174)	.280 (.150)	.014 (.728)

Note. Bayesian information criterion = 4,775.3.

direct effect on the repetitive variable, is retained. As a final adaptation, a direct effect from the group-level latent variable on the creative variable is added, following the BVR$_{group}$ value, and the reasoning that the structure of a team and the overall packet of tasks it realizes may have a direct effect on the creativity an employee has in accomplishing their share of the teamwork.

In Table 9, the BVR, BVR$_{group}$, and BVR$_{pair}$ residuals for the final model are presented. Further attempts to make this model more parsimonious result in models in which significant residuals are reintroduced. Note that the model chosen has a higher BIC value than the previous model (4,768.9 compared with 4,775.3), but given the focus on model fit and misfit, we opt for the less parsimonious model. This choice depends on the goal of the model specification. If the goal is to obtain high posterior probabilities, the model for which the residuals

Table 10. Profile for the Three-class, Three-group-class Model, with Covariance between Variation and Diverse and Direct Effects on Repetitive and Creative

	Group Class 1 Repetitive	Group Class 2 Defined	Group Class 3 Nonrepetitive	Class 1 Diverse	Class 2 Structure	Class 3 Creative	Overall
Nonrepetitive	.301	.316	.613	.554	.130	.233	.400
Creative	.660	.348	.674	.731	.077	.844	.557
Diverse	.822	.521	.754	.953	.209	.526	.698
Capacity	.753	.590	.707	.851	.444	.342	.683
Variation	.786	.506	.704	.962	.263	.000[a]	.665
Class 1	.784	.382	.678				
Class 2	.122	.529	.195				
Class 3	.095	.090	.127				
Prevalence	.352	.345	.302	.613	.284	.103	

Note. Bayesian information criterion = 4,775.3.
[a]Boundary solution.

are presented in Table 8 would be preferred (Burnham and Anderson 2002; Hamaker et al. 2011).

The profile of this final model is presented in Table 10. Comparing these results with those in Table 4, it becomes clear that the individual-level classification is practically identical to that obtained in the model with two group-level LCs and three individual-level LCs. On the group level, the additions to the model, an extra LC and two direct effects, led to splitting up the large first class from the initial solution. The second group-level class in this model is similar to the second class in the model presented in Table 4. The first class from Table 4, however, is split up into two classes. These two classes are rather similar when compared with each other, as they are when compared with the class from the first model, but with a large difference in degree of task repetition reported by the team members.

The results from Table 10 clearly show the difficulty in capturing team differences using team-level classes, as the first and third class differ only with respect to the degree of task repetition. Given that the group-level classes in the initial model are affecting the indicators only indirectly through the lower level LC, such a relatively small difference between teams may become obscured between other characteristics that the teams do have in common. That is, detecting these specific characteristics on the team level in a model without direct effects from the team-level latent variable also requires more classes on the lower level. Such an addition of LCs on either level is not warranted when inspecting the BIC values for these models, which are known to favor model parsimony. However, through the proposed BVR_{group} and BVR_{pair}, this lack of a direct effect between the group-level LC and the repetitiveness variable could be detected, as well as the subsequent need for an additional class on the group level.

Maybe more important, because of the improved fit and the possibility to test assumptions, the model arrives at different substantive results. In this instance, the added group-level class causes a separation based primarily on task repetitiveness. Given that the interest lies in relating the classes to job satisfaction or turnover intent as an outcome, the results may differ between the original model as depicted in Table 4, and the better fitting model arrived at in Table 10. When, for example, task repetitiveness on the team level is detrimental to employee job satisfaction, it would have been hard to distinguish as an important factor in the model with two group-level classes. It would, however, be visible in the model with three group-level classes in which a comparison between the first and third group-level classes would identify repetitiveness as an important factor.

Using the residuals as additional guidance now results in a model with substantial better fit that would likely not have been found when relying only on the BIC or comparable criteria. Both the proposed BVR_{group} and BVR_{pair}, in combination with the BVR, allow the detection of the initial assumption violations, and they identify not only which part of the model but also which specific parameters may prove problematic. Misfit can be pinpointed and tested, allowing far more informed and directed model adjustments, which may lead to different, more thoroughly tested, substantive results.

It must be pointed out that in the application, the residuals were used as guidance for illustration. However, comparable with many residual measures as well as modification indices, the measures are by no means tied to a certain solution and indicate only badness of fit and assumption violations. That is to say, model adjustments should be theoretically driven, and blind adjustments to the model with the mere goal of improving the fit should be discouraged as a poor research practice that may, for example, lead to capitalization on chance (e.g., see Kaplan 1990; MacCallum, Roznowski, and Necowitz 1992).

Table 11. Misspecified Model: Proportion of Replicates with Bootstrap p Values $< .05$

	Nonrepetitive	Creative	Diverse	Variation	Capacities
BVR_{group}	.788	.064	.050	.376	.052
BVR_{pair}	.872	.048	.032	.378	.042

5. SIMULATION

As a proof of concept, a small simulation study is presented in this section. The final model from the application is used as the population model in the two conditions presented, which contains three classes on both levels, a direct effect on the creative and nonrepetitive variables, as well as a covariance between the diverse and variation variables. The exact logit parameters for this model can be found in Appendix C, available in the online journal. Latent GOLD 5.0 is used to generate 500 replicate data sets, which are subsequently analyzed using a misspecified model that does not contain the covariance and direct effects, and it has only two group-level classes as well as the correct model. Identical to the application, bootstrapped p values are obtained on the basis of 500 iterations.

For the misspecified model, the expectation is that the BVR_{group} detects the absence of the two direct effects. Table 11 shows that the power to detect one of these misspecifications indeed turns out to be high (.788). The second direct effect is not detected. However, the logit parameters of the direct effect on the creative variable are minute (effect coded -0.186 and -0.005). In addition, the type I error for the three other variables does not differ significantly from the alpha level. The power to detect the missing class through the BVR_{pair} is equally high. Tables 4 and 10 show that the major change between the two- and three-class models is a separation of classes solely on the basis of the nonrepetitive variable. When the classes are not separated, BVR_{pair} detects the residual dependency between respondents on this particular variable. That the values for the variation variable are higher than the nominal alpha levels can be explained by the fact that the logit parameter in the population model is extremely high. As a result, in an attempt to explain maximum group-level dependency, the model underestimates the dependency resulting from the variation

Table 12. Correct Model: Proportion of Replicates with Bootstrap p Values $< .05$

	Nonrepetitive	Creative	Diverse	Variation	Capacities
BVR_{group}	.034	.028	.044	.034	.040
BVR_{pair}	.046	.052	.076	.046	.048

variable to be able to explain the group-level dependency that results from the nonrepetitive and creative variables.

Table 12 shows the proportion of rejections under the correctly specified model, that is, the type I error. Satisfactory error rates should be close to the nominal .05 level. This is the case in most instances, but BVR_{pair} for the diverse variable, as well as BVR_{group} for the creative variable, differs significantly from .05. Whether this is the result of the additional direct effect and covariance for these variables and is a systematic issue requires further study and a more extensive simulation study. The absolute differences appear to be small, however.

To summarize, although the simulation presented here is necessarily limited in scope, the power of the introduced measures is high, and the type I error rates are close to their nominal levels. This simulation therefore demonstrates that our measures' performance is satisfactory in the case of the application discussed, and it provides proof of concept from which future investigations may depart.

6. DISCUSSION

Several problems occur when using only global fit statistics or information criteria for model selection in multilevel LC analysis. Because of the lack of local fit statistics, potential model misfit may go unnoticed, and there is no information available regarding how a model might be adjusted and improved. Therefore two new local fit statistics, BVR_{group} and BVR_{pair}, are proposed, which test individual areas of the model and as such help in determining which areas of the model are problematic and how a model can best be improved. In conjunction with the standard BVR, they also allow the two local independence assumptions central to multilevel LC models to be inspected and tested. Computation of both the BVR_{group} and BVR_{pair} is already implemented in the user-friendly Latent GOLD 5.0 software package.

By using the BVR_{group} and BVR_{pair} as additional guidance to test and improve a multilevel LC model, it is shown that they enhance the ease with which fruitful model adjustments can be found. The model obtained by relying on the two residuals has better global fit and is known to better adhere to the local independence assumptions. The usefulness of the residuals is further emphasized by the change in substantive results between the initial model selected through the BIC and the latter model as improved through the use of the proposed statistics. That is, the misfit that is detected in this instance is not a mere misspecification against which the model is robust but actually distorts model-based conclusions.

That these model improvements can be found using a stepwise approach, and that such an approach may lead to finding relations and effects that would otherwise go unnoticed, does not, however, mean that these improvements lead to the true population model. It should be noted that there is a substantial risk for capitalization on chance, and in practice, such an approach should be used in conjunction with a form of replication such as cross-validation. That is, the residuals merely indicate local misfit and do not point to a given solution for that particular misfit.

Nonetheless, in this case important sources of misfit that affect the results have been picked up by the two residuals. Still, this paper serves as an introduction, and a more in-depth simulation study is lacking. Such a future study would not focus as much on the type I error rates, as the process of p value bootstrapping is identical to that of the BVR for which it has been extensively tested (Oberski et al. 2013). Rather, it would focus on the consistency with which misspecification is detected under different circumstances and in more complex models, such as models incorporating covariates.

An additional extension that does require future work is to develop a similar residual for LC models for longitudinal data, in which dependencies can be assumed to take on the form of autocorrelation structures. Furthermore, the use of the BVR_{group} and BVR_{pair} may be studied for different methods and models that could also benefit from these statistics (e.g., see Varriale and Vermunt 2012). The residuals are developed with the aim of testing the local fit of multilevel LC models, but they can be applied to all cases in which categorical multilevel data are used. The observed frequencies would be identical when applying the residuals to other methods, and only the expected frequencies would need to be obtained from the alternative approach.

Acknowledgments

We thank the reviewers and the editor for their helpful comments.

Funding

This work was supported by The Netherlands Organization for Scientific Research (grant 453-10-002). Dr. Oberski was also supported by The Netherlands Organization for Scientific Research (grant 451-14-017).

Note

1. The Latent GOLD 5.0 software package does not estimate these probabilities directly but estimates the logit parameters for the logistic parameterization of the model. See Vermunt (2003).

References

Bartholomew, David J. and Shing On Leung. 2002. "A Goodness of Fit Test for Sparse 2^P Contingency Tables." *British Journal of Mathematical and Statistical Psychology* 55(1):1–15.

Bennink, Margot, Marcel A. Croon, Jos Keuning, and Jeroen K. Vermunt. 2014. "Measuring Student Ability, Classifying Schools, and Detecting Item Bias at School Level Based on Student-level Dichotomous Items." *Journal of Educational and Behavioral Statistics* 39(3):180–201.

Bishop, Yvonne M. M., Stephen E. Fienberg, and Paul W. Holland. 1975. *Discrete Multivariate Analysis: Theory and Practice*. Cambridge, MA: MIT Press.

Burnham, Kenneth P. and David R. Anderson. 2002. *Model Selection and Multimodel Inference: A Practical Information-theoretic Approach*. 2nd ed. New York: Springer.

Clogg, Clifford C. and Leo A. Goodman. 1984. "Latent Structure Analysis of a Set of Multidimensional Contingency Tables." *Journal of the American Statistical Association* 79(388):762–71.

Fila, Marcus J., Lisa S. Paik, Rodger W. Griffeth, and David Allen. 2014. "Disaggregating Job Satisfaction: Effects of Perceived Demands, Control, and Support." *Journal of Business and Psychology* 29(4):639–49.

Finch, W. Holmes and Kendall C. Bronk. 2011. "Conducting Confirmatory Latent Class Analysis Using Mplus." *Structural Equation Modeling: A Multidisciplinary Journal* 18(1):132–51.

Glas, Cees A. W. 1999. "Modification Indices for the 2-PL and the Nominal Response Model." *Psychometrika* 64(3):273–94.

Goodman, Leo A. 2002. "Latent Class Analysis: The Empirical Study of Latent Types, Latent Variables, and Latent Structures." Pp. 3–56 in *Applied Latent Class Analysis*, edited by J. A. Hagenaars and A. L. McCutcheon. Cambridge, UK: Cambridge University Press.

Gunter, Barrie and Adrian Furnham. 1996. "Biographical and Climate Predictors of Job Satisfaction and Pride in Organization." *Journal of Psychology: Interdisciplinary and Applied* 130(2):193–208.

Hamaker, Ellen L., Pascal van Hattum, Rebecca M. Kuiper, and Herbert Hoijtink. 2011. "Model Selection Based on Information Criteria in Multilevel Modeling." Pp. 231–56 in *Handbook of Advanced Multilevel Analysis*, edited by J. J. Hox and J. K. Roberts. New York: Routledge.

Harrell, Paul T., Brent E. Mancha, Hanno Petras, Rebecca C. Trenz, and William W. Latimer. 2012. "Latent Classes of Heroin and Cocaine Users Predict Unique HIV/ HCV Risk Factors." *Drug and Alcohol Dependence* 122(3):220–27.

Hox, Joop J. 2010. *Multilevel Analysis: Techniques and Applications*. New York: Routledge.

Kaplan, David. 1990. "Evaluating and Modifying Covariance Structure Models: A Review and Recommendation." *Multivariate Behavioral Research* 25(2):137–55.

Lambert, Eric G., Nancy Lynne Hogan, and Shannon M. Barton. 2001. "The Impact of Job Satisfaction on Turnover Intent: A Test of a Structural Measurement Model Using a National Sample of Workers." *Social Science Journal* 38(2):233–50.

Langeheine, Rolf, Jeroen Pannekoek, and Frank van de Pol. 1996. "Bootstrapping Goodness-of-fit Measures in Categorical Data Analysis." *Sociological Methods and Research* 24(4):492–516.

Laudy, Olav, Mark Zoccolillo, Raymond H. Baillargeon, Jan Boom, Richard E. Tremblay, and Herbert J. A. Hoijtink. 2005. "Applications of Confirmatory Latent Class Analysis in Developmental Psychology." *European Journal of Developmental Psychology* 2(1):1–15.

Liu, Dong, Terence Mitchell, Thomas Lee, Brooks Holtom, and Timothy Hinkin. 2012. "When Employees Are out of Step with Coworkers: How Job Satisfaction Trajectory and Dispersion Influence Individual- and Unit-level Voluntary Turnover." *Academy of Management Journal* 55(6):1360–80.

Lukočiené, Olga, Roberta Varriale, and Jeroen K. Vermunt. 2010. "The Simultaneous Decision(s) about the Number of Lower- and Higher-level Classes in Multilevel Latent Class Analysis." Pp. 247–83 in *Sociological Methodology*, vol. 40, edited by Tim Futing Liao. Hoboken, NJ: Wiley-Blackwell.

Lukočiené, Olga and Jeroen K. Vermunt. 2010. "Determining the Number of Components in Mixture Models for Hierarchical Data." Pp. 241–49 in *Advances in Data Analysis, Data Handling and Business Intelligence*, edited by A. Fink, B. Lausen, W. Seidel, and A. Ultsch. Heidelberg, Germany: Springer.

MacCallum, Robert C., Mary Roznowski, and Lawrence B. Necowitz. 1992. "Model Modifications in Covariance Structure Analysis: The Problem of Capitalization on Chance." *Psychological Bulletin* 111(3):490–504.

Mavridis, Dimitrios, Irini Moustaki, and Martin Knott. 2007. "Goodness-of-fit Measures for Latent Variable Models for Binary Data." Pp. 135–61 in *Handbook of Latent*

Variable and Related Models, edited by S. Lee. Amsterdam, the Netherlands: Elsevier.

Muthén, Bengt O. and Tihomir Asparouhov. 2009. "Multilevel Regression Mixture Analysis." *Journal of the Royal Statistical Society, Series A* 172(3):639–57.

Oberski, Daniel L., Geert H. van Kollenburg, and Jeroen K. Vermunt. 2013. "A Monte Carlo Evaluation of Three Methods to Detect Local Dependence in Binary Data Latent Class Models." *Advances in Data Analysis and Classification* 7(3):267–79.

Ryu, Ehri and Stephen G. West. 2009. "Level-specific Evaluation of Model Fit in Multilevel Structural Equation Modeling." *Structural Equation Modeling: A Multidisciplinary Journal* 16(4):583–601.

Snijders, Tom A. B. and Johannes Berkhof. 2008. "Diagnostic Checks for Multilevel Models." Pp. 141–75 in *Handbook of Multilevel Analysis*, edited by J. de Leeuw and E. Meijer. New York: Springer.

Snijders, Tom A. B. and Roel J. Bosker. 2012. *Multilevel Analysis: An Introduction to Basic and Advanced Multilevel Modeling*. Thousand Oaks, CA: Sage.

van Mierlo, Heleen. 2003. "Self-managing Teamwork and Psychological Well-being." PhD dissertation, Department of Technology Management, Eindhoven University of Technology, Eindhoven, the Netherlands.

van Mierlo, Heleen, Christel G. Rutte, Michiel A. J. Kompier, and Hans A.C.M. Doorewaard. 2005. "Self-managing Teamwork and Psychological Well-being: Review of a Multilevel Research Domain." *Group and Organizational Management* 30(2):211–35.

van Veldhoven, Marc, Theodorus F. Meijman, Jacobus P. J. Broersen, and R. J. Fortuin. 1997. *Handleiding VBBA: Onderzoek naar de Beleving van Psychosociale Arbeidsbelasting en Werkstress met Behulp van de Vragenlijst Beleving en Beoordeling van Arbeid* [VBBA Manual: An Investigation of Perceptions of Psychosocial Workload and Work Stress by Means of the Dutch Questionnaire on the Experience and Evaluation of Work]. Amsterdam, the Netherlands: SKB.

Varriale, Roberta and Jeroen K. Vermunt. 2012. "Multilevel Mixture Factor Models." *Multivariate Behavioral Research* 47(2):247–75.

Vasdekis, Vassilis, Silvia Cagnone, and Irini Moustaki. 2012. "A Composite Likelihood Inference in Latent Variable Models for Ordinal Longitudinal Responses." *Psychometrika* 77(3):425–41.

Vermunt, Jeroen K. 2003. "Multilevel Latent Class Models." Pp. 213–39 in *Sociological Methodology*, vol. 33, edited by Ross M. Stolzenberg. Boston, MA: Blackwell.

Vermunt, Jeroen K. 2004. "An EM Algorithm for the Estimation of Parametric and Nonparametric Hierarchical Nonlinear Models." *Statistica Neerlandica* 58(2): 220–33.

Vermunt, J. K. 2005. "Mixed-effects Logistic Regression Models for Indirectly Observed Outcome Variables." *Multivariate Behavioral Research* 40(3):281–301.

Vermunt, Jeroen K. 2008. "Latent Class and Finite Mixture Models for Multilevel Data Sets." *Statistical Methods in Medical Research* 17(1):33–51.

Vermunt, Jeroen K. and Jay Magidson. 2005. *Technical Guide for Latent Gold 4.0: Basic and Advanced*. Belmont, MA: Statistical Innovations.

Vermunt, Jeroen K. and Jay Magidson. 2013. *Technical Guide for Latent Gold 5.0: Basic, Advanced and Syntax*. Belmont, MA: Statistical Innovations.

Yuan, Ke-Hai and Peter M. Bentler. 2007. "Multilevel Covariance Structure Analysis by Fitting Multiple Single-level Models." Pp. 53–82 in *Sociological Methodology*, vol. 37, edited by Yu Xie. Boston, MA: Blackwell.

Author Biographies

Erwin Nagelkerke is a PhD candidate in the Department of Methodology and Statistics at Tilburg University. He has a bachelor's degree in sociology and a master's degree in social and behavioral sciences from Tilburg University. His research focuses on fit evaluation of latent class models.

Daniel L. Oberski is an assistant professor in the Department of Methodology and Statistics at Tilburg University. His research focuses on model fit evaluation in latent variable model and the application of latent variable models to estimate measurement quality of survey questions and administrative data.

Jeroen K. Vermunt is a full professor in the Department of Methodology and Statistics at Tilburg University. He holds a PhD in social sciences from Tilburg University. He has published extensively on categorical data techniques, methods for the analysis of longitudinal and event history data, latent class and finite mixture models, and latent trait models. He is the codeveloper (with Jay Magidson) of the Latent GOLD software package.

Sociological Methodology
2016, Vol. 46(1) 283–318
© American Sociological Association 2015
DOI: 10.1177/0081175015602484
http://sm.sagepub.com

ॐ 9 ॐ

MODELING INCIDENCE OF NUPTIALITY

Juha Alho*

Abstract

The statistical description of the formation of marriages is hampered by the fact that the intensity of marriage of one sex depends on the available supply of potential spouses from the other. Unlike the situation that occurs in the study of fertility, there is no reason to give a preference to either of the sexes. To address the fundamental problem caused by the two-sex nature of the process, a solution is proposed that considers the sexes jointly. The solution relies on a novel use of generalized averages. The model is formulated in stochastic terms, and it is parametrized in terms of the overall level of nuptiality, the relative propensity of nuptiality by age, and the mutual relative attraction of spouses at different ages. Although national statistics collection relies on data aggregated by age groups, the models are formulated at individual levels to show how estimation could also be carried out in small populations. In particular, examples of how maximum likelihood estimation can be carried out are given for specific parametric models. The methods are illustrated by both monthly and annual nuptiality data from Finland.

Keywords

Finland, generalized averages, maximum likelihood, power transforms, two-sex problem

*University of Helsinki, Finland

Corresponding Author:
Juha Alho, Unioninkatu 35, 00014 University of Helsinki, Finland
Email: juha.alho@helsinki.fi

1. INTRODUCTION

The so-called occurrence/exposure rates form the basis of many of the most frequently used demographic measures. For example, the survival probabilities and life expectancies displayed in a life table are computed from age-specific mortality rates. These are obtained by dividing the number of deaths by an estimate of the person-years lived in each age group. Similarly, age-specific fertility rates are summed to form the total fertility rate, or averaged with chosen weights to get age-standardized fertility measures. In principle, the occurrence of marriages, or *nuptiality*, can also be described in terms of age-specific marriage rates.

Mortality rates are typically computed separately for men and women, or else the sexes are combined. Fertility rates are almost exclusively computed for women. This is probably due to the ease of data collection: a child's mother is nearly always known, but the father may not be. The case of nuptiality is fundamentally different, however. In human populations there is no reason to give preference to either sex, and an analysis from the point of view of either sex is interesting. But the consideration of the two sexes jointly leads to surprising problems of coherence that form the so-called two-sex problem of demography.

Similar problems of coherence appear in other fields of sociological interest, but their practical significance depends on the context. For instance, despite the difficulties of data collection, the male point of view has recently aroused interest in the study of fertility (cf. Li 2011). When total fertility rates are considered separately for women and men, similar problems of coherence arise as in the case of nuptiality, and they can be tackled with the methods proposed in this paper. Survey data on heterosexual behavior are a related example (Smith 2006). In a closed population, the average number of partners for women and men should lead to the same total number of reported partners for the female and male populations, but a major source of discrepancies involve misreporting that can only partially be rectified with our methods. In the study of household trends, one can consider household status (such as living alone, married, or cohabiting) as a characteristic of an individual. The numbers of those living in unions should then match between women and men (cf. Alho and Keilman 2010). But because the prevalence rates or proportions needed in such analyses are conceptually different from the incidences (or rates) considered in this paper, the methods proposed here need to be adjusted. Finally, some similar-sounding problems are only ostensibly related to this work. In enterprise demography, one studies mergers and splits of firms in addition to their births and deaths (e.g., Carroll and Hannan 2000). Yet there is no notion comparable with gender, and the two-sex aspect is missing. In fact, mergers are more akin to same-sex marriages that are also different from the heterosexual marriages considered here.

The essence of the two-sex problem does not involve age. Consider a toy example. Suppose that there are three women and four men, and one marriage occurs. Then, the female marriage rate is 1 in 3, and the male marriage rate is 1 in 4. We are left with two women and three men. For the purpose of our argument, this would suffice. But to make explicit that the populations at risk can also change for other reasons, suppose that one man migrates into the population, which now has two women and four men. Suppose that the incidence of marriage does not change. Then, from the female point of view, the expected number of marriages during the next period would be 2 in 3, but from the male point of view it would be 1. Or, the sex-specific marriage rates are not coherent. The purposes of the paper are to contribute to a better understanding of what the options are for computing coherent nuptiality rates and to propose a practical solution that has some merits over other proposals in the context of producing statistical estimates of nuptiality trends.

There is a large body of literature on the two-sex problem (e.g., Caswell and Weeks 1986; Choo and Siow 2006a, 2006b; McFarland 1972, 1975; Pollard 1975; Pollard and Höhn 1993; Schoen 1981), but no proposal for its solution has found universal acceptance. I develop a rather large class of marriage models for age-structured populations in which the two sexes are considered jointly, and the problem of incoherence does not arise. Different choices for the fundamental measure of nuptiality lead to different definitions of the population exposed, and vice versa. An important empirical finding is that the solution is not very sensitive to the particular choice.

I argue in favor of a particular solution (in the toy example, it results in 0.78 as the expectation) on the grounds of its relative simplicity. In addition to the choice of the fundamental measure of nuptiality, there is an age-specific measure of relative propensity of marriage for both sexes, and there is a measure of preference across age-sex groups. The three measures are logically independent.

The models developed here are *not* behavioral, in the sense that they would be based on social, economic, or biological theories of actual behavior (cf. Bergstrom and Lam 1988; Bozon and Heran 1989; Choo and Siow 2006a, 2006b; Dagsvik, Brunborg, and Flaatten 2001; Henry 1972; Lee, Engen, and Sæther 2008). In particular, there is no notion of utility or equilibrium in the models. Yet, the models developed here do have competition aspects of the kind one would expect in human populations (e.g., Pollak 1990; Pollard and Höhn 1993). In this respect, this model differs from the model of Schoen (1981:206), which assumes "zero-spillover mating" in the terminology of Pollak (1990). On the other hand, the proposed approach has many similarities to the recent proposals of Matthews and Garenne (2013a, 2013b). In particular, both approaches involve marginal age-specific marriage intensities, there is a bivariate function to describe the mutual preferences across age, and an iterative numerical solution for generating marriages when populations at risk change.

The primary difference is that the model proposed here is stochastic, defined in terms of rates and in continuous time, whereas that of Matthews and Garenne (2013a) is deterministic, given in terms of counts and in discrete time. The present model gives an exact probability mechanism of how marriages occur and how the marriage market is changed by a removal of a pair, at an individual level. The second difference is that I provide a whole family of alternative measures of a population at risk, and I can thereby compare their performance in practice. The third difference is that I have access to individual-level data on pair formation, which allows me to empirically estimate both parametric and nonparametric models for the incidence of new marriages using maximum likelihood. Both types of models provide guidelines for finite population microsimulation. On the other hand, the models considered in this paper involve only the nuptiality processes, so the complex considerations one must tackle when the whole process of population renewal is involved (e.g., Kesten 1970a, 1970b; Matthews and Garenne 2013b) are avoided.

The source of randomness in our models is not sampling variability. In fact, the empirical data I use come from population registries, and as such they correspond to a census rather than a probability sample. As is the case in the Bayesian formulations of sampling theory, one can think of the demographic data as being a sample from a superpopulation. Inference is then carried out under the assumed *model* rather than on the basis of the randomization mechanism underlying probability samples (cf. Särndal 1978). This is the approach usually adopted in statistical demography (cf. Alho and Spencer 2005).

I develop the model stepwise in sections 2 to 4. In section 2, the connection between the fundamental measure and the definition of the exposed population is considered in a population with no age. In section 3, the model is extended to involve age-specific propensities of marriage. Section 4 presents a stochastic matching mechanism that involves the mutual relative preferences of the sexes by age, with section 4.2 arriving at an explicit mechanism of how marriages are formed. Section 5 considers maximum likelihood estimation. Section 6 discusses the nature of competition effects that these models entail. Section 7 applies the model to monthly nuptiality data from Finland in 2012 and to annual data from 1988 through 2012. Section 8 concludes the paper by considering the limitations and possible extensions of the findings. Mathematical details, kept to a minimum in the paper, are given in the four appendices available in the online journal.

2. LEVEL OF NUPTIALITY AND THE POPULATION AT RISK

2.1. *Fundamental Measure of Nuptiality and Generalized Averages*

The source of inconsistency in the toy example in section 1 was that the notion of the level of nuptiality was not explicitly defined. Because this choice is independent of age, I now maintain simple notation and consider a population consisting of $V_j > 0$ individuals of the two sexes, whose intensities of marriage are correspondingly $\Lambda_j > 0, j = 1, 2$. The assumption I make is that the marriages occur only among the members of this population (i.e., for estimation purposes, I consider marriages to those living outside the population as censorings), so the intensities of marriage must satisfy the identity

$$\Lambda_1 V_1 = \Lambda_2 V_2. \tag{1}$$

Thus, the two intensities cannot be independently estimated. Instead, I assume that both are functions of a single fundamental measure.

I define the fundamental level of nuptiality Λ in an *implicit* manner, as a *generalized average* of the two intensities. This, together with equation (1), determines the values $\Lambda_j > 0, j = 1, 2$ that will be considered derived measures. The details depend on the choice of the generalized average, but in section 2.4, I provide explicit examples of how the Λ_j values depend on Λ and the V_j values.

More formally, suppose a function g has the inverse function g^{-1}. Let $0 \leq q \leq 1$ be a weight. I then assume that the *fundamental measure of nuptiality* satisfies the relationship

$$\Lambda = g^{-1}(qg(\Lambda_1) + (1 - q)g(\Lambda_2)). \tag{2}$$

That is, to get Λ, first map the Λ_j values to some other scale, then take the average, and map the result back to the original scale. Because the fundamental measure determines the overall (expected) number of marriages, it is the counterpart of the latter number in Matthews and Garenne (2013a).

In nuptiality applications, I use unweighted averages with $q = 1/2$, but there are examples of the use of uneven weights. For example, in ecological studies it is frequently the case that all females mate, but only one male, or only a few of them mate. In this case the number of offspring is primarily limited by fluctuations of the female population.[1] On the other hand, the allowance of uneven weighting is helpful in understanding the nature of the generalized averages.

The intuitive interpretation of the fundamental measure is a bit tricky. I return to it in section 3.1.

There are infinitely many values of g that one could use in equation (2), but I show that a broad class exists that yields tractable solutions (cf. Hadeler, Waldstätter, and Wörz-Busekros 1988:639).

2.2. General Properties

Assume that $g : \mathbb{R}_+ \to \mathbb{R}$ is a strictly monotone, increasing or decreasing function, with inverse $g^{-1} : R_g \to \mathbb{R}_+$, where R_g is the range of g. Consider any two real numbers $x_1 > 0$, $x_2 > 0$, and define their generalized average as $\bar{x}_g = g^{-1}(qg(x_1) + (1 - q)g(x_2))$. In the applications presented here, they will be either intensities of marriage or counts of population at risk. For any such g, if $x_1 = x_2 = \mathrm{x}$, then $\bar{x}_g = x$. Similarly, if $x_1 < x_2$, then $x_1 \leq \bar{x}_g \leq x_2$. Furthermore, $\bar{x}_g$ is an increasing function of both of its arguments. Thus, it has the basic properties expected from a mean value.

The generalized average is also invariant with respect to affine transformations. To see this, suppose that there is another function

$G = a + bg$, where a and b are arbitrary real numbers with $b \neq 0$. If $y = G(x)$, then $y = a + bg(x)$, so $G^{-1}(y) = g^{-1}[(y-a)/b]$. By substituting in $y = qG(x_1) + (1 - q)G(x_2)$, we find that $\bar{x}_G = \bar{x}_g$.

One qualitative implication of the invariance property is that the possible convexity or concavity of g is not important, because if, say, g is convex, then $-g$ is concave, yet both lead to the same generalized average.

2.3. *Averages Based on Power Transforms*

As noted in Hadeler et al. (1988:639–40), most of the earlier proposals for solving the two-sex problem can be seen as different choices of generalized averages based on the *power transform*.

Consider a class of functions $f_p : \mathbb{R}_+ \to \mathbb{R}$, defined as $f_p(x) = (x^p - 1)/p$ for $p \neq 0$ and, by continuity, as $p \to 0$, we define $f_p(x) = \log(x)$ for $p = 0$. Statisticians will recognize this as the so-called Box-Cox transformation that is often used in regression analysis (Box and Cox 1964). Economists, on the other hand, may see this as the form of the utility function displaying *constant relative risk aversion* (cf. Arrow 1971).

The invariance property implies that for the purpose of understanding generalized averages that are induced by the power transform, one needs to look at only two cases: $g_0(x) = \log(x)$ (a natural logarithm; a direct calculation shows that the logarithm to any other base will yield the same generalized average) and $g_p(x) = x^p$, $p \neq 0$.

When $y = g_0(x) = \log(x)$, the inverse is $g_0^{-1}(y) = \exp(y)$ and $\bar{x}_{g_0} = x_1^q x_2^{1-q}$. For $q = 1/2$, this is the usual *geometric mean*, but, for example, Goodman (1968) considered weighted averages.

When $y = g_p(x) = x^p$, the inverse is $g_p^{-1}(y) = y^{1/p}$ and $\bar{x}_{g_p} = (qx_1^p + (1 - q)x_2^p)^{1/p}$ for all $p \neq 0$. (For $p \geq 1$, this is an "L^p norm" of the vector $[x_1, x_2]$. On the other hand, Hadeler et al. [1988:639–40] discussed the case $p \leq 0$ only. As seen below, all real values, including $\pm\infty$, are meaningful for the problem at hand.) For $p = 1$, we obtain $\bar{x}_{g_1} = qx_1 + (1 - q)x_2$, which means that for $q = 1/2$ this is the *arithmetic mean*. For $p = -1$, we obtain $\bar{x}_{g_{-1}} = x_1 x_2/(qx_2 + (1 - q)x_1)$, which means that for $q = 1/2$ this is the *harmonic mean*.[2]

Furthermore, let us fix x_1 and x_2, and consider the function $H(p) = \bar{x}_{g_p}$, where g_p is a member of the power function family. I postpone the calculations to Appendix A in the online journal, but note here that this is an increasing function of p with limits $\bar{x}_{g_p} \to max\{x_1, x_2\}$ as

$p \to +\infty$, and $\bar{x}_{g_p} \to min\{x_1, x_2\}$ as $p \to -\infty$. This result shows that the power transformation family is capable of producing the whole range of reasonable mean values for different choices of p. It also proves the well-known result that harmonic mean $\leq$ geometric mean $\leq$ arithmetic mean, as these correspond to $p = -1, 0,$ and 1, respectively.

2.4. *Relationship between Fundamental Measure and Population at Risk*

Armed with the family of generalized averages based on the power transform, let us turn again to equation (1). Assuming that $p = 0$, then $g(x) = \log(x)$ and $\Lambda = \Lambda_1^q \Lambda_2^{1-q}$. It follows, for example, that $\Lambda_2 = \Lambda^{1/(1-q)} / \Lambda_1^{q/(1-q)}$. Inserting this into equation (1), we see after a short calculation that $\Lambda_1 = \Lambda(V_2/V_1)^{1-q}$. Similarly, we see that $\Lambda_2 = \Lambda(V_1/V_2)^q$. This shows that the implicit construction I have used to define Λ actually does define the derived measures $\Lambda_j, j = 1, 2$.

Furthermore, the overall intensity (equation 1) can be written as ΛV, where the population at risk is

$$V = V_1^q V_2^{1-q}. \tag{3}$$

Thus, for $q = 1/2$, the population at risk is the geometric mean of the population sizes. Under this model, if Λ does not change over time, but the populations at risk change in such a way that their geometric mean remains constant, then the overall intensity of marriage stays at ΛV, although the Λ_j values can change with the relative sizes of the populations. In this sense the Λ_j values are indeed derived measures, and Λ is fundamental.

The result for geometric means is a special case of a general relationship. Suppose that $g(x) = x^p$ for some $p \neq 0$. Equation (2) can equivalently be written as $g(\Lambda) = qg(\Lambda_1) + (1 - q)g(\Lambda_2)$. From equation (1), one can solve $\Lambda_2 = \Lambda_1 V_1/V_2$. Substituting into the equation for $g(\Lambda)$ obtains

$$\Lambda_1 = \Lambda/(q + (1 - q)V_1^p/V_2^p)^{1/p}, \quad \Lambda_2 = \Lambda/(1 - q + qV_2^p/V_1^p)^{1/p}. \tag{4}$$

This makes explicit the way in which the implicit construction for the fundamental measure determines the sex-specific total nuptiality rates for both sexes, when $p \neq 0$.

From equation (1), the result $\Lambda_1 V_1 = \Lambda V$ is then obtained for the intensity of marriage, where

$$V = \frac{V_1 V_2}{((1-q)V_1^p + qV_2^p)^{1/p}} \tag{5}$$

is the size of the population at risk. Emulating the notation above, we write $\bar{\Lambda}_{g_p}$ and $\bar{V}_{g_p}$ for the generalized averages of Λ_j values and V_j values, respectively. The result contained in equations (3) and (5) can then succinctly be stated as a duality:

$$\Lambda = \bar{\Lambda}_{g_p} \Leftrightarrow V = \bar{V}_{g_{-p}} \tag{6}$$

for all $p \in \mathbb{R}$. Thus, the choice of a generalized average parameter p for the intensity is equivalent to choosing the generalized average with parameter $-p$ for the population at risk.

For illustration, if $p = 1$ and $q = 1/2$, the fundamental measure of nuptiality is the arithmetic mean and the size of the population at risk is the harmonic mean, and vice versa.

When applied to the toy example, the generalized average with $p = 1$ and $q = 1/2$ leads to the estimate $\Lambda = (1/3 + 1/4)/2 = 7/24 \approx 0.2917$, the population at risk during the second period is $2 \times 2 \times 4/(2 + 4) = 8/3 \approx 2.667$, and if the fundamental measure does not change, the expected number of marriages in the next period is $7/24 \times 8/3 = 56/72 \approx 0.778$. On the other hand, if $p = 0$ and $q = 1/2$, the fundamental measure is $\Lambda = \sqrt{1/12} \approx 0.2887$, the population at risk is $\sqrt{8} \approx 2.828$, and the expected number of marriages becomes $\sqrt{8/12} \approx 0.816$. From the perspective of demographic applications, the difference in the relative size of the exposed male and female populations in the toy example is very large. Yet the difference in the expected count is rather small.

Recalling the monotonicity properties of $\bar{x}_{g_p}$, we see that $V \to \min\{V_1, V_2\}$ as $p \to +\infty$ and $V \to \max\{V_1, V_2\}$ as $p \to -\infty$.[3] Continuing with the toy example, for $p = +\infty$ we have $\Lambda = 1/3$, and the population at risk during the next period is 2, so the expected number of marriages is $2/3 \approx 0.667$, and for $p = -\infty$ we have $\Lambda = 1/4$, and the expected number of marriages is $4/4 = 1$. Even though the populations at risk can, in principle, change in an arbitrary way during the two occasions, it can be shown that the expected marriages are always given by the extreme values of p. Thus, the expected numbers are within $[2/3, 1]$ for *all* generalized averages in this case.

3. RELATIVE PROPENSITY OF MARRIAGE BY AGE

In human populations, the propensity of marriage depends heavily on age. In industrialized countries, most marriages occur between the ages of 20 and 40 years. To account for the age dependence, I consider continuous age and assume that marriages can occur in ages $0 \leq x \leq \omega < +\infty$.[4] The assumption of a finite upper bound avoids mathematical considerations that are not important for our discussion.

3.1. *Modeling Relative Propensity*

To represent the relative propensity of marriage, I define functions $\phi_j(x_j)$, for $0 \leq x_j \leq \omega, j = 1, 2$, and assume that these are probability densities, with

$$\int_0^\omega \phi_j(x_j)dx_j = 1, \quad j = 1, 2. \tag{7}$$

The normalization has no implications on the empirical analysis, and it is merely used for identifiability. To avoid unnecessary mathematical considerations, I also assume that the densities are continuous. The functions ϕ_j will represent the *relative propensity of marriage* in the sense that if there are two individuals of sex j that are in ages x_j and y_j, and one of them is known to marry, the probability that is the one with age x_j is $\phi_j(x_j)/[\phi_j(x_j) + \phi_j(y_j)]$. These probabilities do not depend on others present in the population, and the functions ϕ_j will be basic parameters in our model, in addition to the overall level Λ. For comparison, Matthews and Garenne (2013a) specified corresponding age-specific reference counts that determine the marginal marriage counts for both sexes by age.

I take time to be continuous and assume that at exact time t, the total marriage intensity of sex $j = 1, 2$ is $\Lambda_j(t) > 0$, and the *age-specific intensity of marriage* is $\Lambda_j(x_j, t) = \Lambda_j(t)\phi_j(x_j)$. This is the second interpretation of the functions ϕ_j, and for this interpretation the normalization (equation 7) is necessary.

> *Example 1:* A possible parametric model for the propensities could be the lognormal density,
>
> $$\phi_j(x_j) \propto \exp\left(-\frac{1}{2\sigma_j^2}\left(\log(x_j) - \mu_j\right)^2\right)/x_j, j = 1, 2, \text{ where } \mu_j \text{ and } \sigma_j \text{ are}$$

the median and the spread of the propensities, respectively. This density has the limit $\phi_j(x_j) \rightarrow 0$, as $x_j \rightarrow 0$, so it remains bounded at the lower boundary.[5]

Now that age is included in the model, the fundamental measure Λ can be characterized. Because Λ_j values are sums (or integrals) of age-specific nuptiality rates, and these are average numbers of marriages per person-year, Λ_j/ω is the average number of marriages of sex j per person-year in a population whose age distribution in ages $[0, \omega]$ is uniform. Such populations are sometimes called *rectangular*. As a further average across sexes, Λ/ω has the same interpretation.[6]

3.2. *Hazard of a New Marriage in an Age-structured Population*

I define $V_j(x_j, t)$ to be the size of the nonmarried population, in ages $\leq x_j$, at exact time t, for $j = 1, 2$. Because the populations will be finite, $V_j(\omega, t) < +\infty$. To ensure coherence, the hazard $h(t)$ of a new marriage in the population must be the same for both sexes. Using the so-called Stieltjes integral notation,[7] I generalize equation (1) to the form

$$h(t) = \Lambda_1(t)W_1(t) = \Lambda_2(t)W_2(t), \tag{8}$$

where

$$W_j(t) = \int_0^\omega \phi_j(x_j)dV_j(x_j, t), j = 1, 2. \tag{9}$$

I now assume that the total marriage intensities are related to the fundamental measure of nuptiality via a generalized average, $\Lambda = \bar{\Lambda}(t)$, where $\bar{\Lambda}(t) \equiv (q\Lambda_1(t)^p + (1 - q)\Lambda_2(t)^p)^{1/p}$ for some $p \neq 0$, or else $\Lambda = \bar{\Lambda}(t)$, where $\bar{\Lambda}(t) \equiv \Lambda_1(t)^q \Lambda_2(t)^{1-q}$ when $p = 0$. As in equation (5), the overall hazard of a new marriage in equation (8) is then of the form $h(t) = \Lambda V(t)$, where for $p \neq 0$, for example

$$V(t) = \frac{W_1(t)W_2(t)}{((1 - q)W_1(t)^p + qW_2(t)^p)^{1/p}}. \tag{10}$$

According to the intended meaning of the model, if the level of nuptiality ($=\Lambda$) does not change over time, then, as in the toy example, the total marriage rates of the two sexes may well change due to the changing age compositions of the two populations that are exposed to marriage. The way the $\Lambda_j(t)$ values change is exactly matched by the change in the population at risk (equation 10).

4. MUTUAL PREFERENCES ACROSS AGE AND SEX

I assume that marriages occur randomly as a stochastic point process (or arrival process) with intensity (equation 8). The functions $\phi_j(x_j), j = 1, 2$ determine for both sexes separately the relative likelihood of marriage between any that are present in the exposed population at time t. What remains to be specified, is how they are matched. In demographic contexts the simplest rule of "random mating" is, of course, not realistic. Spouses tend to come from similar ages.

4.1. *Dependencies via Odds Ratios*

To formulate a model for mutual preferences, I define a function $\gamma(x_1, x_2)$ such that

$$\int_0^\omega \gamma(x_1, z)dz = \int_0^\omega \gamma(y, x_2)dy = 0 \tag{11}$$

for all $0 \leq x_j \leq \omega, j = 1, 2$. Again, the specific identification condition does not influence empirical estimations. The interpretation will be that if one marriage is known to occur such that the spouse with sex $j = 1$ has either age x_1 or y_1, and the age of the spouse of sex $j = 2$ has either age x_2 or y_2, then the odds that the ages are (x_1, x_2) rather than (x_1, y_2) relative to the odds of the ages being (y_1, x_2) rather than (y_1, y_2) is given by

$$\exp\left(\gamma(x_1, x_2) - \gamma(x_1, y_2) - \gamma(y_1, x_2) + \gamma(y_1, y_2)\right). \tag{12}$$

In other words, the function $\gamma(x_1, x_2)$ determines the *odds ratios of the probabilities of marriage* between different possible pairs. The function γ is the third basic parameter in this system, in addition to Λ and $\phi_j, j = 1, 2$.

It may not seem obvious at first glance that such a choice can be made independently of the propensity parameters introduced in section 3. However, it has long been understood that a numerical solution is available (e.g., McFarland 1975). Consider the function

$$\psi_t(x_1, x_2) = \exp\left(\delta_t + \alpha_t(x_1) + \beta_t(x_2) + \gamma(x_1, x_2)\right). \tag{13}$$

It is well known, and can be verified by a direct calculation, that under this matching model, odds ratios $\psi_t(x_1, x_2)/\psi_t(x_1, y_2)/.\psi_t(y_1, x_2)/\psi_t(y_1, y_2)$ are given by equation (12) for all t. The key assumption is that equation (13) is chosen so that it satisfies

$$\int_0^\omega \psi_t(x_1, z)dV_2(z, t) = \Lambda_1(t)\phi_1(x_1),$$
$$\int_0^\omega \psi_t(y, x_2)dV_1(y, t) = \Lambda_2(t)\phi_2(x_2),$$
(14)

for ages x_j with $dV(x_j, t) = 1, j = 1, 2$. With this choice, equation (13) is compatible with the propensity assumptions made in section 3. In particular, if we integrate the two right-hand sides of equation (14) with respect to $V_1(x_1, t)$, and $V_2(x_2, t)$, then both equal $h(t)$, and the correct overall intensity is guaranteed.

Finding δ_t, $\alpha_t(x_1)$, and $\beta_t(x_2)$ that satisfy equation (14) is equivalent to finding a contingency table that matches given marginals but maintains given odds ratios. In the present case, the "contingency table" consists of "cells" formed by all pairs of potential spouses that can be formed between the two sexes at time t. As there are no "zero cells," such a table can always be found (cf. Fienberg 1970). The solution can be calculated using the *iterative proportional fitting* of Deming and Stephan (1940), for example.[8]

4.2. Stochastic Marriage Formation

These considerations provide two equivalent views of how marriages are formed. The first is that the overall intensity of a new marriage at any time $t > 0$ is $\Lambda V(t)$, where $V(t)$ is as given in equation (10). Once a new marriage "arrives," it is between a pair of individuals with ages (x_1, x_2), with probability

$$\frac{\psi_t(x_1, x_2)dV_1(x_1, t)dV_2(x_2, t)}{\int_0^\omega \int_0^\omega \psi_t(y, z)dV_1(y, t)dV_2(z, t)}.$$
(15)

The second view is that we may interpret equation (13) as an intensity that a marriage occurs between any pair of individuals with ages (x_1, x_2) that can be formed between the two sexes at time t. As soon as a marriage occurs, the married couple is removed from the population at risk, and a new $\psi_t(x_1, x_2)$ is formed that satisfies the updated marginals (equation 14).

> *Example 2:* (a) In the so-called log-bilinear models, the functions determining mutual preference are of the form $\gamma(x_1, y_1) = \gamma_1(x_1)\gamma_2(y_1)$, where the functions γ_j integrate to zero. One can see that in this case the odds ratio (equation 12) is of the form $\exp\{[\gamma_1(x_1) -\gamma_1(x_2)][\gamma_2(y_1) -\gamma_2(y_2)]\}$. (b) Every two-dimensional, strictly positive continuous density on $[0, \omega] \times [0, \omega]$ gives rise to an odds ratio specification (equation 12). For example, the bivariate lognormal density gives rise to such a specification. As shown in Appendix B in the online journal, for this model the odds ratio (equation 12) can be written in the form $\exp\{\xi[\log(x_1) - \log(x_2)][\log(y_1) - \log(y_2)]\}$, where $\xi \in \mathbb{R}$ is a parameter that is essentially determined by the correlation coefficient ρ of the bivariate normal distribution that underlies the bivariate lognormal. For example, $\xi \gtrless 0$ as $\rho \gtrless 0$.

As discussed by Deville and Särndal (1992:378, 381), the above solution that preserves odds ratios as measures of mutual attraction is related to a certain minimization problem. Other minimization problems yield other parametrizations, but those details are not pursued in this paper. Yet, another potential approach is given by Cohen and Rothblum (1993).

5. MAXIMUM LIKELIHOOD ESTIMATION

Both individual-level and grouped data can be used as a basis for estimation, and both nonparametric and parametric models can be entertained. A useful feature is that under certain assumptions, the parameters can be estimated separately for the two sexes.

5.1. *Waiting Times to Events*

The generic representation of the events in our model is that if something happens at time $0 < \tau < 1$, then the waiting time until the next event (which could be a marriage, entry to marriage age, death, or entry/ exit via migration or marriage outside the population being considered) is $T > 0$, such that

$$P(T>t) = \exp\left(-\int_0^t \mu(\tau+u) + \Lambda V(\tau+u)du\right), \tag{16}$$

where $V(t)$ is as defined in equation (10) and $\mu(t)$ is the hazard of other events besides marriage. I do not discuss the structure of the latter hazard here, although, for example, in simulations assumptions about it need to be made, possibly along the lines of Matthews and Garenne (2013b).

The model is assumed to be Markovian: whenever a new event arrives, the whole marriage market is assumed to start afresh. If the next event occurs at $\tau + t$, then the probability that it is marriage is $\Lambda V(\tau + t)/[\mu(\tau + t) + \Lambda V(\tau + t)]$. The contribution of a marriage at $\tau + t$ to the likelihood is then

$$f(\tau+t) = \Lambda V(t+\tau) \exp\left(-\int_0^t \mu(\tau+u) + \Lambda V(\tau+u)du\right). \tag{17}$$

5.2. Overall Level of Marriages

5.2.1. Individual-level Data.
I first consider individual-level data, and assume that N events occur during $0 \le t \le 1$, with $M \le N$ marriages. Let $I \subset \{1, \ldots, N\}$ be the subset of indices that correspond to marriages. For any $i = 1, \ldots, N$, we define τ_i as the time of the previous event (with $\tau_1 = 0$), and t_i the corresponding waiting time. To account for the time after the last event, we define $\tau_{N+1} = \tau_N + \tau_N$ and $\tau_{N+1} = 1$ $-\tau_{N+1}$. By Markovianity, equation (17) implies that the log-likelihood equals (up to an additive constant that does not depend on the parameters of interest)

$$\ell = M \log \Lambda - \sum_{i=1}^{N+1} \int_0^{t_i} \mu(\tau_i + u) + \Lambda V(\tau_i + u)du. \tag{18}$$

Differentiating equation (18) with respect to Λ, and setting the derivative to zero, leads to the maximum likelihood estimator (MLE) $\hat{\Lambda} = M/V$, where

$$V = \int_0^1 V(t)dt. \tag{19}$$

This is a classical occurrence/exposure rate, but with "exposure" given by equation (19). If the time unit is one year, then in practical calculations, one might approximate $V \neq [V(0) + V(1)]/2$, where $V(t)$ is defined in equation (10). Here, $V(t)$ depends on the propensities ϕ_j. Their estimation is considered in section 5.3.

5.2.2. *Grouped Data.* Suppose now that grouped data are available by single years of age, and no parametric model is assumed. As is common in demography, we may use a piecewise constant approximation $\Lambda_j(x_j, t) = \Lambda_{ij}$, for $i \leq x_j < i + 1, i = 0, \ldots, \omega - 1$, for $j = 1, 2$, and $0 < t < 1$. Let the corresponding person-years be V_{ij}, and suppose that the corresponding number of marriages is M_{ij}. Using the customary Poisson approximation (cf. Alho and Spencer 2005) $M_{ij} \sim Po(\Lambda_{ij}V_{ij})$, we obtain the MLE $\hat{\Lambda}_{ij} = M_{ij}/V_{ij}$. The estimates of the total nuptiality rate are then $\hat{\Lambda}_j = \sum_{i=0}^{\omega-1} \hat{\Lambda}_{ij}, j = 1, 2$. The estimate of the overall level of nuptiality for the period $0 < t < 1$ is simply the generalized average on the basis of the chosen mapping g, in the simplest case just $\hat{\Lambda} = (\hat{\Lambda}_1 + \hat{\Lambda}_2)/2$. This would correspond to the harmonic mean for the populations at risk.

The grouped data example shows that when a saturated model is assumed, the usual sex-specific marriage intensities are recovered. For statistical description, their arithmetic average would seem to be the preferred measure.

5.3. *Who Marries?*

5.3.1. *Individual-level Data.* To estimate the age- and sex-specific propensity parameters, suppose that we have individual-level data of the form discussed in section 5.2. Let the ages of those who marry be $x_{ij}, i \in I_j$ for $j = 1, 2$. Suppose $\phi_j(x_{ij} \mid \theta_j)$ is a probability density that depends on a parameter θ_j. For example, in the lognormal model of example 1, we would have $\theta_j = (\mu_j, \sigma_j^2)^T, j = 1, 2$. Let R_{ij} be the set of individuals of sex j who were at risk for marriage, when the ith marriage occurred. Under the assumption of Markovianity, the partial likelihood that the age at marriage was x_{ij} given the ages of those in R_{ij} that were at risk, for $i \in I_j$, is (cf. Cox 1975)

$$L_{ij} = \phi_j(x_{ij}|\theta)/ \sum_{k \in R_{ij}} \phi_j(x_{kj}|\theta) \tag{20}$$

for $j = 1, 2$. Then, the likelihoods are $L_j = \prod_{i \in I_j} L_{ij}, j = 1, 2$. I show in Appendix C in the online journal how these can be maximized using Newton's method. The lognormal model of example 1 is used for illustration in section 5.4.

5.3.2. *Grouped data.* For aggregated data, we would use the piecewise constant assumption of section 5.2. In this case, we can condition on the total number of marriages that have occurred, and the multinomial model gives us directly the MLEs $\hat{\phi}_{ij} = \hat{\Lambda}_{ij} / \sum_{k=0}^{\omega-1} \hat{\Lambda}_{kj}, j = 1, 2$, where estimates of age-specific intensity obtained in section 5.2 are used.

5.4. *Which Pairs Are Formed?*

5.4.1. *Individual-level Data.* For individual-level data, a marriage that occurs at time t_i gives us an experiment in which the pair is picked from a multinomial model for all possible pairs at time t_i that have probabilities proportional to $\phi_{t_i}(x_1, x_2)$, as given in equation (13).

Under the bivariate lognormal model mentioned in example 2b, we may substitute $\gamma(x_1, x_2) := \xi \log(x_1)\log(x_2)$, where ξ is a parameter to be estimated. Note that the estimates of marginal parameters, $\hat{\mu}_j, \hat{\sigma}_j, j = 1, 2$ can be obtained as explained in section 5.3, so for any value of $\xi \in \mathbb{R}$ one can determine the functions α_{t_i} and β_{t_i}, and compute the probability of the pair that was actually formed at time t_i. A search over the values ξ will then yield the MLE.

5.4.2. *Individual-level Data Set.* Appendix D in the online journal provides a small artificial data set consisting of nine marriages. It gives the ages of those at risk and records who happened to marry. Appendix C online displays the log-likelihood function for the data in the bivariate lognormal case, and it derives the necessary derivatives for maximizing the likelihood using Newton's method. A reparameterization is found to be useful for the convenience of programming. Although not of substantive interest, I note that the MLEs are $(\hat{\mu}_1, \hat{\mu}_2) = (1.308, 0888), (\hat{\sigma}_1, \hat{\sigma}_2) = (0.553, 0361)$, and $\rho = 0.48$. The computation of the correlation parameter is of some theoretical interest, because it is most easily done by the inspection of the profile log-likelihood. Figure 1 gives its graph as a function ρ, and the maximizing ρ is easily found by a numerical search.

5.4.3. *Grouped Data.* For aggregated data, we continue to assume piecewise constancy, so $\psi_t(x_1, x_2) = \exp(\delta + \alpha_i + \beta_k + \gamma_{ik}) \equiv \phi_{ikt}$ for $i \leq x_1 < i + 1$, $k \leq x_2 < k + 1$, and $0 < t < 1$. The goal is to estimate the odds ratio parameters γ_{ik}. But these are independent of the marginal parameters, so observed odds ratios provide the MLEs via

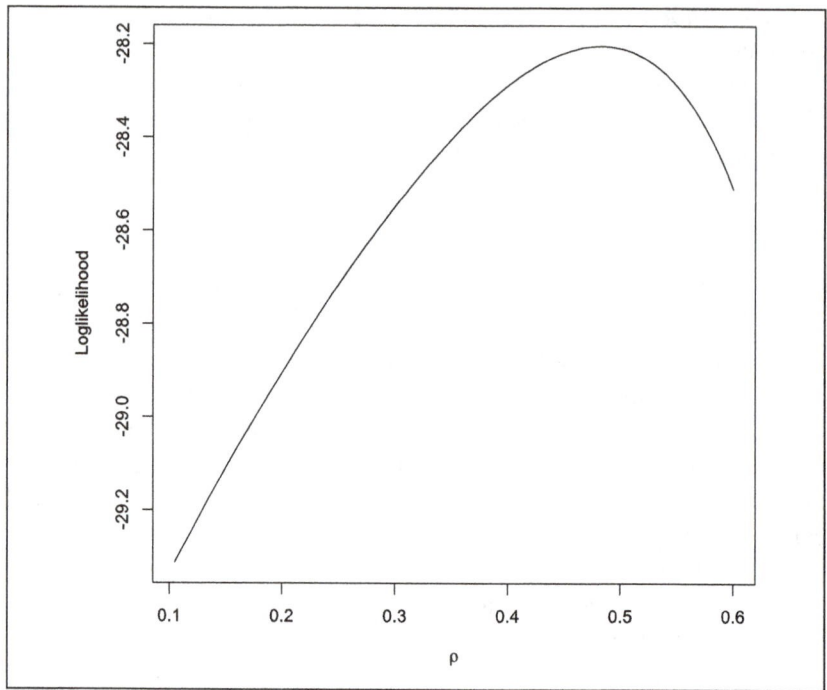

Figure 1. Profile log-likelihood as a function of ρ.

$$\log\{\frac{M_{i,k}}{M_{i',k}} / \frac{M_{i,k'}}{M_{i',k'}}\} = \gamma_{i,k} - \gamma_{i',k} - \gamma_{i,k'} + \gamma_{i',k'}, \qquad (21)$$

where M_{ik} is the count of marriages for which the spouses' ages are (i, k). Together with the marginal estimates, this is equivalent to fitting a saturated model to the counts M_{ik} when the person-years of exposure are given. Such estimates replicate the observed counts exactly.

6. COMPETITION

Although the present models are intended primarily for descriptive demographic purposes, they do entail competition effects of an expected kind (e.g., Dagsvik et al. 2001). This is useful in microsimulation, for example. I formulate the results for the generalized average on the basis of a power transformation with $p \neq 0$ and $0 < q < 1$.

Consider a general situation in which we add to the populations $V_j(x_j, t)$ additional members described by cumulative functions $V_j(x_j)$. Define

$$\phi_j = \int_0^\omega \phi_j(x_j)dV_j(x_j), \quad j=1,2. \tag{22}$$

Thus, if a single individual of sex j and age x_j is added, then $\phi_j = \phi_j(x_j)$, for example.

Suppose that equation (8) held prior to the introduction of new members at t. Denote the total marriage rates after the introduction by Λ_j, for short. We then first obtain

$$\Lambda_1(W_1(t)+\phi_1) = \Lambda_2(W_2(t)+\phi_2). \tag{23}$$

As in equation (4), we also have

$$\Lambda_1 = \Lambda/(q+(1-q)\{(W_1(t)+\phi_1)/(W_2(t)+\phi_2)\}^p)^{1/p}. \tag{24}$$

Suppose, for example, that $\phi_1 > 0$, but $\phi_2 = 0$. In other words, there are new entrants to sex $j = 1$ but none to sex $j = 2$. Then, the ratio in brackets is reduced. When $p > 0$, the whole denominator is increased, and $\Lambda_1 < \Lambda_1(t)$. Conversely, because the overall intensity remains at Λ, we must have $\Lambda_2 > \Lambda_2(t)$. A reader can check that the same happens when $p < 0$, as two ratios are flipped. The intuition is that new entrants increase competition among those of the same sex but reduce competition among the other sex.

Suppose the increase is due to the addition of a single person. Then, $\phi_1 = \phi_1(x_1)$ depends on the age of the entrant. The higher the relative propensity of the entrant, the larger the two effects on total marriage rate are, which is as one would expect. The same is true for the expected number of marriages, as can be seen from equation (10), a generalized average that is an increasing function of both of its arguments. The effects of the new entrants on those already present are proportional to the propensity parameters.[9]

The effect of any other form of change in population composition can be similarly deduced. For example, we see directly from equation (24) that any change satisfying $\phi_1/\phi_2 = W_1(t)/W_2(t)$ leaves the total intensities unchanged, although the expected number of marriages typically does change.

7. ANALYSIS OF NUPTIALITY DATA

For illustration, I first consider monthly nuptiality data from Finland in 2012 and estimate all three sets of nuptiality parameters. As a second illustration, I consider annual data from 1988 through 2012. Here, the emphasis is on the estimation of trends.[10]

7.1. *Basic Characteristics of the Monthly Data*

The occurrences of marriage are classified by single years of ages 17 to 70 years, for both men and women. (A small number of marriages at ages < 17 and at ages > 70 are left out of the data set.) I take 17 to correspond to $i = 0$ and 70 to correspond to $i = 53$, so there are $\omega = 54$ ages in all. The data are for the 12 months, January to December. They are denoted by $t = 1, \ldots, T$, where $T = 12$. The total number of marriages is 28,674.

I consider all marriages irrespective of order, so the population at risk consists of single, divorced, and widowed. It also contains those cohabiting, as they are also at risk for marriage. Although these subgroups are known to display different behaviors with respect to marriage, I take here a descriptive point of view and ignore such heterogeneity. The number of exposure years during calendar year 2012 was 987,246 for women and 1,048,107 for men. Thus, about 2.9 percent of those at risk did marry in 2012. In a rectangular population with 54 ages and a constant marriage intensity, the total nuptiality rate would then be $\Lambda = 54 \times 0.029 = 1.57$.

There are about 6 percent more men than women at risk. Thus, the competition for marriage is slightly higher for men than for women. Figure 2 indicates that this aspect is the strongest at ages 25 to 35.[11] The switch after age 55 appears to be due to higher male mortality. Monthly changes in the population at risk are necessarily very small, so a very similar picture holds for each month.

A remarkable aspect of the Finnish marriage patterns is the strong seasonal variation. As shown in Figure 3, June to August and December are the most popular times to marry. The summer months have long been favored, but in recent years other factors, such as "cute" dates, have had an influence. For example, December 20, 2012 is written "20.12.2012," and December 12, 2012 might be abbreviated "12.12.12.," thus increasing the popularity of December. To account for the seasonality, I allow the fundamental intensity to vary from month to month.

7.1.1. *Estimation of Propensities.* The model I have constructed separates the three aspects: total intensity, propensities by age, and mutual preferences by ages of potential spouses. By construction, the propensities are the most fundamental, and are considered now.

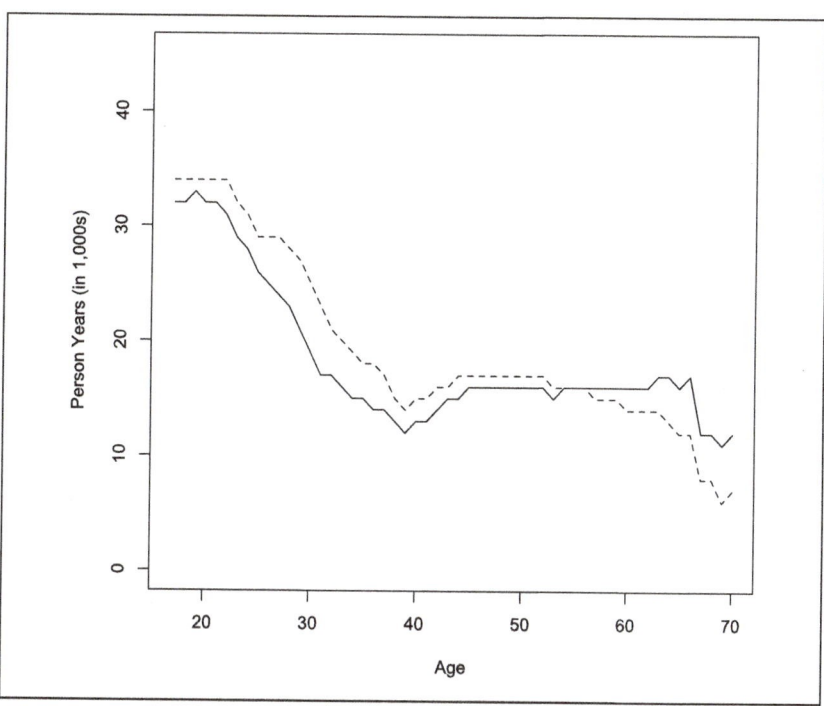

Figure 2. Population at risk for marriage in Finland, in 2012, for women (solid line) and men (dashed line).

I first assume that the propensities are the same throughout the year. Thus, the model is *not* a saturated one. I use a Poisson approximation to the likelihood, so the numbers of marriages for sex j, at age i, in month t are of the form

$$M_{ijt} \sim Po(\Lambda_{jt}\phi_{ij}V_{ijt}), \tag{25}$$

where Λ_{jt} is the total nuptiality rate, V_{ijt} is the person-years of exposure, and the parameters ϕ_{ij} are the propensity parameters that satisfy the condition $\sum_{i=0}^{\omega-1} \phi_{ij} = 1$. I assume that the counts are independent over age and time. Let $M_{\cdot jt} = \sum_{i=0}^{\omega-1} M_{ijt}$. It follows from well-known properties of independent Poisson distributions that for each month $t = 1, \ldots, T$, by conditioning on $M_{\cdot jt}$, we have the multinomial distributions

$$(M_{0jt}, \ldots, M_{\omega-1,jt}) \sim Mult(M_{\cdot jt}; p_{0jt}, \ldots, p_{\omega-1,jt}), \tag{26}$$

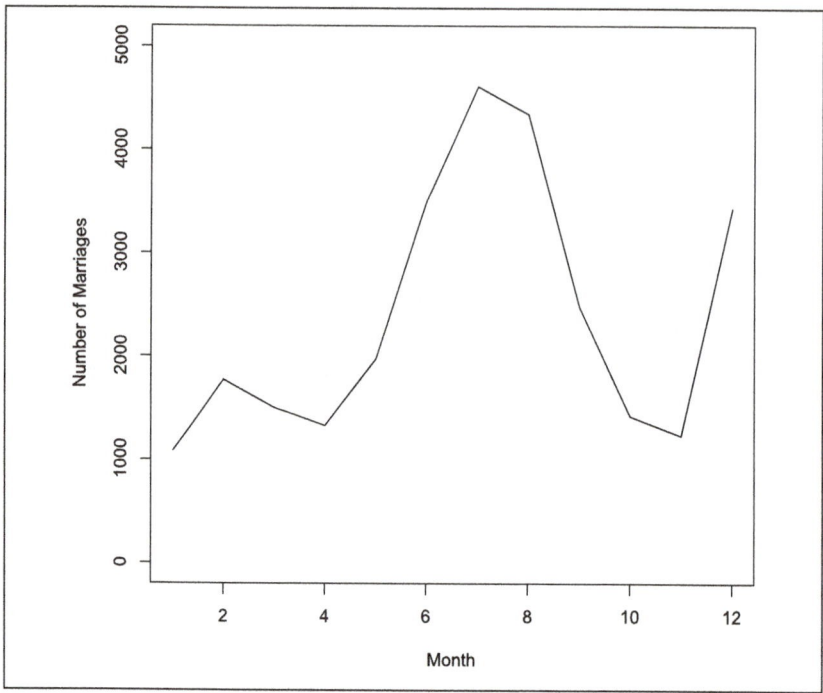

Figure 3. Marriages in Finland, in 2012, by month.

where $p_{ijt} = \phi_{ij} V_{ijt} / \sum_{k=0}^{\omega-1} \phi_{kj} V_{kjt}$, for $i = 1, \ldots, \omega - 1$ and $j = 1, 2$. We see that the overall level does not enter.[12] Setting the derivative of the log-likelihood with respect to the ϕ_{ij} equal to zero, we get the nonlinear equations

$$\phi_{ij} = \frac{M_{ij\cdot}}{\sum_{t=1}^{T} M_{\cdot jt} V_{ijt} / \sum_{i=0}^{\omega-1} \phi_{ij} V_{ijt}}. \tag{27}$$

This system of equations can be solved simply by iteration. For example, we can start by setting $\phi_{ij} = 1/\omega$, on the right-hand side, and then compute the left-hand side, normalize to sum to 1, and iterate. Half a dozen iterations, or fewer, appear to suffice. Figure 4 displays the estimates. Women's propensity for marriage is clearly higher until age 32 or so than that of men of the same age. Presumably, men prefer younger women, and women prefer men who are older than they are.

A notable feature of the propensity curves is the peak at age 49. This appears to be caused by a piece of legislation that entitles a widowed

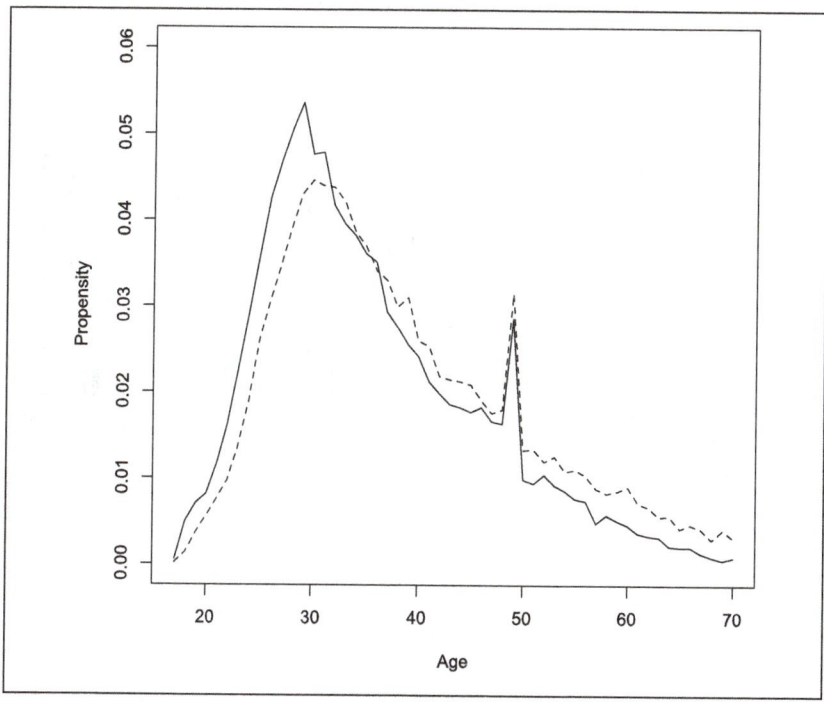

Figure 4. Marriage propensity in Finland, in 2012, for women (solid line) and men (dashed line).

spouse three years of survivor's benefits in the case of the spouse's death, if marriage is formed before age 50.

The lognormal density I have used for illustration captures the overall shape of the propensity patterns, but it has a less peaked mode, and being unimodal, it cannot capture the second local mode at age 49. As suggested later in the paper, in earlier years the second mode at age 49 had not yet appeared, and the overall shape was close to the lognormal.

In large samples the Poisson assumption would almost never hold (cf. Alho and Spencer 2005; Pollard 1975), as there typically are omitted covariates that cause deviations. A routine alternative is to use a negative binomial likelihood, or to fit a mixed Poisson regression model, in which there is a normally distributed random effect for each observation that can account for the possible overdispersion.

The latter was done here with the glmer function of the R package lme4 (http://cran.r-project.org/web/packages/lme4/index.html).[13] The

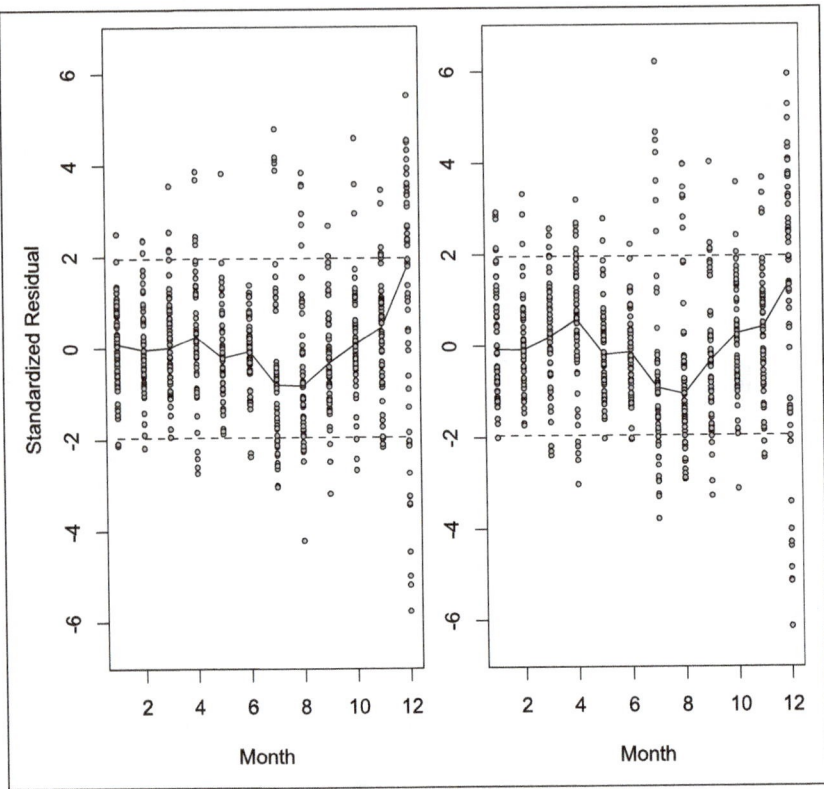

Figure 5. Distribution of standardized residuals and their median by month, for women (right) and men (left).

point estimates were essentially the same as those obtained from the main effects Poisson model.

However, Figure 5 shows the monthly distributions of the standardized residuals from the Poisson model.[14] We see that those marrying in December are different from the rest, especially from those marrying in July and August. In fact, Figure 6 (for men; the figure for women is very similar and is not displayed) shows that it is the *older* couples that have opted disproportionately for December, whereas the young have favored the summer months. The scatterplot also shows that the residuals are autocorrelated over age. Thus, statistical modeling that tries to account for all observable deviations from the standard Poisson assumption needs to take into account both the extra-Poisson variability and its autocorrelation over age and time.

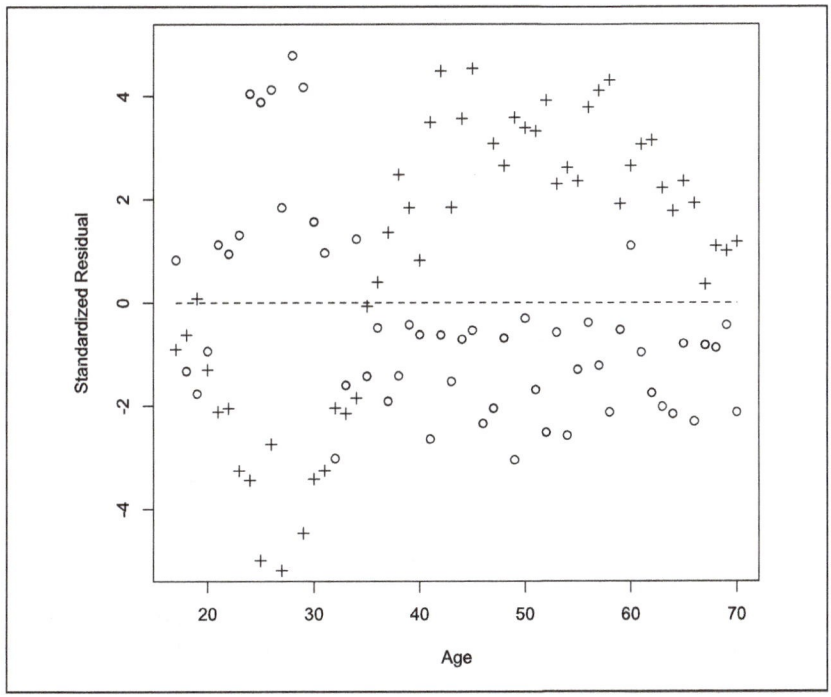

Figure 6. Standardized residuals by age for men marrying in July (o) and in December (+).

7.1.2. Estimation of Total Nuptiality. Given that we now have estimates of the propensities ϕ_{ij} available, we can use discrete-time versions of equations (8) to (10) to estimate the overall level of nuptiality. Aggregating the marriage counts over age for each month $t = 1, \ldots, T$, the monthly numbers of marriages have the approximate Poisson distributions, $M_{\cdot jt} \sim Po(\Lambda_t V_t)$, where the population at risk is $V_t = W_{1t} W_{2t}/((1-q)W_{1t}^p + qW_{2t}^p)^{1/p}$, for $p \neq 0$, and

$$W_{jt} = \sum_{i=o}^{\omega-1} \phi_{ij} V_{ijt}, j = 1, 2. \tag{28}$$

This leads to the MLEs,

$$\hat{\Lambda}_t = M_{\cdot jt}/V_t. \tag{29}$$

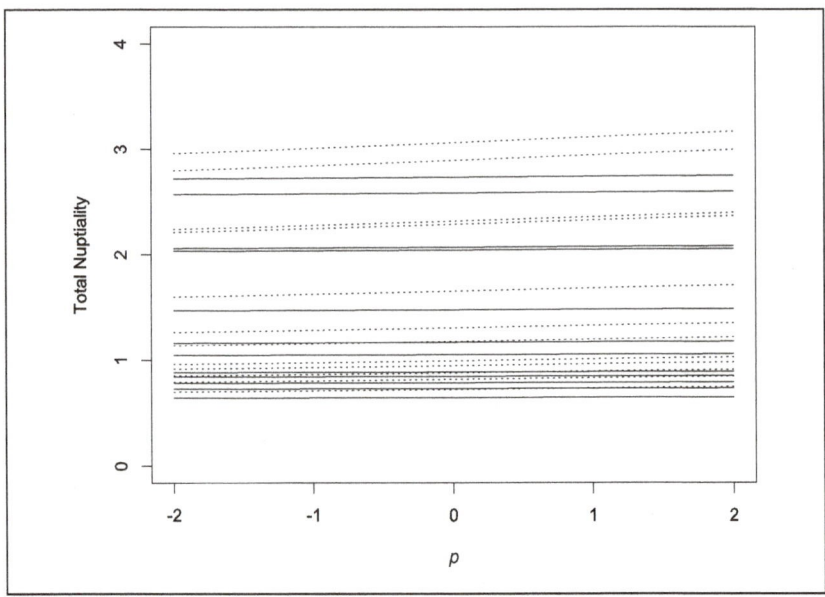

Figure 7. Monthly total nuptiality rates in 2012, in Finland, for different values of the parameter p, actual data (solid line) and modified data (dotted line).

Independently of the estimates of the propensity parameters, or of the values of q and p, all models considered here are saturated, so the MLEs provide a perfect fit for the total number, each month. Therefore, no comparable problem of overdispersion can arise here, as in the previous section, in which unsaturated models were considered.

I assume that $q = 1/2$, but it is of interest to see if the choice of p makes a difference. For illustration, consider a range of values $-2 \leq p \leq 2$. This easily includes all values that are commonly proposed. Recall that we view p as the index used in the generalized average of the rates, so by the duality relationship (equation 6), the sign changes for the population at risk. Therefore, for $p = -1$ the population at risk is the arithmetic mean, for $p = 0$ it is the geometric mean, and for $p = 1$ it is the harmonic mean of the relevant population counts. The solid curves in Figure 7 provides the estimates. The finding could not be clearer. While all lines go gently up (as they must; the increase in p reduces the generalized average of the population counts), the choice of

p hardly matters in the setting of Finland in 2012, where the male population was only about 6 percent larger than the female population.

An obvious question then arises: Does the choice of p matter? A simple modification of the data can give a clue.

> *Example 3:* The sex ratio at birth in Finland is 1.05. Suppose that number is 1.20, as it appears to have been in China around 2000 (e.g., Shuzhuo 2007). At that time, the number of women coming to the marriage age of 17 would have been 87.5 percent of the current value. Suppose their number is depleted by marriage and other forces as before, so it is even less than 87.5 percent of the current value in higher ages. In order not to complicate the calculation by details that are hard to support empirically, let us simply assume that the population at risk gradually reduces to 65 percent of the current values, by age 70. We keep everything else the same (i.e., the numbers of marriages and the estimated propensities). The number of women is then reduced by as much as 23 percent. As shown by the dotted curves of Figure 7, the levels of all curves go up (as they should, when the population of one sex becomes smaller while the counts remain the same), but the slopes increase, notably but not dramatically.

We see that for the purpose of analyzing the very large monthly variations, it does not much matter what value of p is used.

7.1.3. *Estimation of Mutual Preferences.*

The use of data by single years of age becomes inconvenient in the discussion of pair formation because there are numerous age pairs with zero events. These are not structural zeros but simply random occurrences. Parametric models provide one option, but I present here the simplest estimates that are obtainable from aggregate data.

I have restricted my attention to 5-year age groups 18–23, . . ., 53–58. Or, I have considered eight ages for both sexes. Fully saturated models are used that replicate exactly the observed counts. As noted earlier, this means that we can directly use the observed marriage counts M_{ik}, $i,k = 1, . . ., 8$ in estimation. Writing their logarithms as $m_{ik} = \log(M_{ik})$, I first subtract the row means $\bar{m}_{i\cdot}$ (averages over k), and subsequently the column means $\bar{m}_{\cdot k}$ of the row centered values (averages over i) to obtain the estimates $\hat{\gamma}_{ik} = m_{ik} - \bar{m}_{i\cdot} - \bar{m}_{\cdot k}$. These define the odds ratios via equation (21).

Figure 8. The gamma function for ages 18 to 57, on the basis of nuptiality rates for 2012, in Finland, as aggregated in 5-year age groups.

The estimates are depicted in Figure 8. (The odds ratios themselves are difficult to illustrate directly as they are defined on a four-dimensional space.) The higher the surface for a pair of ages, the more they are attracted to each other. The finding is very clear: Both the relatively young and the relatively old prefer potential spouses in their own age ranges, whereas those in the middle are somewhat more open to other choices, in this respect. The estimates in the corners corresponding to the pairing of "old and young," are based on just a handful of cases and cannot be seriously interpreted.

7.2. *Analysis of Annual Trends*

The annual data I use were given for the two sexes separately. In the monthly data, the events were defined so that both spouses were in Finland, and so pair formation by age could be observed. The 1988–2012 historical data have annual counts of women's marriages by single years of ages 17 to 70. This includes (a relatively small number of) marriages with partners outside the country. Because men marry more frequently outside the country, their counts were defined as men's marriages inside the country. The two counts still differ because some marriages occur outside the age range considered here. The difference does not influence the estimation of the propensities, as this is carried out separately for men and women. In the analysis of the annual total counts, the difference is relevant. However, the difference is small and the average of the two counts was used. We do not make this distinction in the notation, however.

The nonmarried male population increased by 40 percent and the female population by 31 percent from 1988 to 2012. The actual number of marriages increased by only 11 percent. This suggests that the level of nuptiality would have decreased by about a fifth. The conclusion is *not* correct, however.

I assume again an approximate Poisson likelihood, $M_{ijt} \sim Po$ $(\Lambda_{jt}\phi_{ijt}V_{ijt})$ for the annual marriage counts in age $i = 0, \ldots, 53$, for sex j $= 1,2$, in year $t = 1, \ldots, 25$, where Λ_{jt} is the total nuptiality rate, ϕ_{ijt} is the propensity function, and V_{ijt} is the person-years of exposure. In other words, the model is a saturated one and the MLEs exactly replicate all observed counts.

Finland had its peak year of the baby boom in 1947–1948. During that period, the total fertility rate was 3.48. Total fertility fell, first slowly and then increasingly rapidly to a low of 1.50 in 1973. After that there was a recovery and fertility fluctuated, mostly in the 1.7 to 1.9 range. The resulting change in age distribution is shown in Figure 9. We see that in the beginning of the present observation period, the baby boomers were at ages with a high propensity of marriage (say, 23- to 35-year-olds), but during the 25 years I consider, they had moved to ages with a lower propensity to marry. Thus, the rapidly changing age structure has tended to *slow down* the increase one might have expected on the basis of the increase in the total nonmarried population alone.

Another factor to consider is propensities. Under the saturated model, they are simply the age-specific intensities of nuptiality, as normalized to sum to one. Figure 10 shows that the propensities have also changed in a major way: High propensities have come down and shifted to older ages. This has tempered the dramatic change in the age structure.[15]

Given that the numerical effect of the choice of *p* appears not to matter much, it seems that the arithmetic average is the preferred measure for the general level of nuptiality. The finding from Figure 11 is that there was initially a small decline in the level of nuptiality, but after that there was a slight *increase*. The net effect is that the level of nuptiality has remained approximately the same. The reason that the numbers of marriages have,

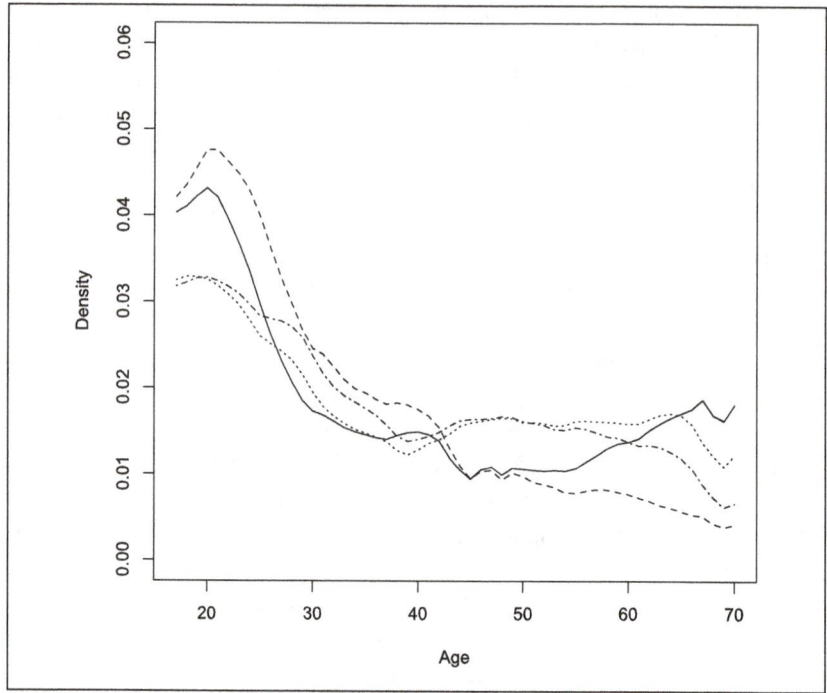

Figure 9. Density of nonmarried population in Finland in 1988 and 2012, for women (solid, dotted lines) and men (dashed lines, dash-dotted lines).

nevertheless, not kept up with the increase in populations at risk is that these populations are older, and in older populations the propensity to marry, even if increased, is still not high enough to compensate for the lower percentage of those in high propensity ages. In other words, we have an example of the classical confounding effect of changing age structure.

8. REMARKS

In this paper I have developed a stochastic model of marriage formation. The principle of maximum likelihood leads to natural measures of intensity. The proposed model solves the two-sex problem by providing a class of models that one can entertain. A particular choice for the general measure of nuptiality determines the appropriate measure for the population at risk, and vice versa.

Happily, the choice of a particular model is not numerically important. Therefore, the choice can be based on the ease of communication.

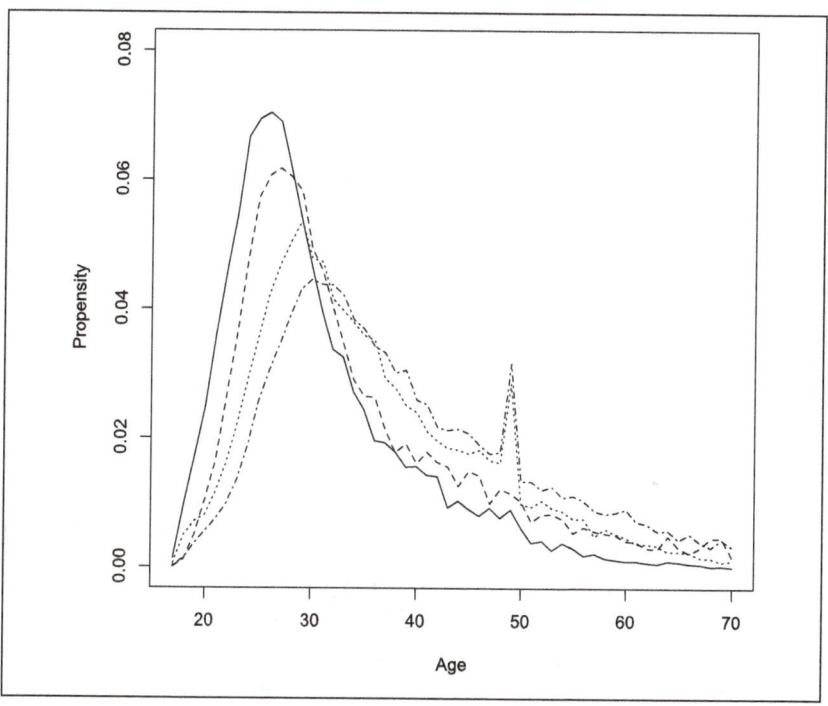

Figure 10. Marriage propensity in Finland, in 1988 and 2012, for women (solid lines, dotted lines) and men (dashed lines, dash-dotted lines).

I propose that the arithmetic average of the total nuptiality measures of the two sexes be adopted. This corresponds to the use of the harmonic mean as the risk population.

An intriguing question may now be asked: Can we base the choice of the generalized average (or p) on data alone? The answer is, it depends. If the models are not saturated with respect to time, then such an estimate can be obtained.

Consider, for example, the annual data from Finland discussed earlier. Suppose we assume that the overall level of nuptiality does not change over time, and suppose that the propensities are assumed to be the same all through the 25-year period. Then, the best fit to the data turns out to be given by $p = +\infty$ (details omitted). This is not very satisfactory, however. First, the difference in the fit provided by all values $p \geq 0$ (recall again that p here refers to the generalized average regarding the rates, not the populations at risk) is so small as to be of no practical consequence. Second, although the assumption of a constant level of

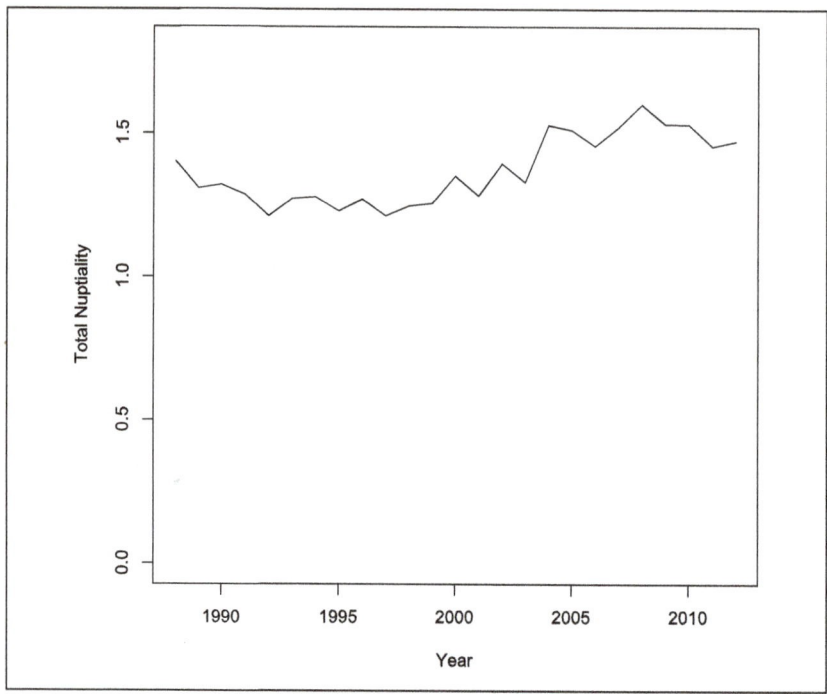

Figure 11. Total nuptiality in Finland, 1988 to 2012.

nuptiality is not unreasonable in view of the analyses given above, the assumption of constant propensities is not. Third, the solution $p = +\infty$ corresponds to the minimum of the two populations, as the size of the population at risk. Between 1990 and 2012, the minimum is given by the female population, but in 1988 and 1989 it is given by the male population. It would be difficult to explain to the general public why sometimes the population at risk is the male population, and sometimes it is the female population.

I have emphasized the role of the fundamental level of nuptiality primarily because its handling allows me to overcome the long-standing two-sex problem. However, for an economist or a sociologist, the estimated propensities and the odds ratios may be of even greater importance, as reasons for their change over time can be of considerable substantive interest.

Similarly, changes in legislation and mores that are occurring in many industrialized countries will lead to further issues that will influence the interpretation of the parameters of the system. As cohabitation has received, in many countries, a position comparable with that of marriage, and same-sex marriages have gained acceptance, it may become useful to reconsider the question of what is the relevant event of pair formation to measure and model. In the current formulation, same-sex marriages would be interpreted as censorings that remove individuals from the risk for a two-sex marriage. However, a new marriage market is arising within each of the two sexes. It seems that such markets can be modeled on age-specific propensities and mutual preferences as in the examples included here, but the logically more difficult two-sex aspect does not arise.

The results given here are for marriages, irrespective of order. Restriction to first marriages along the same lines appears both feasible and useful. This would provide refined versions of the currently used total marriage rates that are sometimes used to approximate probabilities of ever marrying. A similar extension seems feasible in other areas. For example, in Alho and Keilman (2010), a stochastic model for household shares was developed. In past data, the empirical counts of couples living together necessarily agree for men and women, but for forecasts such agreement is not guaranteed. A solution along the lines developed here seems feasible. Finally, and as pointed out above, the fact that we have an explicit probability mechanism for generating the marriages, it is straightforward to simulate such processes, whether on an individual or a group level.

Acknowledgments

An early version of the paper was presented at the Keyfitz Centennial Symposium on Mathematical Demography in Columbus, Ohio, in June 2013. The author would like to thank H. Caswell, J. Cohen, N. Keiding, N. Keilman, R. Lee, and S. Tuljapurkar for useful suggestions at various stages of the work. Comments from the editor and two reviewers helped to improve the presentation.

Funding

Part of this research was carried out when the author participated in the international research group Changing Family Patterns in Norway and Other Industrialized Countries at the Centre for Advanced Study, the Norwegian Academy of Science and Letters, Oslo, during the academic year 2006–2007, and as part of the project AGHON in 2010–2012, funded by NORFACE.

Notes

1. And indeed, the practice of calculating human fertility rates for women only is an example of a similar assumption of *female dominance*, a usage that may sound a bit unintuitive.

2. Note how the weighting operates in the denominator. There is a typographical error in the corresponding formula in Hadeler et al. (1988:639).

3. I am grateful to Hal Caswell for pointing out that min$\{V_1, V_2\}$ is a relevant measure for population at risk in ecology. Emperor penguins are strictly monogamous during the breeding season because the feeding of chicks requires both parents (Jenouvrier et al. 2010:743).

4. Age could be replaced by, or combined with, other cardinal characteristics and possibly crossed with nominal types, such as race or geographic location.

5. The lognormal is given here because of its analytical tractability. A relevant alternative might be the log gamma density of Kaneko (2003). In Matthews and Garenne (2013a), the so-called Picrate function is proposed for the parametric description of the first marriage intensities (Matthews, Leclerc, and Garenne 2009).

6. I am not restricting attention here to first marriages, so my concept of Λ should not be confused with the "total marriage rate" that is sometimes used in official statistics as a proxy for the probability of ever marrying. Note also that in this respect, marriage differs from fertility. Both are repeatable events, but marriage typically removes a person from the population at risk for several years, whereas childbirth only removes a woman from the population at risk for less than one year at a time. This is usually not of practical consequence, so the total fertility rate (that corresponds to Λ) can be understood as the expected number of children a woman will have in her lifetime, provided that she survives to the end of childbearing age.

7. The cumulative functions have a jump of size 1, or $dV_j(x_j, t) = 1$, for each individual who is at exact age x_j at exact time t, otherwise $dV_j(x_j, t) = 0$. Thus, the integrals in equation (8) represent finite sums over the individuals who belong to the exposed populations at time t.

8. In Matthews and Garenne (2013a) this iteration is described as a "heuristic bidding process."

9. This is in contradiction with requirement A7 in Dagsvik et al. (2001) that the effects will be the greatest for those whose ages are the closest to those of the entrants. In the present model, the introduction of new individuals does not alter the relative propensity of those already present. This choice is deliberate, as it permits the separate estimation of the propensities of the two sexes.

10. Thanks are due to T. Nikander of Statistics Finland, who kindly provided both data sets as well as insight into the seasonality observed in the monthly data.

11. For example, at age 31 there are nearly 30 percent more men at risk than women.

12. Note also that when time is measured in years, V_{ijt} is approximately 1/12 of the average population size during the month, but this factor of proportionality also cancels each month.

13. A corresponding Bayesian analysis of the hierarchical model is available via the R package MCMCglmm (http://cran.r-project.org/web/packages/MCMCglmm/index.html).

14. Here $(M_{ijt} - \hat{M}_{ijt})/\sqrt{\hat{M}_{ijt}}$, for months $t = 1, \ldots, 12$, ages $i = 17, \ldots, 70$ for the two sexes $j = 1, 2$, where $\hat{M}_{ijt} = M_{ijt}\,\hat{p}_{ij}$ as defined in equation (26).

15. Note also that the peak at age 49 has appeared (gradually) during the latter part of the period.

References

Alho, Juha, and Nico Keilman. 2010. "On Future Household Structure." *Journal of the Royal Statistical Society, Series A* 173:117–43.

Alho, Juha M., and Bruce D. Spencer. 2005. *Statistical Demography and Forecasting*. New York: Springer.

Arrow, Kenneth J. 1971. *Lectures on Decision Making under Uncertainty*. Helsinki, Finland: ETLA.

Bergstrom, Ted, and David Lam. 1988. "The Effects of Cohort Size on Marriage Markets in Twentieth Century Sweden." Presented at the IUSSP Seminar on the Family, the Market, and the State in Ageing Societies, Sendai, Japan.

Box, George E. P., and David R. Cox. 1964. "An Analysis of Transformations." *Journal of the Royal Statistical Society, Series B* 26:211–52.

Bozon, Michel, and François Heran. 1989. "Finding a Spouse: A Survey of How French Couples Meet." *Population: English Edition* 44:90–121.

Carroll, Glenn R., and Michael T. Hannan. 2000. *The Demography of Corporations and Industries*. Princeton, NJ: Princeton University Press.

Caswell, Hal, and Daniel E. Weeks. 1986. "Two-sex Models: Chaos, Extinction, and Other Dynamic Consequences of Sex." *American Naturalist* 128:707–35.

Choo, Eugene, and Aloysius Siow. 2006a[b]. "Estimating a Marriage Matching Model with Spillover Effects." *Demography* 43:463–90.

Choo, Eugene, and Aloysius Siow. 2006b[a]. "Who Marries Whom and Why." *Journal of Political Economy* 114:175–201.

Chung, Kai L. 1974. *A Course in Probability Theory*. 2nd ed. New York: Academic Press.

Cohen, Joel E., and Uriel G. Rothblum. 1993. "Nonnegative Ranks, Decompositions, and Factorizations of Nonnegative Matrices." *Linear Algebra and Its Applications* 190:149–68.

Cox, David R. 1975. "Partial Likelihood." *Biometrika* 62:269–76.

Dagsvik, John K., Helge Brunborg, and Ane S. Flaatten. 2001. "A Behavioral Two-sex Marriage Model." *Mathematical Population Studies* 9:97–121.

Deming, W. Edwards, and Frederic C. Stephan. 1940. "On a Least Squares Adjustment of a Sampled Frequency Table When the Specified Marginal Totals Are Known." *Annals of Mathematical Statistics* 11:427–44.

Deville, Jean-Claude, and Carl-Erik Särndal. 1992. "Calibration Estimators in Survey Sampling." *Journal of the American Statistical Association* 87:376–82.

Fienberg, Stephen E. 1970. "An Iterative Procedure for Estimation in Contingency Tables." *Annals of Mathematical Statistics* 41:907–17.

Goodman, Leo A. 1968. "Stochastic Models for the Population Growth of the Sexes." *Biometrika* 55:469–87.

Hadeler, K. P., R. Waldstätter, and A. Wörz-Busekros. 1988. "Models for Pair Formation in Bisexual Populations." *Journal of Mathematical Biology* 26:635–49.

Henry, Louis. 1972. *Démographie. Analyse et Modèles*. Paris: Larousse.

Jenouvrier, Stèphanie, Hal Caswell, Christophe Barbraud, and Henri Weimerskirch. 2010. "Mating Behavior, Population Growth, and the Operational Sex Ratio: A Periodic Two-sex Model Approach." *American Naturalist* 175:739–52.

Kaneko, Ryuichi. 2003. "Elaboration of the Coale-McNeil Nuptiality Model as the Generalized Log Gamma Distribution: A New Identity and Empirical Enhancements." *Demographic Research* 9:223–62.

Kesten, Harry. 1970a. "Quadratic Transformations: A Model for Population Growth I." *Advances in Applied Probability* 2:1–82.

Kesten, Harry. 1970b. "Quadratic Transformations: A Model for Population Growth II." *Advances in Applied Probability* 2:179–228.

Lee, Aline, Steinar Engen, and Bernt-Erik Sæther. 2008. "Understanding Mating Systems: A Mathematical Model of the Pair Formation Process." *Theoretical Population Biology* 73:112–24.

Li, Zhang. 2011. *Male Fertility and Its Determinants*. Dordrecht, the Netherlands: Springer.

Matthews, Alan P., and Michel L. Garenne. 2013a. "A Dynamic Model of the Marriage Market—Part 1: Matching Algorthm Based on Age Preference and Availability." *Theoretical Population Biology* 88:78–85.

Matthews, Alan P., and Michel L. Garenne. 2013b. "A Dynamic Model of the Marriage Market—Part 2: Simulation of Marital States and Application to Empirical Data." *Theoretical Population Biology* 88:78–85.

Matthews, Alan P., Pauline M. Leclerc, and Michel L. Garenne. 2009. "The Picrate Model for Fitting the Age Pattern of First Marriage." *Mathematics and Social Sciences* 47:17–28.

McFarland, David D. 1972. "Comparison of Alternative Marriage Models." Pp. 89–106 in *Population Dynamics*, edited by T.N.E. Greville. New York: Academic Press.

McFarland, David D. 1975. "Models of Marriage and Fertility." *Social Forces* 54:66–83.

Pollak, Robert A. 1990. "Two-sex Demographic Models." *Journal of Political Economy* 98:399–420.

Pollard, John. 1975. "Modelling Human Populations for Projection Purposes—Some of the Problems and Challenges." *Australian Journal of Statistics* 17:63–76.

Pollard, John, and Charlotte Höhn. 1993. "The Interaction between the Sexes." *Zeithschrift für Bevölkerungswissenschaft* 19:203–28.

Särndal, Carl-Erik. 1978. "Design-based and Model-based Inference in Survey Sampling." *Scandinavian Journal of Statistics* 5:27–52.

Schoen, Robert. 1981. "The Harmonic Mean as the Basis of a Realistic Two-sex Marriage Model." *Demography* 18:201–16.

Shuzhuo, Li. 2007. "Imbalanced Sex Ratio at Birth and Comprehensive Intervention in China." Presented at 4th Asia Pacific Conference on Reproductive and Sexual Health and Rights, October 29–31, Hyderabad, India.

Smith, Tom W. 2006. "American Sexual Behavior." GSS Topical Report No. 25. Chicago: National Opinion Research Center.

Author Biography

Juha Alho received his PhD from Northwestern University and is now a professor of social statistics at the University of Helsinki, Finland. He is a former editor of the *Scandinavian Journal of Statistics* and an associate editor of the *International Journal of Forecasting*. He is the author, with B. D. Spencer, of *Statistical Demography and Forecasting*, and he has edited, with S. Hougaard Jensen and J. Lassila, *Demographic Uncertainty and Fiscal Sustainability*.

Sociological Methodology
2016, Vol. 46(1) 319–344
© American Sociological Association 2016
DOI: 10.1177/0081175016654737
http://sm.sagepub.com
⑤SAGE

✌ 10 ✌

ADDRESSING MEASUREMENT ERROR BIAS IN GDP WITH NIGHTTIME LIGHTS AND AN APPLICATION TO INFANT MORTALITY WITH CHINESE COUNTY DATA

Xi Chen*

Abstract

As an emerging research area, application of satellite-based nighttime lights data in the social sciences has increased rapidly in recent years. This study, building on the recent surge in the use of satellite-based lights data, explores whether information provided by such data can be used to address attenuation bias in the estimated coefficient when the regressor variable, Gross Domestic Product (GDP), is measured with large error. Using an example of a study on infant mortality rates (IMRs) in the People's Republic of China (PRC), this paper compares four models with different indicators of GDP as the regressor of IMR: (1) observed GDP alone, (2) lights variable as a substitute, (3) a synthetic measure based on weighted observed GDP and lights, and (4) GDP with lights as an instrumental variable. The results show that the inclusion of nighttime lights can reduce the bias in coefficient estimates compared with the model using observed GDP. Among the three approaches discussed, the instrumental-variable approach proves to be the best approach

*Quinnipiac University, Hamden, CT, USA

Corresponding Author:
Xi Chen, Quinnipiac University, 275 Mount Carmel Avenue, Hamden, CT 06518, USA
Email: xi.chen@quinnipiac.edu

*in correcting the bias caused by GDP measurement error and estimates the
effect of GDP much higher than do the models using observed GDP. The
study concludes that beyond the topic of this study, nighttime lights data have
great potential to be used in other sociological research areas facing estimation bias problems due to measurement errors in economic indicators. The
potential is especially great for those focusing on developing regions or
small areas lacking high-quality measures of economic and demographic
variables.*

Keywords

measurement error, nighttime lights, GDP, demography

1. INTRODUCTION

Conducting empirical studies with local-level economic or demographic
data is often problematic for developing countries, which tend to have
relatively poor statistical systems. As a result, observed indexes created
by using conventional methods tend to have large systematic errors at
the local level. Even if some regional economic and demographic statistics are collected regularly by national bureaus or agencies, the measurement errors in these indicators or indexes generally are not reported or
are unknown. In empirical analysis, using variables with large measurement errors as a regressor can produce attenuation bias in coefficient
estimates. In sociological studies, scholars often use economic indicators, such as income or Gross Domestic Product (GDP), as a regressor
or control variable in hypotheses testing. Attenuation bias caused by
large measurement error in such variables can lead scholars to underestimate the true effect of economic conditions on the subject of interest.
As a result, researchers often focus on analyzing data at the country or
province/state level, as national information can be obtained easily from
international agencies (Babones 2013). A proxy variable or instrumental
variable can provide a potential solution for attenuation bias problems.
However, valid proxy or instrumental variables themselves often cannot
be found at the local level either.

Take China as an example. Economists have been uncertain of the
true value of China's GDP, and it has recently been reported that the
real value of its GDP per capita was likely 30% higher in 2005 than the
World Bank estimated (Feenstra et al. 2013). Among the reasons that
cause the real GDP value to be underestimated, Feenstra and colleagues

emphasize the discrepancy in price data provided by China for urban and rural areas, different index number methods, and different concepts of real GDP that were adopted in GDP calculations. If at the national level the conventional method of collecting and calculating GDP is problematic, it is not hard to imagine that at the local level the data involve even more significant problems of estimation errors. Often, the true extent of such GDP error at the county or city level in China is unknown, as no literature has been identified that has investigated such topics. It is assumed in this research that data collected on the local level has greater measurement errors than data collected at the national level. This inevitably leads to problems in empirical analysis that uses Chinese GDP as a regressor to estimate the impact of economic conditions on other social phenomena in China. In China's case, a valid proxy measure or instrumental variable may not be available, as the People's Republic of China (PRC) publishes few or no local data. Thus, for studies that use local-level data, such as county or city level data, the true effect theorized may be considerably underestimated.

To address this issue related to a lack of high-quality data, the use of information collected through unconventional, nonsurvey methods, including remote sensing information, has been suggested as a possible solution (Babones 2013; Goodchild and Janelle 2004). One such source that is emerging as a promising alternative to conventional measures is nighttime lights data. This source provides great potential in estimating regional, local development, and demographic changes, as the data have been shown to highly correlate with main economic indicators (Chen and Nordhaus 2011; Henderson, Storeygard, and Weil 2011). The conceptual explanation of the close association between lights and economic development is quite straightforward and self-evident. The degree of luminosity captured from space depends on artificial light on the ground, which is generated by electricity (Mellander et al. 2015). Regions with high economic activities—especially business-dense and population-dense commercial and industrial areas—tend to use more electricity and therefore are brighter at night. The advantage of lights data collected from space is that such information covers almost the entire globe, it is updated almost instantly by satellites, and it can provide information at very small scales and is less likely to be affected by reporting or coding errors due to discrepancies in definition, collection, and calculation across regions or countries—all problems often encountered with conventional data.

This article is a first attempt to investigate whether nighttime lights can be used to address the miscalculation in conventional GDP by providing additional information or can be used as an instrumental variable in a model that predicts infant mortality rate (IMR). In theory, IMR is associated with many development indicators at the local level, such as income level, income inequality, per capita GDP, or urban population (Frisbie 2006). With China's rapid economic growth since the 1970s, IMR declined from 63 in the 1970s to 18 during the period between 2005 and 2010 (United Nations 2013). Yet, the economic gains have been geographically uneven, occurring largely in coastal areas and larger cities. Much less is known about China's regional local economic development over the last 20 years and its links to IMR at the county or city level. Data limitation has hindered research efforts in such studies. An exception to this was work conducted by Poston (1996) about 20 years ago. Using 1,441 cases from Census 1980, the author found that the development index is one of the strongest predictors of local IMR. Despite the availability of more recent China census data, no other empirical work has been identified to further investigate this topic.

The primary goal of this article is to explore how lights data can be included in a regression model that reduces the bias in coefficient estimates of GDP. Different statistical approaches are introduced and discussed, and then the empirical analyses based on these approaches are conducted with Chinese county-level data. Through evaluating the results of the analyses, I recommend the use of nighttime lights data as a general method potentially applicable to many fields of sociology in addressing the attenuation bias due to measurement error in observed economic indicators.

2. THE USE OF NIGHTTIME LIGHTS IN SOCIAL SCIENCE STUDIES

Nighttime lights data are imagery data collected by the U.S. Department of Defense, and they are processed from satellite-based remote sensing information. The publicly accessible lights data were developed by the Defense Meteorological Satellite Program—Operational Linescan System (DMSP-OLS 2015). The earliest satellite-based information has existed since the 1970s, but its application within social science research dates back only to the 1990s.

Most early studies using nighttime lights were conducted and published by geoscience and remote sensing field scholars to estimate a wide range of demographic and economic indicators, including population density, GDP, income per capita, wealth, urbanization, and energy use (Doll, Muller, and Elvidge 2000; Ebener et al. 2005; Elvidge et al. 1997; Elvidge et al. 2001; Elvidge et al. 2007; Elvidge et al. 2009; Noor et al. 2008; Sanderson et al. 2000; Sutton, Elvidge, and Ghosh 2007). Over the last decade, the number of studies exploring the usefulness of lights has increased, and among these studies the topics have expanded to include electricity usage rates (Elvidge et al. 2011), global distribution of economic activity (Ghosh et al. 2010a), and fossil fuel carbon dioxide emissions (Ghosh et al. 2010b; Oda and Maksyutov 2011).

Social scientists have only recently begun paying attention to the potential of nighttime lights data, especially in the field of economics. Over the last five years, increasing numbers of economists have been looking to the lights data for economic output estimates at both national and subnational levels with formal statistical approaches (Chen and Nordhaus 2011; Henderson et al. 2011; Henderson, Storeygard, and Weil 2012; Nordhaus and Chen 2015). Nordhaus and Chen's analyses show that lights can provide more additional information in estimating GDP for poor countries than for middle-income and rich countries, as the poor countries often report GDP with large errors. Lights data are also found to closely correlate with small area poverty for many poor and developing countries (Chen 2015, 2016). Likely due to China's increasing importance on the world stage and concurrently the lack of high-quality measures on some variables, China has also been the focus of at least a handful of studies using lights to estimate GDP, regional economic development, and electric power consumption (Li et al. 2013; Shi et al. 2014; Zhao, Currit, and Samson 2011). It is estimated that lights data have been used in almost 3,000 studies to investigate economic phenomena alone (Nordhaus and Chen 2015). These studies, establishing the strong association between lights and economic indicators, undoubtedly have now formed a foundation for further applications of lights in disciplines beyond economics.

In terms of different approaches to using lights, there are largely three main categories. The first primarily explores the association between lights and other variables. Most early work on nighttime lights belongs to this category. Often in these studies, correlation statistics are used to demonstrate the close association between lights and other variables.

The second category focuses on how to use lights data information to create new indicators, such as the human footprint index (Sanderson et al. 2000) and Night Light Development Index (NLDI) (Elvidge et al. 2012), or to improve a current measurement of indicators, such as GDP and grid cell output (Chen and Nordhaus 2011; Nordhaus and Chen 2015) and national income (Henderson et al. 2012; Henderson et al. 2011). The final category uses lights data as a regressor in hypothesis testing. The rationale for using lights data in regression is the large body of existing literature that has demonstrated a strong correlation between lights and demographic and economic indicators. For example, Bharti et al. (2011) used lights directly as a population density measure in a model that predicted seasonal fluctuations of measles in urban areas in Niger, where direct population measures are not available. The lights variable is also used as a proxy for economic shocks to predict civil conflict in African countries (Hodler and Raschky 2014), and it has been used as a development indicator to predict steel stock for buildings and civil engineering infrastructure in China (Liang et al. 2014).

Although using lights as a regressor in a model is innovative and may indeed be able to address the poor data quality and measurement error issues in many research areas, the strength and weakness of the different methods of using lights in hypothesis-testing have not been carefully explored. To address issues regarding measurement error bias, there are three potential approaches, among which two have been previously suggested or implemented in the literature reviewed above. The first approach uses lights directly as a regressor to substitute the variable that researchers are interested in testing but that is often unavailable or measured with a large error. The variable being substituted can include a population and economic development index, as lights are indeed highly correlated with these variables. The second approach is to replace the theoretical variable with a modified, improved measure based on additional information from lights. Although the formal statistical procedure of how to use lights to improve current measures of GDP has been thoroughly discussed (Chen and Nordhaus 2011; Nordhaus and Chen 2015), its further applications in hypothesis testing have not yet been explored. Finally, the conventional solution for measurement error bias is the instrumental variable approach, and lights can potentially be a valid instrumental variable, as these data are highly correlated with GDP; in addition, because lights data are collected by NASA satellites instead of by local or regional government agencies, they do not suffer the same problems of conventional GDP measures. To simplify the terms in the analysis that follows, I refer to the first approach as the *direct substitution approach*, the second as the *synthetic measure approach*, and the third as the *instrumental variable approach*.

Section 3 discusses each approach in detail in terms of whether it can reduce the measurement error bias in regression coefficient estimates. In the context of predicting county IMR with GDP values, four models are specified: (1) those using observed GDP values published by the PRC, (2) those using the lights variable as a direct substitute for GDP, (3) those using synthetic GDP measures based on weighted lights and observed GDP, and (4) those using lights as an instrumental variable for GDP. The primary goal is to see whether lights-based measures, in one way or another, can reduce the bias in GDP coefficient estimators. The results of four regression models using a Chinese county sample in the year 2000 are compared to show the improvement of estimated GDP coefficients with the different methods.

3. METHODS

The main question in the analysis is how GDP affects IMR. To simplify the expressions, we consider only the case of regression with a single regressor:

$$w = c + Y\mu + u, \tag{1}$$

where w is IMR. We can assume that it is measured without errors, c denotes a constant, u is an error term, and Y is the theoretical variable, GDP. The true GDP, Y, is measured with error ε:

$$y = Y + \varepsilon. \tag{2}$$

The ordinary least squares (OLS) estimator via the regression of w on y yields a biased estimator of μ:

$$w = c_{gdp} + y\mu_{gdp} + u_{gdp}$$

and

$$\hat{\mu}_{gdp} = \left(\frac{\sigma_Y^2}{\sigma_Y^2 + \sigma_\varepsilon^2}\right)\mu. \tag{3}$$

In the hope that lights can provide a solution to reduce bias in the OLS estimator, the three different approaches are presented and discussed below.

3.1. *Direct Substitution Approach*

The direct substitution approach uses the lights variable as a simple substitute for the GDP measure in the equation. In such a case, the relationship between lights and GDP is overlooked. Although lights data are highly correlated with many variables, including GDP, they can also be influenced by other physical and social factors not related to GDP. Thus, the correlation coefficient between lights and GDP is not equal to 1, and using lights directly as a regressor can still cause a bias in estimating the true effect of GDP on the dependent variable, IMR.

The equation for such an approach can be expressed as

$$w = c_{light} + m\mu_{light} + u_{light}, \qquad (4)$$

where m is an observed light measure. The relationship between true GDP and the observed lights variable can be expressed with the following equation:

$$Y = m\gamma + v. \qquad (5)$$

Here, $\hat{\mu}_{lights}$ is still a biased estimator of true GDP, as the relationship between $\hat{\mu}_{lights}$ and μ can be expressed with the equation

$$\hat{\mu}_{lights} = \mu * \gamma. \qquad (6)$$

In equation (5), γ is the standardized coefficient. Unless γ is equal to 1, the direct substitution approach will lead to a downward bias, which is determined by γ.

3.2. *Synthetic Measure Approach*

This approach first combines information from lights and observed GDP to create an improved, synthetic measure of GDP that is then used as a regressor in the model. The study adopts Nordhaus and Chen's (2015) method, as the authors have specified a statistical procedure to use lights to improve current measures of GDP. According to this method, the precise amount of information provided by lights can be calculated. Theoretically, the synthetic measure created through optimal weights of lights and observed GDP is the best estimate of true GDP. However, its further implementation in regression models has not been explored in detail. One caveat related to this method is that the value of synthetic GDP is determined by three parameters that can potentially cause bias in the final GDP coefficient estimate in IMR regression models, although it may be better than the OLS estimator based on observed GDP values.

The equation to be estimated with a synthetic value can be expressed as the following:

$$w = c_{syn} + s\mu_{syn} + u_{syn}. \tag{7}$$

A synthetic GDP measure, s, can be calculated based on weighted information from both observed GDP, y, and the observed lights variable m. The section that follows illustrates how to calculate synthetic GDP, according to Chen and Nordhaus (2011) and Nordhaus and Chen (2015). We can begin with an assumption that there is a structural relationship between observed lights and true GDP:

$$m = \alpha + Y * \beta + \tau. \tag{8}$$

A proxy measure of GDP can be calculated by inverting the coefficient β. However, because the true values of Y are unknown, the true value of β is unknown. We can use observed GDP and lights to estimate β in equation (8), with the resultant estimation referred to as $\hat{\beta}$. This coefficient is biased due to measurement error in y, ε. To get the error-corrected estimates of the structural coefficient, $\tilde{\beta}$, we can use the classic errors-in-variable correction, as shown below:

$$\tilde{\beta} = \left(\frac{\sigma_y^2 + \sigma_\varepsilon^2}{\sigma_y^2} \right) \hat{\beta}.$$

Here, $\hat{\beta}$ is the regression coefficient based on observed data. σ_y^2 and σ_ε^2 can be determined if ε is known.

The proxy GDP variable z is calculated by inverting $\tilde{\beta}$:

$$\hat{z} = (1/\tilde{\beta})m.$$

The proxy measure of GDP, z, provides an alternative measure to the observed GDP value. To improve the current observed GDP measure, we need information from both observed value y and the lights-based proxy measure z. A synthetic measure of Y can then be calculated by taking weighted averages of y and $\hat{z}$:

$$\hat{s} = (1 - \theta)y + \theta\hat{z}. \tag{9}$$

The optimal weight, θ, is important as it indicates how much information is from lights-based GDP (see the detailed discussion in Nordhaus and Chen [2015] and Appendix A in the online journal for θ estimation):

$$\theta = \frac{\beta^2\sigma_\varepsilon^2}{\beta^2\sigma_\varepsilon^2 + \sigma_\tau^2}.$$

Here σ_τ^2 and β can be consistently estimated with equation (8). The value of σ_ε^2 is based on a prior estimate of measurement error in Chinese county-level GDP. Let us assume it is 40%, as it is the value used for a 1 degree latitude by 1 degree longitude grid cell of economic output in Chen and Nordhaus (2011), and the measurement error at the national level can be up to 30% (Feenstra et al. 2013). The synthetic measure of GDP, s, can be calculated once θ is known, and it can then be used as a regressor in equation (7). The μ_{syn} in equation (7) is an OLS estimator, and it can be expressed as a function of μ (see details in Appendix B in the online journal):

$$\mu_{syn} = \frac{(1 - \theta + \theta * \frac{\beta}{\beta}) * \sigma_Y^2}{(1 - \theta + \theta * \frac{\beta}{\beta}) * \sigma_Y^2 + (1 - \theta)^2\sigma_\varepsilon^2 + (\frac{\theta}{\beta})^2\sigma_\tau^2} * \mu. \tag{10}$$

According to the above equation, μ_{syn} has the following three properties: First, when θ equals 0—that is, when no weight is provided by lights—μ_{syn} equals μ_{gdp} in equation (3), and again μ_{syn} is a downward-biased estimator of μ. Second, under the assumption that $\beta = \hat{\beta}$, the equation can be written as

$$\mu_{syn} = \frac{\sigma_Y^2}{\sigma_Y^2 + (1 - \theta)^2\sigma_\varepsilon^2 + (\frac{\theta}{\beta})^2\sigma_\tau^2} * \mu.$$

Here, the estimation bias is not determined by the error in y, σ_ε, as in equation (3), but rather is determined by the weighted error terms of σ_ε and σ_τ. In addition, because the final synthetic value is a product of weighted observed and lights-based measure of GDP, the extent of bias in μ_{syn} is also determined by the weight, θ. Third, unless σ_ε is zero, μ_{syn} will always be a downward-biased estimator of μ, because a part of the denominator $(1 - \theta)^2 \sigma_\varepsilon^2 + (\frac{\theta}{\beta})^2 \sigma_\tau^2$ is always larger than zero. However, this bias can be smaller than the model using observed GDP, particularly when $\sigma_\varepsilon^2 < \frac{\theta \sigma_\tau^2}{\beta(2-\theta)}$. Furthermore, when assumptions of the two-stage least squares (2SLS) method are not met, the synthetic approach probably is a better solution than the substitution approach, as it considers the structural relationship between lights and GDP (σ_τ^2 and β).

3.3. Instrumental Variable Approach

The final possible way of using lights information is to treat lights as an instrumental variable and estimate the GDP coefficient with a 2SLS regression. The benefit of using lights as an instrumental variable is that it not only meets the two criteria (explained below) of being a valid instrumental variable for GDP but it is also available almost anywhere in most time periods and such data can be aggregated to various scales. The lights variable is highly correlated with observed GDP (criteria 1), as shown in Table 1 and Figure 1, and it is less likely to correlate with the error term (criteria 2) in the original regression equation using observed GDP, or u_{gdp} in equation (3).

The measurement error in GDP provided by the Chinese government is more likely to correlate with the error term in the regression model, because measurement error in GDP, ε, is strongly influenced by the incompetency of local statistical systems (Feenstra et al. 2013), which can also relate to u_{gdp}, the error term in the OLS regression. Satellite lights data collected and processed by U.S. agencies are not influenced by the data collection procedures of the Chinese government, and they are therefore less likely to have the error that is correlated with u_{gdp}. In the first-stage regression, the observed GDP, y, is regressed on lights variable m. The OLS estimate of the lights coefficient can then be used to project y. The projected value, $\hat{Y}_{2sls}$, is used as the regressor in the second-stage regression:

$$w = c_{2sls} + \hat{Y}_{2sls}\mu_{2sls} + u_{2sls}$$

and

Table 1. Correlation Coefficients of Variables

	IMR	GDP	Synthetic GDP	Lights	Total Population	Illiterate Females (%)	Female Workers (%)	Agricultural Population (%)	General Hospital	Maternal and Child Health Service
GDP	−.5195									
Synthetic GDP	−.5481	.9965								
Lights	−.6301	.8779	.9149							
Total population	−.3378	.7045	.7007	.6106						
Illiterate females (%)	.4032	−.3911	−.4107	−.4635	−.3172					
Female workers (%)	.1231	.0304	.0157	−.0568	.1409	.2059				
Agricultural population (%)	.1089	−.0265	−.0475	−.1438	.1697	.2363	.4393			
General hospital	.1628	−.3617	−.3571	−.2986	−.478	.0942	−.1223	−.2501		
Maternal and child health service	.1175	−.3428	−.3373	−.2765	−.4355	.1157	−.0519	−.1037	.2919	
Family planning service	.2258	−.319	−.328	−.3377	−.3017	.0957	.0262	.0301	.0769	.1865

Note: GDP = Gross Domestic Product; IMR = infant mortality rate.

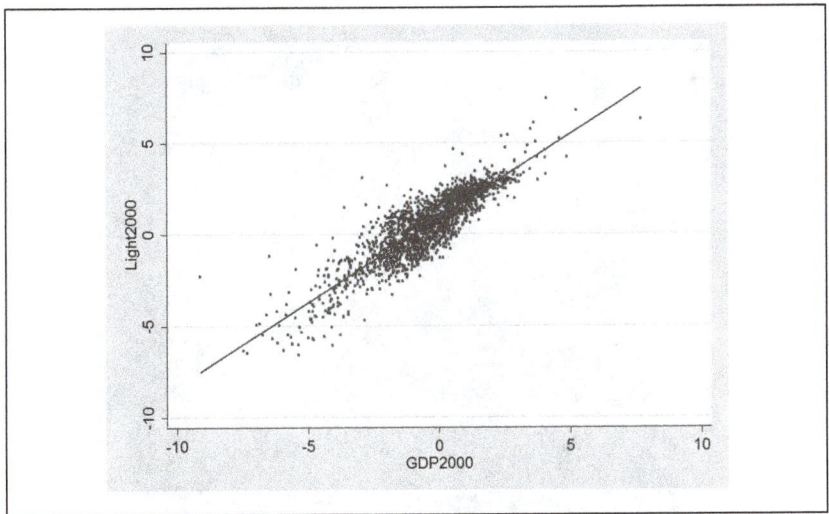

Figure 1. Scatterplot of log Gross Domestic Product (GDP) and log lights for Chinese counties, 2000.

$$\mu_{2sls} = \mu + \frac{Corr(m, u_{2sls})}{Corr(m, y)} * \frac{\sigma_{u_{2sls}}}{\sigma_y}. \tag{11}$$

The instrumental variable (IV) estimator is consistent if $Corr(m, u_{2sls}) = 0$. The bias occurs when $Corr(m, u_{2sls})$ is large and $Corr(m, y)$ is very small. In this study, $Corr(m, y)$ is .88. Thus, if $Corr(m, u_{2sls})$ is zero or very small, the IV estimator will have a very small bias. In the specific case of China, air pollution in Chinese cities has increased with GDP growth over the last decade and could correlate with model errors in the IMR model. Air pollution could also have a dimming effect on lights. However, no studies have been identified investigating to what extent air pollution influences nighttime lights quality in China. Considering that the data are from the year 2000, a period when air pollution in China was not as severe as it is today, we can assume it does not pose a serious problem for the IV approach here. In the above three approaches, estimators are derived in the case of a simple linear regression with one regressor. In the case of multiple regression, the conditional variances and covariance of variables should be considered. In Section 4, the results of three approaches using Chinese county-level data are compared and the difference in outcomes is further explored.

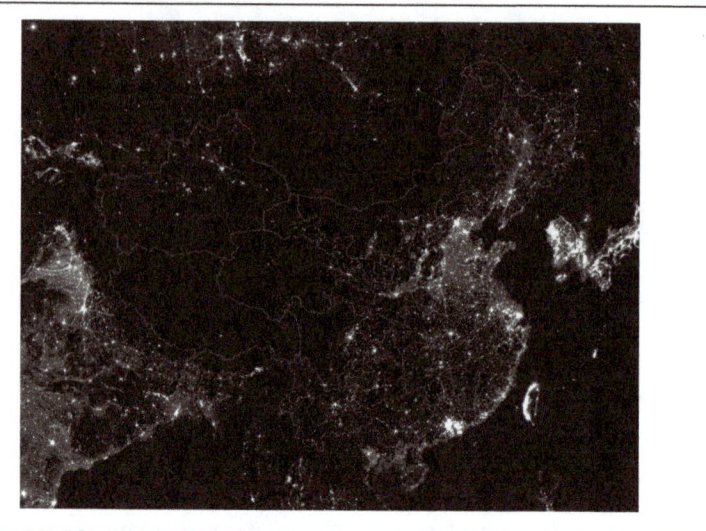

Figure 2. Radiance (calibrated) nighttime light map of China. *Data source:* DMSP-OLS 2014.

4. DATA AND ANALYSIS

4.1. *Lights Data*

The nighttime lights data used in this study are obtained from the National Geophysical Data Center (NGDC) of the National Oceanic and Atmospheric Administration (http://ngdc.noaa.gov/eog/). NGDC provides multiple lights data sets, among which stable and raw lights are annual averaged lights stored in 30 arc-second grids and available from 1992 to 2013 (DMSP-OLS 2015). Because stable lights are processed from raw lights with multiple clean-up steps, they are considered a better quality and are used most often in the lights studies mentioned in Section 2. Most lights data cover the entire globe, except areas with high latitude—that is, beyond 75°N and 65°S latitude. In addition to time series data, DMSP also generates "radiance" or "calibrated" lights (Figure 2). In comparison to stable lights, calibrated lights introduce less measurement error caused by saturation and overglowing, but they are only available for certain years. Because the calibrated lights are available for the year 2000, which is the same year that China county and city IMR was measured, the analysis that follows uses calibrated lights. Log lights density is the main variable used below. It is measured with the logarithmic value of average digital numbers (DNs) of lights per square kilometer, capturing the average intensity of lighting at night. (Additional information on digital numbers can be found at the NOAA DMSP website, http://ngdc.noaa.gov/eog/.)

4.2. *China County/City Data*

The small area IMR data are obtained from China Data Online (chinadataonline.org). The IMR is a crude measure, and it is defined as 1 minus the percentage of the number of living children to live births in the year 2000. There are 2,869 valid observations for this variable, including almost all counties defined by the China 2000 census. The corresponding administrative boundary digital map can also be downloaded from China Data Online. Other county variables downloaded from this website include total population, percentage of illiterate female to total population of 15 and over, percentage of employed female population to total employed population, percentage of agricultural population over total population, general hospitals per 100,000 persons, maternal and child health services per 100,000 persons, and family planning services per 100,000 persons. Population density is defined as the total population residing in a county area divided by the area size in square kilometers. The general economy indicator, GDP, is downloaded from "County Statistics" from China Data Online. The data provide county-level GDP (in 100 million yuan) in 2000 for 2,065 observations. Based on GB (Guobiao) code and county names, only 1,964 observations are merged into IMR data. Then all China demographic and GDP data are merged with lights data. The sample, without missing values, is 1,915. All variables are expressed in natural logarithms, except three local health condition control variables—general hospital, maternal and child health services, and family planning services, as many counties have none of these facilities.

Four models are analyzed: model 1 uses observed GDP as the regressor in OLS; model 2 uses lights data as a substitute for a GDP measure; model 3 uses a lights-based synthetic measure of GDP; and model 4 uses a lights variable as an instrumental variable and 2SLS regression. A set of covariates are also included in all four models. The comparison of the results can help determine whether inclusion of lights data can improve the GDP coefficient estimates compared with the model using observed values of GDP, and it can determine which model produces the estimates with the least attenuation bias in GDP coefficient estimates. The standard errors of estimated coefficients are adjusted with provincial clusters in all models to address potential heteroscedasticity problems. In model 4 (IV method), the standard error reported is adjusted by taking into account original GDP measures.

5. RESULTS

As shown earlier, Table 1 reports correlation coefficients for all variables used in regression models. Observed GDP and lights are strongly correlated ($r = .878$). This is also shown in the scatterplot of log GDP and log lights in Figure 1. The strong association between GDP and lights does not seem to vary across more or less developed counties. The scatterplot (Figure 1) shows that their association is quite consistent across all counties, and correlation coefficients for counties with GDP above and below the GDP mean are .75 and .76, respectively. Differences in correlation coefficients are also small between the poorest county group ($p = .75$ for counties in lowest quantile of GDP) and the most developed county group ($p = .71$ for counties in the highest quantile of GDP). Thus, although economic conditions of Chinese counties are highly uneven, the lights and GDP relationship is quite consistent across all counties.

The synthetic GDP using an optimal weight method is also calculated based on equation (9). The value of θ is predicted based on sample estimates and the prior estimate of measurement error in y. Specifically, for the sample data, the regression coefficient of lights on observed y, $\hat{\beta}$ is .923 ($p < .0001$). The measurement error in reported Chinese county-level GDP in 2000 is estimated to be 40%, and the estimated value of $\tilde{\beta}$ is therefore equal to .973. According to equation (9), the synthetic measure of GDP can be calculated as

$$\text{GDP}_{syn} = .839 * \text{GDP} + .161 * light * 1.028.$$

Here, .839 is the weight for observed GDP, and .161 is the weight for lights-based proxy GDP. The error-corrected, inverted coefficient in the GDP-light structural model is 1.028. In other words, in the synthetic measure of GDP, about 84% of information is from the observed GDP and 16% of information is from lights-based estimates of GDP. Table 1 shows that the synthetic measure is strongly correlated with both observed GDP ($r = .997$) and lights ($r = .915$).

Table 2 reports regression results for four models that predict IMR at the county level. Again, all variables are converted to logarithmic values, except for general hospital, maternal and child health, and family planning services. Because both the theoretical regressor and dependent variable are log-transformed, we can compare the coefficients across models, as they reflect the ratio of the percentage change in IMR to the percentage change in GDP measures. The coefficient value of GDP varies substantially across the four models. It is the lowest in model 1 ($-.195$), which uses observed GDP. It is the highest in model 4, which uses lights as an instrumental variable of GDP ($-.354$). The values of the coefficients in model 2 (using direct substitution) and model 3 (using synthetic GDP) are very close and in between the values in model 1 and model 4. Comparing the goodness-of-model-fit index, R-squared, model 1 also has the lowest value, while model 4 has the highest.

Table 3 reports the Wald tests for equal coefficients across any two models. The tests reject the null hypothesis in all pairs, except for the pair in models 2 and 3. These results indicate significant improvement in coefficient estimates in all lights-based models (models 2, 3, and 4) over the model using observed GDP (model 1). Furthermore, the coefficient of the 2SLS approach is significantly higher than the coefficients in the other three models.

These results suggest that the measurement error in current Chinese county GDP leads to substantial downward bias in OLS estimates. Most likely such error is caused by mistakes or inconsistency in coding, reporting, and calculating in Chinese local or national statistical data. Subsequently, when lights information is included, the downward bias is reduced significantly in models 2, 3, and 4, most likely due to the fact that lights data are not collected by conventional methods. However, comparing models that use lights information, model 4 shows the largest increase in both GDP coefficient and IMR variance explained. The estimation bias in model 2 is probably caused by the imperfect association between lights and true GDP, as lights can be influenced by other factors that are unrelated to GDP, as illustrated in equation (5). We can expect that when the correlation between true GDP and lights is very high, the downward bias in the coefficient estimator (model 2) could be relatively small. The estimator in model 3 is also a biased one, as the

Table 2. Regression Results on Infant Mortality Rate

	Model 1: OLS Using Observed GDP	Model 2: OLS Using Lights as GDP Measure	Model 3: OLS Using Synthetic GDP	Model 4: 2SLS
GDP	-.195**			-.354***
	(-3.55)			(-5.58)
Lights		-.247***		
		(-5.99)		
Synthetic GDP			-.220***	
			(-4.03)	
Total population	.048	.066	.072	.237*
	(.50)	(.87)	(.76)	(2.28)
Illiterate females (%)	.286*	.186+	.263*	.208*
	(2.50)	(1.79)	(2.38)	(1.99)
Female workers (%)	.304	.265	.302	.337
	(1.11)	(1.10)	(1.14)	(1.33)
Agricultural population (%)	-.014	-.180	-.052	-.140
	(-.07)	(-.88)	(-.25)	(-.60)
General hospital	.007	.002	.005	-.002
	(.54)	(.20)	(.42)	(-.14)
Maternal and child health service	-.051*	-.040+	-.051*	-.063*
	(-2.41)	(-1.87)	(-2.35)	(-2.26)
Family planning service	.017	.007	.014	.008
	(1.32)	(.71)	(1.20)	(.80)
Constant	4.792**	5.669***	4.748**	3.230*
	(3.20)	(3.97)	(3.21)	(2.00)
N	1915	1915	1915	1915
Adjusted R-squared	.332	.422	.354	.425

Note: GDP = Gross Domestic Product; OLS = ordinary least squares; 2SLS = two-stage least squares.
+$p < .10$. *$p < .05$. **$p < .01$. ***$p < .001$ in two-tailed tests.

coefficient, μ_{syn}, is determined by the error term in observed GDP (ε) as well as the structural parameters β and τ, as shown in equation (10). In comparison with other estimated coefficients, model 4, using lights as an instrumental variable, produces a better result. In this particular study, the Durbin and Wu-Hausman test rejects the null hypothesis (both statistics are significant at the .0001 level) that the observed GDP is uncorrelated with error terms in the structural model. Therefore, an instrumental-variables estimator is really needed here. The method section also proves that if assumptions of a valid instrumental variable are satisfied, the 2SLS estimator is unbiased. For Chinese county data, the

Table 3. The Values of Chi-square in the Wald Tests of Equal Coefficients across Models

	Model 1: OLS Using Observed GDP	Model 2: OLS Using Lights as GDP Measure	Model 3: OLS Using Synthetic GDP	Model 4: 2SLS
Unstandardized coefficient	−.195	−.247	−.22	−.354
Robust standard error	.055	.041	.055	.063

Wald test for equal coefficient. The null hypothesis is $b1 - b2 = 0$.

Model 2	3.45[+]		
Model 3	17.71***	1.31	
Model 4	28.96***	36.02***	29.31***

Note: The above test considers the covariance of the estimators in addition to standard errors of estimators, as the models are fit on the same sample, and therefore the estimators are not stochastically dependent. GDP = Gross Domestic Product; OLS = ordinary least squares; 2SLS = two-stage least squares.

$+p < .10$. $*p < .05$. $**p < .01$. $***p < .001$ in two-tailed tests.

test on a weak instrumental variable reports the minimum eigenvalue statistic of 2895.01, which rejects the null hypothesis that the nighttime lights variable is a weak instrument.

In short, to address GDP measurement error issues in regression analysis, the instrumental variable approach seems to provide a better solution, if assumptions for the IV regressions are met. The changes in the values of coefficient and goodness-of-fit statistics of IMR regressions illustrate this. Using observed GDP measures, the model 1 results indicate that for a 1% of increase in GDP, the IMR is reduced by .196%, while with lights as an instrumental variable for GDP, the results in model 4 actually show that county-level GDP has a much larger impact on reducing IMR: A 1% increase in GDP leads to a .354% decrease in IMR. The effect estimated with the input of lights information almost doubles the effect estimated with observed GDP data. The IV method can potentially address multiple sources of endogeneity bias including measurement error, omitted variables, and reverse causation. In this particular analysis, additional explanatory variables are added to reduce bias caused by omitted variables, and reverse causation seems counterintuitive, so the IV approach here most likely corrects the bias due to measurement error in GDP.

The coefficients of other independent variables are also shown in Table 2, among which female education level and maternal and child health services are significant. Specifically, the percentage of illiterate women is positively associated with IMR and is significant in all four models. The number of maternal and child health services per 1,000 persons is negatively associated with local IMR, and the coefficient signs are consistent in all models and significant at $\alpha = .05$ in three models. Because the primary goal of this study is to determine the most appropriate approach of using lights in addressing estimation bias, literature and results related to other predictors will not be discussed in detail here. However, more studies on these topics are needed because there is only a small body of literature on *local* IMR and its determinants.

6. DISCUSSION

The primary goal of this study is to demonstrate how nighttime lights can be used to address measurement error issues in economic indicators, particularly in GDP, when economic indicators are used to predict other social patterns or changes. The Chinese IMR study presented here is used as an empirical example for introducing lights data and statistical approaches, which can have wide implications for future sociological studies. First, economic index is crucial to understanding many subjects in sociology, including studies dealing with political systems, demographic processes, health outcomes, cultural adaption, inequality, and other social problems, just to name few. In almost every subfield of sociology, we can identify some midrange theory hypothesizing the effect of economic conditions on the topic that is central to this field, such as economic push-pull factors in immigration studies. In addition, economic conditions are often treated as control variables when other theoretical variables are the subject of research interest. If the index of economic condition is measured with large error, the results of empirical analysis problematize our findings, due to the attenuation bias. That is, the effect of economic conditions may not appear significant or will appear substantially smaller than what is actually the case. The existence of such a problem not only undermines the methodological rigor of the discipline but also provides misleading information to scholars for theory building and empirical forecasting and to governmental agencies for policymaking. Although satellite-based nighttime lights data have made breakthrough methodological contributions in economics over the last 10 years, with more than 3,000 articles discussing their use since 2000 (Nordhaus and Chen 2015), their implication and further contribution to sociology studies have not yet been fully investigated. With formal statistical approaches, this study shows both mathematically and empirically that lights data are able to address attenuation bias when economic measures are used as a regressor.

Furthermore, sociological research on smaller areas, developing regions, or nations with restricted information flow can benefit from findings in this study. The lights data and methodology approaches presented here provide a potential solution for data-quality problems. Data of small areas or from developing regions, such as sub-Saharan Africa, in general have large measurement errors. With appropriate methods, including primary analysis on the reliability of lights,[1] researchers can improve the empirical results substantially. In other cases, lights, as a source for additional information for both economic and demographic indicators, can help researchers working in areas where governments are not willing to publish such information. China is such an example. Fortunately, Chinese local-level information cannot hide from satellites, and the finest resolution lights data published by NOAA (VIIRS lights) can provide information by 15 arc-second (450 meters) geographic grids, and they are not affected by flaws in the collection process or limited by political administrative boundaries. As shown in IMR studies, lights as an instrumental variable for Chinese county GDP, where no other high-quality data are available, improve the estimated coefficient on IMR significantly. Similar approaches can be applied to study other topics in China—for example, labor migration, protests, wealth inequality—or to study those same topics in other developing regions of the world.

Another area that can potentially benefit from the lights data and methods presented here is cross-national comparison research. Measurement issues are crucial in cross-national research (Singh 1995). The attenuation bias caused by measurement error in cross-national indicators can even lead to unexpected results predicted by otherwise valid theoretical propositions (You and Khagram 2005). Measurement errors in such studies commonly exist due to data incompatibility in time frame, different variable definitions and operationalizations, and inconsistent units of analysis across countries or regions. Lights data can potentially provide solutions to these problems, as they cover the entire globe and collect data across regions at the same time, and they can be aggregated to the scale needed by particular research. With rigorous statistical methods, such as the synthetic variables and 2SLS suggested in this study, lights can be used to effectively address attenuation bias problems in cross-national comparison studies, especially in those using economic indicators such as GDP, income, and business establishment. Studies using population density and urbanization can also consider the applicability of lights, as lights data are highly correlated with these variables as well (see Section 2). Additional studies on how to apply lights in cross-national comparison studies are needed. In short, nighttime lights data, coupled with rigorous methods, can address a wide range of methodological problems that are currently unanswered in sociology studies.

7. CONCLUSION

This study is the first to show that nighttime lights data can be used to address estimator bias issues due to measurement error in a regressor. It further points out the wide range of possible applications of nighttime lights in sociological studies. In addition to methodological expositions, the particular methods are explained with an illustration of a GDP and IMR study at the Chinese county level. The methodology section shows that the lights variable can be used to address attenuation bias through three approaches: using lights as a direct substitute measure for GDP, creating a synthetic variable with optimal weights from lights and observed GDP, and using lights as an instrumental variable. The three estimators and their potential biases are discussed thoroughly. Compared with an OLS estimator with observed GDP, the substitution and synthetic measure approaches also cause attenuation bias, but the magnitude of their bias is mostly or partially determined by the parameters of the structural equation between lights and true GDP. Both approaches could potentially reduce the coefficient estimate bias in the regression models in the case that measurement error in observed GDP is very large. However, the degree of reduction is determined by other parameters (see again the methods discussion in Section 3). The analysis using Chinese county-level data in the year 2000 shows that the estimated effects of GDP are higher in substitution and synthetization models than in the model that uses observed GDP. Although using lights as an instrumental variable for economic indicators has not been widely practiced, the methodological and empirical exposition in this study suggests that such an approach can be very effective in addressing bias in coefficient estimates caused by large measurement error in observed economic data. In short, when lights information is added in one form or another, improvement in the estimated economic effect and overall model fit can be substantial, compared with models that use only observed economic data. In the case of Chinese IMR studies, the improvement in the instrumental variable approach is most significant.

The approaches discussed here can be implemented in many other sociology research areas. As widely used predictors or controls in sociology studies, the economic, demographic, and urbanization variables measured with conventional survey methods can cause biased estimation when they are measured with large errors and included in regression models. Theoretically these variables have a great impact on a wide spectrum of subjects of interest to the field. Prior to this study, there were no effective solutions to the empirical problem of attenuation bias across many fields of sociology. Lights data hold a great potential to address such a significant problem. In addition, such data can also meet the needs of researchers with interests in developing regions of the world, small areas, or comparison studies. Coupled with statistical approaches, lights data can improve methodological rigor in these studies, given that data quality is a quite common problem, often becoming an obstacle for advanced empirical analysis.

Even though research into the use of satellite-based nighttime lights data, or other remote-sensing information in the social sciences, is still at an early stage, there now appears little doubt that with appropriate methods, such information can potentially address some limitations or deficiencies in research based on conventional data collection or calculation. With more satellite imagery data made available to researchers over the last several decades, there is a possibility that sociology and other social science disciplines will expand research beyond the limitations imposed by current data problems for some parts of the world. This not only benefits scholars, by opening avenues of new information and hence new studies, but also aids policymakers, by providing more precise regionalized information with which to create and assess policy.

Acknowledgments

I thank the following individuals for feedback on earlier drafts of this paper: Salvatore Babones (University of Sydney), Zhipeng Liao (UCLA), William Nordhaus (Yale University), Dudley Poston (Texas A&M University), and Joel M. Vaughan and Keith Kerr (Quinnipiac University).

Notes

1. Researchers should check the association between lights and the variable of their interest prior to using lights data in a formal statistical model, as such association can vary by country or vary across subsamples, even though this is not the case for GDP and lights for Chinese counties. See Chen and Nordhaus (2011) for more explanation on the variation of the relationship between lights and GDP.

References

Babones, Salvatore J. 2013. *Methods for Quantitative Macro-comparative Research.* Thousand Oaks, CA: Sage Publications.

Bharti, Nita, Andrew J. Tatem, Matthew Ferrari, Rebeeca Grais, Ali Djibo, and Bryan Grenfell. 2011. "Explaining Seasonal Fluctuations of Measles in Niger Using Nighttime Lights Imagery." *Science* 334(6061):1424–27.

Chen, Xi, and William Nordhaus. 2011. "Using Luminosity Data as a Proxy for Economic Statistics." *The Proceedings of National Academy of Sciences* 108(21): 8589–94.

Chen, Xi. 2015. "Explaining Subnational Infant Morality and Poverty Rates: What Can We Learn from Night-time Lights?" *Spatial Demography* 3(1):27–53.

Chen, Xi. 2016. "Using Nighttime Lights Data as a Proxy in Social Scientific Research." Pp. 301–23 in *Recapturing Space: New Middle-range Theory in Spatial Demography*, edited by F. Howell, J. Porter, and S. Matthews. New York: Springer International.

DMSP-OLS. 2015. "National Defense Meteorological Satellite Program, Image and Data processing by NOAA's National Geophysical Data Center. DMSP data collected by the US Air Force Weather Agency." Retrieved April 10, 2015 (http://www.ngdc.noaa.gov/dmsp/dmsp.html).

Doll, Christopher, Jan-Peter Muller, and Christopher Elvidge. 2000. "Nighttime Imagery as a Tool for Global Mapping of Socio-economic Parameters and Greenhouse Gas Emissions." *Ambio* 29(3):157–62.

Ebener, Steve, Christopher Murray, Ajay Tandon, and Christopher Elvidge. 2005. "From Wealth to Health: Modeling the Distribution of Income per Capita at the Sub-national Level Using Night-time Light Imagery." *International Journal Health Geographics* 4(1):5–14.

Elvidge, Christopher D., Kimberly E. Baugh, Eric Kihn, Herbert Kroehl, E. R. Davis, and C. W. Davis. 1997. "Relation Between Satellites Observed Visible-near Infrared Emissions, Population, Economic Activity and Electric Power Consumption." *International Journal of Remote Sensing* 18(6):1373–79.

Elvidge, Christopher D., Marc L. Imhoff, Kimberly E. Baugh, Vinita Ruth Hobson, Ingrid Nelson, Jeff Safran, John B. Dietz, and Benjamin T. Tuttle. 2001. "Night-time Lights of the World: 1994–1995." *ISPRS Journal of Photogrammetry and Remote Sensing* 56(2):81–99.

Elvidge, Christopher D., Jeffrey Safran, Benjamin Tuttle, Paul Sutton, Pierantonio Cinzano, Donald Pettit, John Arvesen, and Christopher Small. 2007. "Potential for Global Mapping of Development via a Nightsat Mission." *GeoJournal* 69 (1-2): 45–53.

Elvidge, Christopher D., Paul C. Sutton, Tilottama Ghosh, Benjamin T. Tuttle, Kimberly E. Baugh, Budhendra Bhaduri, and Edward Bright. 2009. "A Global Poverty Map Derived from Satellite Data." *Computers and Geosciences* 35(8):1652–60.

Elvidge, Christopher, Kimberly Baugh, Paul Sutton, Budhendra Bhaduri, Benjamin Tuttle, Tilottama Ghosh, Daniel Ziskin, and Edward H. Erwin. 2011. "Who's in the Dark: Satellite Based Estimates of Electrification Rates." Pp. 211–24 in *Urban Remote Sensing: Monitoring, Synthesis and Modeling in the Urban Environment*, edited by X. Yang. Chichester, England: Wiley-Blackwell.

Elvidge, Christopher, Kimberly Baugh, Sharolyn J. Anderson, Paul Sutton, and Tilottama Ghosh. 2012. "The Night Light Development Index (NLDI): A Spatially Explicit Measure of Human Development from Satellite Data." *Social Geography* 7(1):23–35.

Feenstra, R. C., H. Ma, J. Peter Neary, and D. S. Prasada Rao. 2013. "Who Shrunk China? Puzzles in the Measurement of Real GDP." *The Economic Journal* 123(573): 1100–29.

Frisbie, W. P. 2006. "Infant Mortality." Pp. 251–82 in *Handbook of Population*, edited by D. L. Poston Jr. and M. Micklin. New York: Springer.

Ghosh, Tilottama, Rebecca Powell, Christopher Elvidge, Kimberly Baugh, Paul Sutton, and S. Anderson. 2010a. "Shedding Light on the Global Distribution of Economic Activity." *The Open Geography Journal* 3(1):148–61.

Ghosh, Tilottama, Christopher Elvidge, Paul Sutton, Kimberly Baugh, Daniel Ziskin, and Benjamin Tuttle. 2010b. "Creating a Global Grid of Distributed Fossil Fuel CO_2 Emissions from Nighttime Satellite Imagery." *Energies* 3(12):1895–1913.

Goodchild, Michael F., and Donald G. Janelle. 2004. *Spatially Integrated Social Science*. New York: Oxford University Press.

Henderson, Vernon, Adam Storeygard, and David N. Weil. 2011. "A Bright Idea for Measuring Economic Growth." *American Economic Review* 101(3):194–99.

Henderson, Vernon, Adam Storeygard, and David N. Weil. 2012. "Measuring Economic Growth from Outer Space." *American Economic Review* 102(2):994–1028.

Hodler, Roland, and Paul A. Raschky. 2014 "Economic Shocks and Civil Conflict at the Regional Level." *Economics Letters* 124(3):530–33.

Li, Xi, Huimin Xu, Xiaoling Chen, and Chang Li. 2013. "Potential of NPP-VIIRS Nighttime Light Imagery for Modeling the Regional Economy of China." *Remote Sensing* 5(6):3057–81.

Liang, Hanwei, Hiroki Tanikawa, Yasunari Matsuno, and Liang Dong. 2014. "Modeling In-use Steel Stock in China's Buildings and Civil Engineering Infrastructure Using Time-series of DMSP/OLS Nighttime Lights." *Remote Sensing* 6(6):4780–4800.

Mellander, Charlotta, José Lobo, Kevin Stolarick, and Zara Matheson. 2015 "Nighttime Light Data: A Good Proxy Measure for Economic Activity?" *PLoS One* 10(10): e0139779.

Noor, Abdisalan M., Victor A. Alegana, Peter W. Gething, Andrew J. Tatem, and Robert W. Snow. 2008. "Using Remotely Sensed Night-time Light as a Proxy for Poverty in Africa." *Population Health Metric* 6(1):1–13.

Nordhaus, William, and Xi Chen. "A Sharper Image? Estimates of the Precision of Nighttime Lights as a Proxy for Economic Statistics." *Journal of Economic Geography* 15(1):217–46.

Oda, Tomohiro, and Shamil Maksyutov. 2011. "A Very High-resolution (1 km × km) Global Fossil Fuel CO_2 Emission Inventory Derived Using a Point Source Database and Satellite Observations of Nighttime Lights." *Atmospheric Chemistry and Physics* 11(2):543–56.

Poston, Dudley. 1996. "Patterns of Infant Mortality." Pp.47–65 in *China: The Many Facets of Demographic Change*, edited by A. Goldstein and W. Feng. Boulder, CO: Westview Press.

Sanderson, Eric W., Malanding Jaiteh, Marc A. Levy, Kent H. Redford, Antoinette V. Wannebo, and Gillian Woolmer. 2000. "The Human Footprint and the Last of the Wild." *Bioscience* 52(10):891–904.

Shi, Kaifang, Bailang Yu, Yixiu Huang, Yingjie Hu, Bing Yin, Zuoqi Chen, Liujia Chen, and Jianping Wu. 2014. "Evaluating the Ability of NPP-VIIRS Nighttime Light Data to Estimate the Gross Domestic Product and the Electric Power Consumption of China at Multiple Scales: A Comparison with DMSP-OLS Data." *Remote Sensing* 6(2):1705–24.

Singh, Jagdip. 1995. "Measurement Issues in Cross-national Research." *Journal of International Business Studies* 26(3):597–619.

Sutton, Paul, Christopher Elvidge, and Tilottama Ghosh. 2007. "Estimation of Gross Domestic Product at Sub-national Scales Using Nighttime Satellite Imagery." *International Journal of Ecological Economics and Statistics* 8(S07):5–21.

United Nations, Department of Economic and Social Affairs, Population Division. 2013. *World Population Prospects: The 2012 Revision*. New York: Population Division of the Department of Economic and Social Affairs of the United Nations Secretariat.

You, Jong-Sung, and Sanjeev Khagram. 2005 "A Comparative Study of Inequality and Corruption." *American Sociological Review* 70(1):136–57.

Zhao, Naizhuo, Nate Currit, and Eric Samson. 2011. "Net Primary Production and Gross Domestic Product in China Derived from Satellite Imagery." *Ecological Economics* 70(5): 921–28.

Author Biography

Xi Chen is an assistant professor of sociology at Quinnipiac University. She was previously a research scientist for the Department of Economics at Yale University before joining Quinnipiac University, and she is currently still managing the Yale G-Econ project (Gecon.yale.edu). Her research specialties include quantitative methods, demography, ethnic population in China, and the application of satellite-based nighttime lights data.

Sociological Methodology
2016, Vol. 46(1) 345–357
© American Sociological Association 2016
DOI: 10.1177/0081175016654736
http://sm.sagepub.com

$SAGE

$\mathcal{S} 11 \mathcal{E}$

MODELING CAUSAL IRRELEVANCE IN EVALUATIONS OF CONFIGURATIONAL COMPARATIVE METHODS

Alrik Thiem*
Michael Baumgartner*

Abstract

In volume 44 of Sociological Methodology, *Lucas and Szatrowski argued that the method of qualitative comparative analysis (QCA) suffers from a built-in confirmation bias due to a proclivity for including conditions in its output models whose corresponding factors are not systematically correlated with the endogenous factor. The authors therefore urged that QCA be abandoned. With this comment, we pursue four related objectives: first, we explain why correlation-based evaluation designs for testing QCA's power of discrimination with respect to causally irrelevant factors are unsuitable; second, we show how appropriate tests must be constructed; third, we offer an* R *function that implements a routine for such tests; and fourth and finally, we conduct three series of comprehensive tests, all of whose results indicate that QCA does not suffer from the kind of confirmation bias criticized by Lucas and Szatrowski.*

*University of Geneva, Geneva, Switzerland

Corresponding Author:
Alrik Thiem, University of Geneva, Department of Philosophy, Rue de Candolle 2/Bât. Landolt, 1211 Geneva, Switzerland
Email: alrik.thiem@unige.ch

Keywords

causal inference, confirmation bias, method evaluation, qualitative compara-
tive analysis (QCA), simulations, configurational comparative methods

1. INTRODUCTION

All procedures of causal inference must pass key tests with simulated
data before they can be relied on to work as intended with real-life data,
whose behavior is no longer under the full control of the researcher (cf.
Pearl 2009; Spirtes, Glymour, and Scheines 2000). Only if the procedure in question passes these tests satisfactorily can it be safely assumed
that whenever the procedure's background assumptions are likely to be
satisfied, it is equally likely the generated output will be correct. Yet, in
the case of qualitative comparative analysis (QCA), a method of empirical data analysis largely popularized by Ragin (1987, 2000, 2008), the
order of this standard protocol has not been adhered to.

Although already part of the methodological repertoire of numerous
scientific communities, QCA has only lately been subjected to some
first evaluations of a few of its properties (Baumgartner 2015; Bowers
2014; Glaesser and Cooper 2014; Hug 2013; Krogslund, Choi, and
Poertner 2015; Lucas and Szatrowski 2014; Seawright 2005, 2014;
Skaaning 2011; Thiem 2014a, 2014b; Thiem, Spöhel, and Duşa 2016).[1]
However, not all of these evaluations suggest that the method's reversal
of the standard protocol for establishing the correctness of a new procedure of causal inference has been inconsequential. In a recent contribution to this journal, Lucas and Szatrowski (2014) (LS) argue nothing
less than that QCA "identifies causal patterns in noncausal data," even
in the most ideal circumstances, whereby it is rendered "a wholly inef-
fective research method, providing a fatal distraction that, far from rea-
lizing the promise of qualitative and comparative historical research,
sabotages such analyses" (p. 3). The implications of this assessment are
a reason for serious concern. If QCA were indeed unfit for causal inference, as LS assert, a quarter century after its introduction and hundreds
of empirical applications, partly in areas as sensitive as human health
research and environmental policy (e.g., Britt et al. 2000; Cragun et al.
2014; Dy et al. 2005; Kaminsky and Javernick-Will 2014; Robinson,
Holland, and Naughton-Treves 2014; Srinivasan et al. 2012), the widespread diffusion the method has achieved to date would mean a scientific disaster of considerable magnitude.

Because of these potentially far-reaching repercussions, it is all the more incumbent on methodologists to carefully scrutinize the evaluation by LS. The one experiment on which these authors base their assessment is simple and straightforward. A single data set of 40 cases under four causally relevant exogenous factors X_1 through X_4, an endogenous factor Y, and one additional factor Z_1 is generated in three steps (Lucas and Szatrowski 2014:17–23). First, Bernoulli trials are performed for X_1 through X_4 to simulate independent observations; second, LS let Y take on the value 1 if, and only if, $X_1 = X_2 = 1$ or $X_3 = X_4 = 1$; and third, the authors add Z_1, which is not systematically correlated with the endogenous factor. Then, LS expect QCA to recover the model $\mathbf{m}(\Delta_{LS})$ in Expression (1),

$$\mathbf{m}(\Delta_{LS}) : X_1^{\{1\}} X_2^{\{1\}} \vee X_3^{\{1\}} X_4^{\{1\}} \Leftrightarrow Y^{\{1\}}; \tag{1}$$

and not to attribute any causal relevance to Z_1.[2] The authors find, however, that "[f]ar from identifying Z_1 as noncausal, QCA deeply embeds the noncausal Z_1 into the causal story" (p. 20), a result that allows of only one conclusion, namely, that "QCA is wrong" (p. 23).

LS add random variables *ex post facto* to their data sets, a test design that representatives of QCA have also implemented in their critique of LS (Fiss, Marx, and Rihoux 2014:97). However, the results of such tests are futile because *QCA analyzes implicational patterns between combinations of factor levels, not correlational patterns between factors*. The recognition of this fundamental distinction is crucial, yet it has, strangely enough, been completely missing so far in all existing debates on QCA (cf. Thiem, Baumgartner, and Bol 2016). In Section 2, we thus lay out, for the first time in the literature, how to construct adequate tests for confirmation bias in QCA that are in accordance with the method's search target.

2. MODELING CAUSAL IRRELEVANCE IN QCA

The theory of causation underlying QCA is based on the concept of an *INUS cause*—an *i*nsufficient but *n*ecessary part of a condition that is itself *u*nnecessary but *s*ufficient for an outcome (Mackie 1965:245, 1974:59–87). Under the theory of INUS causality, some condition $A^{\{\cdot\}}$ is causally relevant to another condition $B^{\{\cdot\}}$ within the same causal field $\mathscr{F}$ if, and only if, $A^{\{\cdot\}}$ is a difference maker for $B^{\{\cdot\}}$ in $\mathscr{F}$.[3] It is such a difference maker if there exists a pair of (Boolean) conjunctions $(A^{\{\cdot\}}\Phi, \neg A^{\{\cdot\}}\Phi)$, where Φ denotes another condition of unspecified complexity, for which it is true that $A^{\{\cdot\}}\Phi$ implies $B^{\{\cdot\}}$, while it is not true that $\neg A^{\{\cdot\}}\Phi$ also implies $B^{\{\cdot\}}$.[4]

As QCA searches for INUS conditions, it must be guaranteed in tests of its powers of discrimination with respect to causally irrelevant factors that the data to which QCA is applied do not contain pairs of conjunctions of the aforementioned type if the factor in question is not to be identified as somehow causally relevant to the outcome. Because of the way in which LS have simulated their data, however, the absence of such pairs could not be ensured, as a result of which QCA identified the putatively irrelevant factors as causally relevant. Data simulation strategies by means of which the absence of difference-making scenarios under the framework of INUS causality can be guaranteed through rendering endogenous factors implicationally independent are described in Section 2.1. To also create a transparent basis for our own evaluation in Section 2.2, we embed the introduction of these strategies in a sufficiently comprehensive formalization of QCA's work flow. Because LS have not evaluated multivalue QCA (Cronqvist and Berg-Schlosser 2009; Thiem 2013), we limit ourselves to bivalent (Boolean) factors in the remainder of this article.[5]

2.1. *The Work Flow of QCA and Implicational Independence*

The input to QCA is a data set δ of dimension $n \times k + 1$, where n stands for the number of cases (rows) and k for the number of factors (columns) under the factor frame $\mathbf{F} = \{F_1, F_2, \ldots, F_k, F_o\}$. The output of QCA is a solution $\mathbb{S}(\mathbf{M})$, which is composed of a set of models $\mathbf{M} = \{\mathbf{m}_1, \mathbf{m}_2, \ldots, \mathbf{m}_j\}$, each of which provides an equally well fitting causal account of the outcome $F_o^{\{\cdot\}}$. If QCA's background assumptions are satisfied, a nonempty subset of $\mathbf{M}$, $\mathbf{M}'(\Delta) = \{\mathbf{m}_1, \mathbf{m}_2, \ldots, \mathbf{m}_i\}$, includes only models all of whose properties reflect at least some properties of the data-generating (yet usually unknown) causal structure Δ. If the data are not fragmented, $\mathbf{M}'(\Delta)$ will include the model that reflects all properties of Δ (cf. Baumgartner and Thiem 2015b). Between the input and the output, QCA condenses the information in δ into a truth table $\mathbf{T}(\delta)$ of dimension $d \times (k + 1)$, with $d = 2^k$, before minimizing all implicational dependencies present in $\mathbf{T}(\delta)$ by means of a purpose-built algorithm (cf. Thiem 2015b:727; Thiem and Duşa 2013a:508).

The antecedent Ψ of any model $\mathbf{m} \in \mathbf{M}$ is a minimally necessary disjunction of conjunctions, each of which is minimally sufficient for $F_o^{\{\cdot\}}$. Such conjunctions are called *prime implicants* in QCA. An *implicant* $\bigwedge F_i^{\{\cdot\}} = \Phi$ is a prime implicant of $F_o^{\{\cdot\}}$ if, and only if, Φ is sufficient for $F_o^{\{\cdot\}} (\Phi \Rightarrow F_o^{\{\cdot\}})$, and there exists no proper subset Φ^C of Φ such that $\Phi^C \Rightarrow F_o^{\{\cdot\}}$. A proper subset Φ^C of Φ results from a reduction of Φ by at least one of its conjuncts. A disjunction of prime implicants $\bigvee \Phi_i = \Psi$ is minimally necessary for $F_o^{\{\cdot\}}$ if, and only if, Ψ is necessary for $F_o^{\{\cdot\}} (F_o^{\{\cdot\}} \Rightarrow \Psi)$, and there exists no proper subset Ψ^C of Ψ such that $F_o^{\{\cdot\}} \Rightarrow \Psi^C$. A proper subset Ψ^C of Ψ results from a reduction of Ψ by at least one of its prime implicants. If Ψ is the antecedent in a model $\mathbf{m}$ of $F_o^{\{\cdot\}}$, then it is both minimally sufficient and minimally necessary for $F_o^{\{\cdot\}} (\Psi \Leftrightarrow F_o^{\{\cdot\}})$ (cf. Baumgartner 2008).

For $F_o^{\{\cdot\}}$ to be implicationally independent of F_j, the latter of whose levels are ℓ_1 and ℓ_2, Ψ must neither feature $F_j^{\{\ell_1\}}$ nor $F_j^{\{\ell_2\}}$. This can be operationalized in various ways; implicational independence under INUS causality merely requires F_j not to be a difference maker with respect to $F_o^{\{\cdot\}}$. We introduce six syntactically dissimilar yet logically equivalent possibilities to ensure the causal irrelevance of F_j. They are given in Expressions (2) to (7):

$$\mathbf{m}(\Delta) : \left(\Phi_1 \vee \Phi_2 \vee \cdots \vee \Phi_z \Leftrightarrow F_o^{\{\cdot\}} \right) \wedge \left(F_j^{\{\ell_1\}} \vee F_j^{\{\ell_2\}} \right) \quad (2)$$

$$\mathbf{m}(\Delta) : \left(\Phi_1 \vee \Phi_2 \vee \cdots \vee \Phi_z \right) \wedge \left(F_j^{\{\ell_1\}} \vee F_j^{\{\ell_2\}} \right) \Leftrightarrow F_o^{\{\cdot\}} \quad (3)$$

$$\mathbf{m}(\Delta) : \left(\Phi_1 \wedge \left(F_j^{\{\ell_1\}} \vee F_j^{\{\ell_2\}} \right) \vee \Phi_2 \vee \cdots \vee \Phi_z \right) \Leftrightarrow F_o^{\{\cdot\}} \quad (4)$$

$$\mathbf{m}(\Delta) : \left(\Phi_1 \vee \Phi_2 \wedge \left(F_j^{\{\ell_1\}} \vee F_j^{\{\ell_2\}} \right) \vee \cdots \vee \Phi_z \right) \Leftrightarrow F_o^{\{\cdot\}} \quad (5)$$

$$\mathbf{m}(\Delta) : \left(\Phi_1 \vee \Phi_2 \vee \cdots \wedge \left(F_j^{\{\ell_1\}} \vee F_j^{\{\ell_2\}} \right) \vee \Phi_z \right) \Leftrightarrow F_o^{\{\cdot\}} \quad (6)$$

$$\mathbf{m}(\Delta) : \left(\Phi_1 \vee \Phi_2 \vee \cdots \vee \Phi_z \wedge \left(F_j^{\{\ell_1\}} \vee F_j^{\{\ell_2\}} \right) \right) \Leftrightarrow F_o^{\{\cdot\}} \quad (7)$$

Thus, any evaluation with the aim of testing whether QCA detects the causal irrelevance of F_j necessitates generating its data on the basis of the *complete* set of factors in **F**. Simply appending random variables *ex post facto* to δ, as Fiss et al. (2014), Krogslund et al. (2015), and LS have done, is incorrect. If data are added to δ, the sampling function must resort to **F** in order to prevent the instantiation of conjunctions that are not in accordance with Δ. As this is impossible when further cases are created randomly and independently of Δ, the appropriate way is to take the reverse approach and simulate data on the basis of an output-evaluated truth table, with any of the functions in Expressions (2) to (7) being equally suitable for this purpose.

2.2. Revisiting LS

In this subsection, we attempt to replicate the finding of LS of their Study 1 (Lucas and Szatrowski 2014:17–27) that QCA suffers from confirmation bias. As LS did not provide us with their original data set even after an official request was made through the editorial office of *Sociological Methodology*, we decided to extend their single experiment to three different series of tests: one in which correlations between the irrelevant factor and the outcome factor are closely distributed around zero (mimicking the original design of LS), one in which correlations systematically approach positive unity, and one in which correlations systematically approach negative unity. Each series includes five blocks of 1,000 data experiments each, yielding a total of 15,000 individual data experiments. We vary the size of the data sets for each block and the range of correlations for the two test series in which correlations approach unity. We begin with data sets of 40 cases and double their size four times to 80, 160, 320, and 640 cases.[6] To create the set of ideal circumstances that LS have claimed to be irrelevant to QCA's failure, we assume neither limited empirical diversity nor inconsistencies.[7] The causal structure we seek to recover remains the same as that stipulated by LS, namely, Expression (1).

A scatterplot of correlations between the irrelevant factor and the outcome factor within all 15,000 data experiments is shown in Figure 1. Each field of points represents a block of experiments. The small gray dots plot the correlations of the zero-correlation series, whereas the larger black dots on the right-hand side plot the correlations of the positive-correlation series, and the larger black dots on the left-hand side plot the correlations of the negative-correlation series.

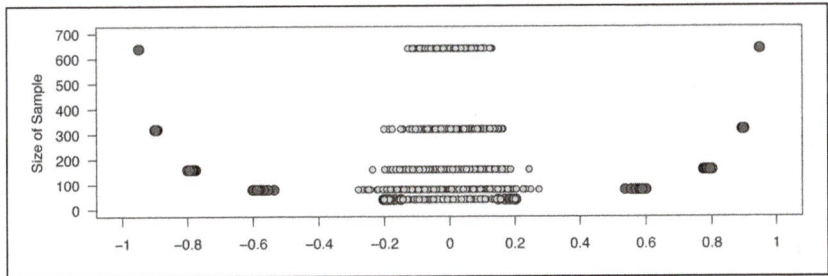

Figure 1. Scatterplot of correlations between irrelevant factor and outcome factor.

Underlying our tests is an *R* function called *implicIndep* (*implicational independence*), which implements Expression (2) for building random data sets on the basis of any Boolean function in disjunctive normal form. This function has been incorporated into the *QCApro* package (Thiem 2016).[8] Besides the possibility of specifying a suitable Boolean function and the number of samples, the correlation between the irrelevant factor and the outcome factor can be controlled via two additional arguments, one for the direction of correlation and one for the sample size of the data sets.[9] For reasons of transparency, the mechanism of *implicIndep* is summarized in the pseudocode in Algorithm 1.

From the Boolean function fed into *implicIndep*, an unevaluated truth table is constructed. All rows that are in accordance with the function are subsequently filtered to constitute an evaluated truth table, which provides the base data set with one case per configuration.[10] Subsequently, as many random data sets of the specified size are built while ensuring the desired direction of correlation between the irrelevant factor and the outcome factor. The larger the size of the data sets, the stronger the correlation. After all QCA solutions have been generated, *implicIndep* tests whether the irrelevant factor is found in any model recorded in the solution that has been derived from a data set.[11] If in at least one of the solutions of a block of data experiments this factor appears, in either of its two levels, the test does not count as passed.

The result of our three series of tests is unambiguous; it can be put into a single sentence: In each run, QCA successfully recovered the data-generating structure in Expression (1) and correctly eliminated the irrelevant factor Z_1. This outcome provides ample evidence that QCA does not suffer from the type of built-in confirmation bias criticized by LS. It also corroborates empirically what we have argued theoretically

```
1 Function implicIndep (Boolean function, number of samples, sample size,
     direction of correlation)
        Input:   A Boolean function in disjunctive normal form
        Output:  A test result of whether the irrelevant factor has been eliminated for all
                 samples, and a vector of correlations between the outcome factor and
                 the irrelevant factor
        /* syntax and consistency checks performed first    */
        /* code proper from here                            */
2       factors ← vector of size k of unique factor names entailed by function;
3       unev.tt ← unevaluated truth table of dimension 2^{k+1} × (k + 1);
4       base.dat ← base data set (rows from unev.tt compatible with function);
5       sol.list ← list of length i (# samples) for QCA solutions of eQMC;
6       if correlation is set to zero then
7           for each i do
8               construct data set from base.dat, add cases randomly chosen from this
                    data set, compute QCA solution, record this solution
9           end
10      end
11      else
12          if correlation is set to positive then
13              record all cases for which outcome factor and irrelevant factor show same
                    levels in vector add.pos
14          else
15              record all cases for which outcome factor and irrelevant factor show not
                    same levels in vector add.neg
16          end
17          for each i do
18              construct base data set from base.dat, add cases randomly chosen from
                    add.pos or add.neg, compute QCA solution, record this solution
19          end
20      end
21      for each i do
22          test whether irrelevant factor contained in Boolean function is found in any
                model of a recorded solution
23      end
24 end
```

Algorithm 1: Pseudo code of test function *implicIndep*

in Section 2.1, namely, that correlations are an inappropriate criterion in evaluations of QCA, which searches for redundancy-free patterns of implications under the theory of INUS causality.

3. CONCLUSIONS

In a recent method evaluation published as the center of a symposium in volume 44 of *Sociological Methodology*, Lucas and Szatrowski (2014) have argued that QCA suffers from confirmation bias and should therefore be abandoned immediately. In this comment, we have shown why the conclusions drawn in that evaluation are uninformative as a result of the test design used by LS, which was not in accordance with the search target of QCA. Instead of ensuring implicational independence in the search for INUS conditions—the adequate way to model causal irrelevance with configurational data and methods—tests were constructed on the basis of unsystematic pairwise correlations. If QCA is to be assessed properly, however, its search target must be respected. Evaluating a procedure of causal inference on the basis of properties of causal structures the procedure was never designed to uncover represents a pointless exercise.[12]

The appropriate standards for evaluating QCA with respect to its power to discriminate between causally relevant and causally irrelevant factors have been introduced and implemented in a flexible R function we have made available together with this article. Using this function, we carried out three test series ourselves and found that in each single run, QCA correctly eliminated the causally irrelevant factor. The method successfully recovered all relevant properties of the data-generating structure, irrespective of the direction or size of the correlation between factors in the data.

Although we have shown that QCA's reversal of the standard protocol for establishing the validity of a new procedure of causal inference has been inconsequential with respect to its evaluation of confirmation bias on the basis of correlation-based designs, we emphasize that our conclusions in no way affect evaluations of other methodological properties of QCA as currently practiced. Further work may well reveal as yet unknown problems of a considerable magnitude that would have been identified already if the order of the standard protocol for introducing procedures of causal inference to applied research had been adhered to. For example, the concealment of model ambiguities and the incorrectness of two solution types have been exposed as two additional, serious problems in QCA (Baumgartner and Thiem 2015a, 2015b; Thiem 2014c). Moreover, it is reasonable to conjecture that given limited diversity and factor set misspecification—two problems of configurational data analysis whose combined effect on the validity of QCA has not yet been evaluated—QCA will produce incorrect results by embedding a causally irrelevant factor in solutions. Irrespective of the issue in question, however, basic rules of method evaluation, such as the recognition of the search target of the procedure on trial, must be abided by in all future tests of QCA if they are to be of scientific value.

Funding

This research received financial support from the Swiss National Science Foundation (project award number PP00P1_144736/1).

Notes

1. See http://www.compasss.org for an overview of QCA-related publications since 1984.
2. Instead of uppercase and lowercase notation, we use value notation to reduce ambiguities in the representation of factors and factor levels. As usual, the conjunction operator, $\wedge$, is dropped whenever there is no risk of ambiguity.
3. The concept of INUS causality has not been well understood in the QCA literature, either (Thiem and Baumgartner 2016).
4. For example, let $A^{\{\cdot\}}$ denote the presence of a sprinkler system in an analysis of a series of arson attacks. If it turned out that a wooden house never burned down when such a system had been present and the fire department was notified within the first 12 minutes, whereas every wooden house without such a system did burn down even when the fire department was notified in time (no house built of stone burned down irrespective of whether it had a sprinkler system or the fire department was notified in time), the presence of sprinklers represents an INUS cause of a wooden house's not burning down.
5. Multivalue QCA is a generalization of crisp-set QCA; see Thiem (2014d, 2015a) for more details.
6. Although we are unsure as to what LS would expect QCA to output when correlations systematically increase in magnitude, we include the two additional test series to further support our argument that correlational relations between factors are no valid criterion for evaluating QCA.
7. The truth table created by LS was not free of limited diversity. However, proponents of QCA support the assumption, contrary to what LS claimed, that ideal recovery contexts for QCA require the absence of limited diversity and inconsistencies (Schneider and Wagemann 2012:119). Space constraints do not allow us to explain why the presence of limited diversity in the truth table created by LS is inconsequential with respect to an evaluation of QCA under ideal conditions.
8. The *QCApro* package is a successor to the *QCA* package (Duşa and Thiem 2014; Thiem and Duşa 2013b). For carrying out the minimization process, *implicIndep* uses the *eQMC* algorithm (Duşa and Thiem 2015). An annotated replication script is available in the online appendix to this article or from the corresponding author on request.
9. As we do not examine the effect of limited empirical diversity, the number of cases is tied to the strength of the correlation for reasons of functional efficiency.
10. As the truth table must not show limited diversity, the size of a data sample can never be smaller than the number of rows in the baseline data set.

11. Demanding the irrelevant factor to be absent from all models in **M** is at least as strict as demanding it to be absent from all models in **M'**(Δ).
12. Such efforts would yield some benefits only if it were found that the procedure *systematically* uncovers properties of causal structures for whose discovery it has not been primarily designed in addition to those for whose discovery it has been explicitly designed.

References

Baumgartner, Michael. 2008. "Regularity Theories Reassessed." *Philosophia* 36(3): 327–54.

Baumgartner, Michael. 2015. "Parsimony and Causality." *Quality and Quantity* 49(2): 839–56.

Baumgartner, Michael, and Alrik Thiem. 2015a. "Model Ambiguities in Configurational Comparative Research." *Sociological Methods and Research*. Retrieved June 8, 2016 (http://smr.sagepub.com/content/early/2015/10/22/0049124115610351).

Baumgartner, Michael, and Alrik Thiem. 2015b. "Often Trusted but Never (Properly) Tested: Evaluating Qualitative Comparative Analysis." Presented at the 12th Conference of the European Sociological Association, August 25–28, Czech Technical University, Prague, Czech Republic.

Bowers, Jake. 2014. "Comment: Method Games—A Proposal for Assessing and Learning about Methods." *Sociological Methodology* 44(1):112–17.

Britt, David W., Samantha T. Risinger, Virginia Miller, Mary K. Mans, Eric L. Krivchenia, and Mark I. Evans. 2000. "Determinants of Parental Decisions after the Prenatal Diagnosis of Down Syndrome: Bringing in Context." *American Journal of Medical Genetics* 93(5):410–16.

Cragun, Deborah, Rita D. DeBate, Susan T. Vadaparampil, Julie Baldwin, Heather Hampel, and Tuya Pal. 2014. "Comparing Universal Lynch Syndrome Tumor-screening Programs to Evaluate Associations between Implementation Strategies and Patient Follow-through." *Genetics in Medicine* 16(10):773–82.

Cronqvist, Lasse, and Dirk Berg-Schlosser. 2009. "Multi-value QCA (mvQCA)." Pp. 69–86 in *Configurational Comparative Methods: Qualitative Comparative Analysis (QCA) and Related Techniques*, edited by Benoît Rihoux and Charles C. Ragin. London: Sage Ltd.

Duşa, Adrian, and Alrik Thiem. 2014. "QCA: A Package for Qualitative Comparative Analysis, R Package Version 1.1-4."

Duşa, Adrian, and Alrik Thiem. 2015. "Enhancing the Minimization of Boolean and Multivalue Output Functions with *eQMC*." *Journal of Mathematical Sociology* 39(2):92–108.

Dy, Sydney M., Pushkal Garg, Dorothy Nyberg, Patricia B. Dawson, Peter J. Pronovost, Laura Morlock, Haya Rubin, and Albert W. Wu. 2005. "Critical Pathway Effectiveness: Assessing the Impact of Patient, Hospital Care, and Pathway Characteristics Using Qualitative Comparative Analysis." *Health Services Research* 40(2):499–516.

Fiss, Peer C., Axel Marx, and Benoît Rihoux. 2014. "Comment: Getting QCA Right." *Sociological Methodology* 44(1):95–100.

Glaesser, Judith, and Barry Cooper. 2014. "Exploring the Consequences of a Recalibration of Causal Conditions When Assessing Sufficiency with Fuzzy Set QCA." *International Journal of Social Research Methodology* 17(4):387–401.

Hug, Simon. 2013. "Qualitative Comparative Analysis: How Inductive Use and Measurement Error Lead to Problematic Inference." *Political Analysis* 21(2):252–65.

Kaminsky, Jessica A., and Amy N. Javernick-Will. 2014. "The Internal Social Sustainability of Sanitation Infrastructure." *Environmental Science and Technology* 48(17):10028–35.

Krogslund, Chris, Donghyun D. Choi, and Mathias Poertner. 2015. "Fuzzy Sets on Shaky Ground: Parameter Sensitivity and Confirmation Bias in fsQCA." *Political Analysis* 23(1):21–41.

Lucas, Samuel R., and Alisa Szatrowski. 2014. "Qualitative Comparative Analysis in Critical Perspective." *Sociological Methodology* 44(1):1–79.

Mackie, John L. 1965. "Causes and Conditions." *American Philosophical Quarterly* 2(4):245–64.

Mackie, John L. 1974. *The Cement of the Universe: A Study of Causation.* Oxford, UK: Oxford University Press.

Pearl, Judea. 2009. *Causality: Models, Reasoning, and Inference.* 2nd ed. Cambridge, UK: Cambridge University Press.

Ragin, Charles C. 1987. *The Comparative Method: Moving beyond Qualitative and Quantitative Strategies.* Berkeley: University of California Press.

Ragin, Charles C. 2000. *Fuzzy-set Social Science.* Chicago: University of Chicago Press.

Ragin, Charles C. 2008. *Redesigning Social Inquiry: Fuzzy Sets and Beyond.* Chicago: University of Chicago Press.

Robinson, Brian E., Margaret B. Holland, and Lisa Naughton-Treves. 2014. "Does Secure Land Tenure Save Forests? A Meta-analysis of the Relationship between Land Tenure and Tropical Deforestation." *Global Environmental Change* 29:281–93.

Schneider, Carsten Q., and Claudius Wagemann. 2012. *Set-theoretic Methods for the Social Sciences: A Guide to Qualitative Comparative Analysis (QCA).* Cambridge, UK: Cambridge University Press.

Seawright, Jason. 2005. "Qualitative Comparative Analysis vis-à-vis Regression." *Studies in Comparative International Development* 40(1):3–26.

Seawright, Jason. 2014. "Comment: Limited Diversity and the Unreliability of QCA." *Sociological Methodology* 44(1):118–21.

Skaaning, Svend-Erik. 2011. "Assessing the Robustness of Crisp-set and Fuzzy-set QCA Results." *Sociological Methods and Research* 40(2):391–408.

Spirtes, Peter, Clark N. Glymour, and Richard Scheines. 2000. *Causation, Prediction, and Search.* 2nd ed. Cambridge, MA: MIT Press.

Srinivasan, V., E. F. Lambin, S. M. Gorelick, B. H. Thompson, and S. Rozelle. 2012. "The Nature and Causes of the Global Water Crisis: Syndromes from a Meta-analysis of Coupled Human-water Studies." *Water Resources Research* 48(10): W10516.

Thiem, Alrik. 2013. "Clearly Crisp, and Not Fuzzy: A Reassessment of the (Putative) Pitfalls of Multi-value QCA." *Field Methods* 25(2):197–207.

Thiem, Alrik. 2014a. "Membership Function Sensitivity of Descriptive Statistics in Fuzzy-set Relations." *International Journal of Social Research Methodology* 17(6):625–42.

Thiem, Alrik. 2014b. "Mill's Methods, Induction and Case Sensitivity in Qualitative Comparative Analysis: A Comment on Hug (2013)." *Qualitative and Multi-Method Research* 12(2):19–24.

Thiem, Alrik. 2014c. "Navigating the Complexities of Qualitative Comparative Analysis: Case Numbers, Necessity Relations, and Model Ambiguities." *Evaluation Review* 38(6):487–513.

Thiem, Alrik. 2014d. "Unifying Configurational Comparative Methods: Generalized-set Qualitative Comparative Analysis." *Sociological Methods and Research* 43(2):313–37.

Thiem, Alrik. 2015a. "Parameters of Fit and Intermediate Solutions in Multi-value Qualitative Comparative Analysis." *Quality and Quantity* 49(2):657–74.

Thiem, Alrik. 2015b. "Using Qualitative Comparative Analysis for Identifying Causal Chains in Configurational Data: A Methodological Commentary on Baumgartner and Epple (2014)." *Sociological Methods and Research* 44(4):723–36.

Thiem, Alrik. 2016. "QCApro: Professional Functionality for Performing and Evaluating Qualitative Comparative Analysis, R Package Version 1.1-0." Retrieved June 8, 2016 (http://www.alrik-thiem.net/software/).

Thiem, Alrik, and Michael Baumgartner. 2016. "Back to Square One: A Reply to Munck, Paine, and Schneider." *Comparative Political Studies* 49(6):801–6.

Thiem, Alrik, Michael Baumgartner, and Damien Bol. 2016. "Still Lost in Translation! A Correction of Three Misunderstandings between Configurational Comparativists and Regressional Analysts." *Comparative Political Studies* 49(6):742–74.

Thiem, Alrik, and Adrian Duşa. 2013a. "Boolean Minimization in Social Science Research: A Review of Current Software for Qualitative Comparative Analysis (QCA)." *Social Science Computer Review* 31(4):505–21.

Thiem, Alrik, and Adrian Duşa. 2013b. "QCA: A Package for Qualitative Comparative Analysis." *R Journal* 5(1):87–97.

Thiem, Alrik, Reto Spöhel, and Adrian Duşa. 2016. "Enhancing Sensitivity Diagnostics for Qualitative Comparative Analysis: A Combinatorial Approach." *Political Analysis* 24(1):104–20.

Author Biographies

Alrik Thiem is a postdoctoral researcher in the Department of Philosophy at the University of Geneva. His work focuses on empirical research and evaluation methods, primarily configurational comparative ones such as coincidence analysis, event structure analysis, and QCA, on which he has published widely in many journals, including *Comparative Political Studies, Evaluation Review, Field Methods, Political Analysis, Sociological Methodology,* and *Sociological Methods and Research.*

Michael Baumgartner is a Swiss National Science Foundation professor in the Department of Philosophy at the University of Geneva, with a specialization in the philosophy of science and logic. He developed the configurational method coincidence analysis and has numerous publications on causation and causal reasoning in general and QCA in particular. Moreover, he is working on mechanistic constitution, cognition, interventionism, determinism, and logical formalization.

Comments

Sociological Methodology
2016, Vol. 46(1) 358–372
© American Sociological Association 2016
DOI: 10.1177/0081175016663592
http://sm.sagepub.com
$SAGE

$ {\cal S}\, 12\, {\cal C} $

MODEL MISSPECIFICATION WHEN ELIMINATING A FACTOR IN AGE-PERIOD-COHORT MULTIPLE CLASSIFICATION MODELS

Robert M. O'Brien*

Abstract

The impossibility of uniquely estimating all of the age, period, and cohort coefficients in age-period-cohort multiple classification (APCMC) models without imposing a constraint on the model is widely recognized. The problem results from a linear dependency in the design matrix, and this dependency involves the linear trends of age effects, period effects, and cohort effects. This article critiques the use of fit statistics to assess the overall importance of the effects of ages, periods, and cohorts in APCMC models. In particular, one proposed strategy to avoid the APCMC model identification problem is to test to see if including only two of the factors in a model (e.g., ages and cohorts) produces a fit that is not significantly different statistically from a model that includes all three factors. If the third factor (in this example periods) does not account for a statistically significant amount of variance, this strategy suggests that one should use the model with only the two factors. This is consistent with model selection approaches. The two-factor model is identified and produces estimates of the individual effects of ages and cohorts. There is, however, a fundamental problem with this approach when used with APCMC models. That

*Department of Sociology, University of Oregon, Eugene, OR, USA

Corresponding Author:
Robert M. O'Brien, HC 64 Box 2604, Castle Valley, UT 84532, USA
Email: bobrien@uoregon.edu

problem results from the complete confounding of the linear effects of the three factors.

Keywords

age-period-cohort models, linear confounding, variable selection, eliminating factors

1. INTRODUCTION

For decades, researchers have known that it is not possible to separate the effects of ages, periods, and cohorts in a straightforward manner since they are intrinsically confounded (Mason et al. 1973; Schaie 1965). Schaie (1965), Mason et al. (1973), and other researchers consider a particular parameterization of this problem: the age-period-cohort multiple classification (APCMC) model. In this parameterization, each of the ages, periods, and cohorts are coded categorically (with a single category for each of these factors serving as reference categories). I will focus on this parameterization, which is widely used in the literature.

The most popular solution to this confounding problem (identification problem) places a single constraint on a model that contains all three factors; for example, the first age-group coefficient equals the second age-group coefficient or the linear trend for the period coefficients is zero. The single just-identifying constraint produces a solution for the age, period, and cohort coefficients under the constraint.[1] Unfortunately, there is an infinite number of these solutions that fits the data equally well, and these solutions can differ considerably depending on the constraint used. Only if the constraint is consistent with the age, period, and cohort parameters that generated the data will the coefficients under the constraint be unbiased estimates of the data-generating parameters. When using this method, we should use theory/substantive knowledge to set a constraint that is more likely to be close to the parameters that generated the data.[2]

The incremental factor fit strategy, which is the focus of this article, does not depend on theory/substantive knowledge. It eliminates one of the factors (ages, periods, or cohorts) if that factor does not make the model fit significantly better statistically when added to a model containing the other two factors. In the ordinary least squares (OLS) situation, we could compute R^2 for the three-factor model ($R^2_{three\ factors}$) and then run a regression model with just two of the factors ($R^2_{two\ factors}$).[3] If the increment in R^2 due to the third factor ($R^2_{increment} = R^2_{three\ factors} - R^2_{two\ factors}$) is not statistically significant, researchers often conclude that the third factor is not important and can be left out of the model with little or no harm. This is a typical variable selection approach. It creates an *atheoretical solution* to the APC identification problem since any just-identifying constrained three-factor solution produces the same $R^2_{three\ factors}$, and if the three-factor model fit is not significantly better statistically than the two-factor model, the two-factor model serves as the solution to the APC identification problem.[4]

The same rationale is used for generalized linear models (GLM), although the criterion differs. Typically, one would calculate the likelihood ratio chi-square: $-2 \times [\ln(\text{likelihood of two factor model}) - \ln(\text{likelihood of the three factor model})]$. This likelihood ratio chi-square (G^2) serves as a significance test for the contribution to the fit of the model of the third factor. If this test does not reject the null hypothesis of no statistically significant improvement in the fit of the model by the third factor, researchers would often drop the third factor.[5]

These incremental fit tests, using OLS or GLM, are employed by many authors working in the APC tradition; for example, see Clayton and Schifflers (1987b); Greenberg and Larkin (1985); Hall, Mairesse, and Turner (2005); Phillips (2014); Shahpar and Li (1999); and Yang and Land (2013). Working with just two of the factors identifies what otherwise (without a constraint) would be an unidentified model.[6]

2. INCREMENTAL FIT TESTS OF FACTORS IN APCMC MODELS

Although it is known that the linear trends of the categorical effect coefficients for ages, periods, and cohorts are linearly dependent (Holford 1983; Luo 2013; O'Brien 2011b), the implications of this dependence are often overlooked when it comes to employing incremental fit tests for the overall importance of ages, periods, and cohorts. Recently Yang and colleagues (Yang et al. 2008; Yang, Fu, and Land 2004; Yang and Land 2013) suggested that such tests should be conducted before deciding to use the intrinsic estimator (IE). Yang and Land (2013:107) state:

> One way to select among models is to conduct model fit tests of whether all three of the A, P, and C effects are present and should be simultaneously

estimated (e.g., see Mason and Smith 1985). That is, analysts should succesively estimate models with A, P, C, AP, AC, PC, and APC sets of effect coefficients and examine the corresponding model fit statistics for improvement as additional sets of combinations of coefficients are added. This gives a sense of the relative importance of A, P, C, effects and the best models that summarize the trends in the observed data.

They conclude: "We reiterate that imposition of a full APC model on data when a reduced model fits the data equally well or better constitutes a model misspecification and should be avoided" (Yang and Land 2013:109).

Yang, Land, and colleagues are not alone in advocating the incremental fit tests for the two- and three-factor models. Clayton and Schifflers (1987a, 1987b) in epidemiology and Hall et al. (2005) in economics provide two other prominent examples. As noted previously, many other authors have used this approach to identify APCMC models.

The problem with such solutions derives from the confounding of the *linear trends* of the three factors: age, period, and cohort. To explain the operational meaning of linear trends in the APCMC context, I use the age-factor as an example. To obtain the linear trend for ages, regress the age effect coefficients (based on one of the just-identifying constrained solutions) from the youngest to the oldest age group on $i = 1$ to I, where I is the number of age groups. A similar procedure can be used to calculate the trend for periods from the earliest to the most recent period and for cohorts from the earliest to the most recent cohort for that same constrained solution. These linear trends differ depending on the constraints used to identify the model. It is important to note that these linear trends are based on the coefficients associated with the APCMC model that is the focus of this article: They might arise from a curvilinear data-generating effect, but the confounding in the model will still exist.[7] Because of the categorical coding of the age, period, and cohort effects in APCMC models, any linear trend of the left-out factor is attributed to the other two factors in a model. Table 1 summarizes this problem.

Table 1 uses R^2 notation to show the amount of variance in the dependent variable that is explicitly accounted for (column 2) by different two-factor models (column 1) and the amount that is actually accounted for (captured) by the model (column 3). We could write analogous notation for GLM models. For the standard APCMC model with cohort as the left-out factor, the *explicit model* specifies just the two main effects

Table 1. Explicit and Estimated Model R^2 and the Assumptions Associated with the Explicit Model for the Standard Age-period-cohort Multiple Classification (APCMC) Approach When Dropping a Factor

APCMC Approach When Dropping a Factor	Explicit Model	Captured by Model	Assumptions about the Left-out Factor When It Is Dropped from the Model
Age-period model	R^2_{AP}	$R^2_{AP(C_L)}$	Linear and nonlinear effects equal zero $coh1 = coh2 = \cdots = cohK = 0$
Age-cohort model	R^2_{AC}	$R^2_{AC(P_L)}$	Linear and nonlinear effects equal zero $per1 = per2 = \cdots perJ = 0$
Period-cohort model	R^2_{PC}	$R^2_{PC(A_L)}$	Linear and nonlinear effects equal zero $age1 = age2 = \cdots = ageI = 0$

(R^2_{AP}) by including the categorically coded variables for ages and periods. This estimated model, however, accounts for the two main effects and any linear effects of the third factor $(R^2_{AP(C_L)})$. The increment in $R^2 F$-test using the standard approach tests only whether the nonlinear effects (deviations from the linear trend) associated with the third factor are statistically significant. The fourth column of Table 1 indicates that when a factor is dropped from the model, the researcher fixes both its linear effects *and* the deviations from linearity to zero. Since the two-factor model already captures the linear effects of the third factor, the difference in the fit of the three-factor model and the two-factor model is that the three-factor model accounts for the deviations of the dropped out factor's effects from its linear trend. In order for the coefficients associated with the two-factor model to be correct in terms of the process that generated the outcome data, the assumption is that the third factor's linear trend is zero and the deviations of its effects from their linear trend are zero.

3. THE APCMC PROBLEM

The APCMC problem is well known. The individual coefficients for ages, periods, and cohorts are linearly dependent, and therefore, unique estimates of these coefficients are not estimable. The problem is that the linear effects of any two of the three factors (age, period, and cohort)

are linearly related to the third (Clayton and Schifflers 1987b; Holford 1983; O'Brien 2014, 2015a; Smith 2004). Models that contain each of these factors are not identified, and the matrix of independent variables does not have an inverse. This is the case whether these factors are linearly coded or whether they are coded using categorical variables. I focus on the situation in which the ages, periods, and cohorts are categorically coded (the APCMC model) since this is the most common situation, but I note the linear coding situation next.

3.1. *Linear Confounding with Linear Coding*

The confounding for the linearly coded variables is easy to describe. If we code age from the youngest age to the oldest age as $1, 2, \ldots, I$, where I is the number of ages; period from the earliest to the most recent period as $1, 2, \ldots, J$, where J is the number of periods; and cohorts from the earliest to the most recent cohort as $1, 2, \ldots, K$, where K is the number of cohorts, then we can identify each of the cohorts as $k = I - i + j$. If we know an observation's age and period, we can determine the cohort. With linear coding, using just two of these factors (variables) fits the data as well as using all three. The information in the third factor is redundant with the information contained in the other two factors. In this situation, researchers are not likely to eliminate a factor because it does not improve the model fit when added to a model that contains the other two factors. As we have seen from the literature cited earlier, however, this temptation is more difficult to resist when the factors are categorically coded.

3.2. *Linear Confounding with Categorical Coding*

Categorical coding of the three factors is far more common in APC analysis. I denote the APCMC model with categorical coding as

$$Y_{ij} = \mu + \alpha_i + \pi_j + \chi_{(I-i+j)} + \epsilon_{ij}, \tag{1}$$

where Y_{ij} is the observed value of the dependent variable in the ith age group and jth period, μ represents the intercept, α_i is the age effect for the ith age group, π_j is the period effect for the jth period, $\chi_{(I-i+j)}$ is the cohort effect for the kth cohort ($k = I - i + j$), and ϵ_{ij} is the residual term for the ijth age-period-specific observation.

Table 2. Relationships between the Linear Effects of Ages, Periods, and Cohorts*

	Period 1 (1)	Period 2 (2)	Period 3 (3)	Period 4 (4)	Period 5 (5)
Age 1 (1)	4	5	6	7	8
Age 2 (2)	3	4	5	6	7
Age 3 (3)	2	3	4	5	6
Age 4 (4)	1	2	3	4	5

*Cohort coding is in the cells corresponding to the observations in each cohort.

For concreteness, Table 2 shows the linear coding in a 4 by 5 age-period table. The linear coding for ages is 1 to 4 on the rows; the linear coding of periods is coded as 1 to 5 on the columns. Using this linear coding and ($k = I - i + j$), we can generate the linear coding for cohorts; for example, the second age group in the second period is in cohort 4 ($= 4 - 2 + 2$), as is the first age group in the first period [$4 (= 4 - 1 + 1)$]. The linear effects of cohorts are linearly dependent on the linear effects of age and period. Thus, when the linear effects of age and period are in the model (when age and period are in the model), this controls for the linear effects of cohorts. Similar linear relationships apply to periods ($j = k - I + i$) and to ages ($i = j - k + I$). *Any two factors in the model account for the linear effect of the third factor.*

4. RELATIONSHIPS BETWEEN THE LINEAR TRENDS FOR AGES, PERIODS, AND COHORTS

Although different constrained solutions can lead to very different estimates of the age, period, and cohort coefficients, the linear trends for ages, periods, and cohorts for different solutions are systematically related. Rodgers (1982:782) shows that the linear trends of the age-effects, period-effects, and cohort-effects are related to each other in the following straightforward manner (presented using my own notation):

$$t_a^* = t_a + k$$

$$t_p^* = t_p - k$$

$$t_c^* = t_c + k. \tag{2}$$

Interpreting the first equation, t_a represents the linear trend in the age coefficients from an "original" constrained solution. If we calculate a different constrained solution and the linear trend in the age coefficients is t_a^* and that trend is k more than the linear trend for the original solution, then the trend for the new solution for periods will be k less than the original trend for periods, and the trend in the cohort coefficients for the new solution will be k greater than the linear trend for the cohort coefficients in the original solution. That is, different constrained solutions produce different trend estimates for the ages, periods, and cohorts, but the differences in the trends for ages, periods, and cohorts based on different solutions are systematically related. The relationships in equation 2 hold exactly when an APC model is based on a single just-identifying constraint. Equation 2 holds approximately when we drop one of the categorically coded factors from the model. Dropping one factor from the model results in more constraints than needed to identify the model. It constrains each of the categories of that factor to have a zero relationship with the dependent variable in the model and, because of this, a zero linear trend. This is why dropping the factor from the model does not exactly reproduce the pattern in equation 2 or produce fit statistics that are the same as when fixing one of the trends to zero.

I focus in the following on two situations. Leaving one of the three factors out of the model is the situation most commonly found in the literature. The other is to constrain the trend of one of the factors to be zero, which is a just-identifying constraint and does follow equation 2 exactly. With categorical coding, the linear trends are similar using either strategy; but, as noted previously, they are not exactly the same. Intuitively, imagine that the trend of periods for the data-generating process is k and k is positive, but we constrain the slope for the period coefficients to be zero. This means that the expected value of the estimated trend under this constraint is not an unbiased estimate of the trend of the period parameters that generated the outcome data. It is too small by k. Given equation 2, this also means that the trend for ages and for cohorts based on the data-generating process is overestimated by k.

The relationships between the slopes of the three factors described in equation 2 provide insight into what happens when we drop one of the factors from the APC model.

1. The left-out factor's effects are constrained to be zero, both its linear trend effects and the effects of its deviations from the linear trend.

2. When the model contains just two factors, those two factors take credit for their own linear trend effects, their effects that involve deviations from their linear trends, *and they take credit for the linear trend effects of the third (left-out) factor.*

3. Since the other two factors get credit for any linear trend effects in the data-generating process due to the left-out factor, this may make the two factors appear to be statistically significant even if their contribution to the data-generating process is not.

4. When the third factor is added to the two-factor model to determine whether it accounts for addition variance in the dependent variable, any variance due to its linear trend has already been "controlled for," and the third factor will not get credit for any effects due to its linear trend.

5. With the test for incremental variance, only effects that are associated with the third factor's deviations from its linear trend will be attributed to it.

6. Leaving the third factor out of the model based on its incremental fit not being statistically significant will too often eliminate a substantively important factor. This elimination affects the coefficient estimates of the two factors in the model.

This unique variance associated with the third factor is an estimable function (O'Brien 2014); it is the same no matter which just-identifying constraint is applied to the three-factor model. The increment in R^2 in this APCMC situation, however, is only a test of the nonlinear effects after any linear trend effect of the third factor has been removed.

5. REGRESSION PROCEDURES AND LINEAR TRENDS FOR AGES, PERIODS, AND COHORTS

The main purpose of this section is to provide a greater intuitive understanding of these relationships: Researchers typically have not (and will not) run these regressions while conducting their research (except for 1 below). I use cohort as an example of the "third" or left-out factor. The analogous procedures can be used with periods or ages as the left-out factor and with any GLM procedure.

1. Run an OLS regression with any single just-identifying constraint to calculate R^2_{apc}. R^2_{apc} is the proportion of the variance in the dependent variable accounted for by the age, period, and cohort factors. The

individual age, period, and cohort effects differ depending on the constraint employed; however, R^2_{apc} is the same for any just-identifying constraint.

2. Run a regression with just the age and period factors (categorically coded variables) and calculate R^2_{ap}. To show that R^2_{ap} captures these linear trend effects, regress the dependent variable on the age and period factors and with cohorts coded linearly from 1 to K for the individual cohorts from earliest to most recent cohorts (as opposed to the categorical coding of cohorts). This model will require that you use one just-identifying constraint since the linear components for cohorts are included in the model. The result is $R^2_{ap(c_L)}$ and $R^2_{ap} = R^2_{ap(c_L)}$ (see again Table 1). The linear trends for the explicit $R^2_{ap(c_L)}$ model (the model that codes the linear trend for cohorts and contains the age and period categorically coded variables) will differ depending on the constraint used to identify the model.

3. The difference between the R^2 for the full APC model and R^2 for the age and period factors $\left(R^2_{apc} - R^2_{ap} \right)$ is due to the deviation of the cohort effects around the linear trend in the cohort effects. To show that this interpretation holds, we find the deviations of the cohort effects from their linear trend in a three-factor model. To calculate these deviations, run any R^2_{apc} model with a single constraint. Calculate the trend in the cohort coefficients, then calculate the deviations in the estimated cohort coefficient from their predicted values based on this trend, then run a model with the age and period categorically coded variables and code each cohort with the deviations of the cohort coefficients from the linear trend in cohorts (these deviations are estimable functions). The result is $R^2_{ap(cohort\ deviations)} = R^2_{apc}$. We can, of course, conduct such analyses with age or period as the third factor.

4. The F-test for the statistical significance of $R^2_{apc} - R^2_{ap}$ is often treated as a test of the importance of a third factor in the generation of the outcome variable. The standard F-test for the increment in variance accounted for does not test for the overall effects of cohorts in the process that generated the outcome data because it does not include any linear trend effects of cohorts. This incremental F-test for $R^2_{apc} - R^2_{ap}$ is a *sufficient test* for the statistical significance of the cohort factor since it indicates whether the cohort variation around its linear trend is statistically significant (without the additional variance in the dependent

variable that might be accounted for by any linear trend in the cohorts). These deviations from the linear trend are estimable; they are the same no matter which just-identifying constraint is used to calculate R^2_{apc}. This test can also be used to test the total effects of deviation from the linear trends for periods $(R^2_{apc} - R^2_{ac})$ and for ages $(R^2_{apc} - R^2_{pc})$. We assume throughout that the correctly specified model is the full APCMC model. Similar tests for GLM models, using likelihood chi-square tests, can be used with the same interpretational caveats.

There is a difference between leaving a factor out of the equation (then it is constrained to have no effect on the solution either from its linear trend or deviations around its linear trend) and constraining the linear trend to be zero. In the latter case, the linear trend effect is zero, but the deviations from the linear trend are allowed to account for variance in the dependent variable. In both cases, however, any linear effect in the factor is absorbed by the other two factors.

6. DISCUSSION

One feature of the APC problem is nearly universally recognized: the impossibility of estimating a model with all three of these factors simultaneously in the model unless some sort of constraint is placed on the model to identify it. Placing a constraint on the model results in a specific set of estimates for the age, period, and cohort effects; a different constraint results in a different set of estimates. Only if the constraint is consistent with the parameters that generated the outcome variables will the solutions under the constraint be an unbiased estimate of those parameters. In fixed models, those constraints are explicit, and in mixed models (Bell and Jones 2014; O'Brien, Hudson, and Stockard 2008; Yang and Land 2006), the constraint is embedded in the model. The linear trends of ages, periods, and cohorts are linearly dependent, and this creates the APC identification problem, while the deviations around the linear trend of these factors are identified/estimable (Clayton and Schifflers 1987a, 1987b; Holford 1983; O'Brien 2014).

What is less well understood are the implications of the complete confounding (linear dependency) of the linear trends estimates of ages, periods, and cohorts on results that compare three-factor and two-factor models. For the categorically coded APC models that drop one of the factors from the model, any linear trend in the dropped out factor is absorbed by the other two factors. The two-factor model takes credit for any linear trend of the effects in the third factor. When the linear trend in the third factor is constrained to zero (rather than dropped from the model), the relationship between the effect of this constraint on the linear trend in the third factor and the linear trends in the other two factors is exact, as shown in equation 2. These insights explain specific shifts between the coefficients of the APCMC model when different factors are dropped from the model and show that the test for incremental variance should not be used as a criterion for dropping a factor from the model. The incremental fit test in the APCMC context is an atheoretical test that is likely to lead to a misspecified model.

What is the researcher to do? The following strategies are not a panacea, but they can provide valuable information about the age, period, and cohort effects.

1. Estimable functions do not depend on the constraint used to identify the APCMC model. Some examples of estimable functions are the second differences of age effects, period effects, and cohort effects: deviations of age effects from the linear trend in the age effects, deviations of period effects from linear trends in the period effects, and deviations of cohort effects from linear trends in the cohort effects and the variance accounted for by the three-factor model. Each of these estimable functions (and others) tells us something about the age, period, and cohort effects (O'Brien 2014).

2. Factor characteristic models can be used to estimate the effects of ages, periods, and cohorts. For example, we can categorically code ages and periods and code cohorts using a proxy variable such as the proportion of the cohort that was born out of wedlock or the relative size of the birth cohort. The effectiveness of this approach depends in part on how well the characteristics capture the effects of cohorts on the dependent variable (O'Brien 2000).

3. We can use a just-identifying constraint but should make sure the constraint is based on theory and/or substantive knowledge. The rationale is that if the theory and/or substantive knowledge are nearly correct and the constraint is based on these, then the constraint is more likely consistent with the data-generating parameters. If the constraint is approximately consistent with the data-generating parameters, it should provide an approximately unbiased estimate of those parameters. Using multiple approaches that reach similar conclusions will build confidence that the

analysis may be getting at the data-generating parameters (O'Brien 2015a).

Notes

1. A just-identifying constraint in this context is a single constraint that just identifies the model (under the constraint) so that with the constraint there is a unique solution.
2. This is not an error-free process, but it is a process that is likely to lead to a defensible constraint and one that hopefully closely matches the parameters that generated the data.
3. To run the three-factor model, one just-identifying constraint is necessary. No matter which just-identifying constraint is used, the R^2 for the three-factor model is the same (Mason et al. 1973). R^2 for the three-factor model is an estimable function (O'Brien 2014).
4. The authors using this method do not justify the constraint used to produce the three-factor solution because no such justification is necessary for the incremental fit test.
5. Information criteria such as Akaike Information Criterion (AIC) and Bayesian Information Criterion (BIC) are also utilized when making this decision.
6. Some approaches examining data varying by ages, periods, and cohorts do not attempt to separate the effects of ages, periods, and cohorts (Firebaugh 1989; Harding and Jencks 2003), and they are not the subject of this article.
7. Bell and Jones (2015) discuss this point in their paper on hierarchical age-period-cohort models.

References

Bell, Andrew, and Kelvyn Jones. 2014. "Another 'Futile Quest'? A Simulation Study of Yang and Land's Hierarchical Age-period-cohort Model." *Demographic Research* 30:333–60.

Bell, Andrew, and Kelvyn Jones. 2015. "Should Age-period-cohort Analyst Accept Innovation without Scrutiny? A Response to Reither, Masters, Yang, Powers, Zheng, and Land." *Social Science and Medicine* 128:331–33.

Clayton, D., and E. Schifflers. 1987a. "Models for Temporal Variation in Cancer Rates I: Age-period and Age-cohort Models." *Statistics in Medicine* 6:449–67.

Clayton, D., and E. Schifflers. 1987b. "Models for Temporal Variation in Cancer Rates II: Age-period-cohort Models." *Statistics in Medicine* 6:468–81.

Firebaugh, Glenn. 1989. "Methods for Estimating Cohort Replacement Effects." Pp. 243–62 in *Sociological Methodology*. Vol. 19, edited by C. C. Clogg. Washington, DC: American Sociological Association.

Greenberg, David F., and Nancy J. Larkin. 1985. "Age-cohort Analysis of Arrest Rates." *Journal of Quantitative Criminology* 1:227–40.

Hall, Bronwyn H., Jacques Mairesse, and Laure Turner. 2005. "Identifying Age, Cohort, and Period Effects in Scientific Research Productivity: Discussions and Illustration

Using Simulated and Actual Data on French Physicists." Working Paper No. 11739, National Bureau of Economic Research, Cambridge, MA.

Harding, David J., and Christopher Jencks. 2003. "Changing Attitudes toward Premarital Sex: Cohort, Period, and Aging Effects." *Public Opinion Research* 67: 211–26.

Holford, Theodore R. 1983. "The Estimation of Age, Period, and Cohort Effects for Vital Rates." *Biometrics* 39:311–24.

Luo, Liying. 2013. "Assessing Validity and Application Scope of the Intrinsic Estimator Approach to the Age-period-cohort Problem." *Demography* 50:1945–67.

Mason, William M., and Herbert L. Smith. 1985. "Age-period-cohort Analysis and the Study of Deaths from Pulmonary Tuberculosis." Pp. 151–228 in *Cohort Analysis in Social Research: Beyond the Identification Problem*, edited by W. M. Mason and S. E. Fienberg. New York: Springer-Verlag.

Mason, Karen O., William M. Mason, H. H. Winsborough, and W. Kenneth Poole. 1973. "Some Methodological Issues in Cohort Analysis of Archival Data." *American Sociological Review* 38:242–58.

O'Brien, Robert M. 2000. "Age Period Cohort Characteristic Models." *Social Science Research* 29:123–39.

O'Brien, Robert M. 2011a. "The Age-period-cohort Conundrum as Two Fundamental Problems." *Quality and Quantity* 45:1429–44.

O'Brien, Robert M. 2014. "Estimable Functions in Age-period-cohort Models: A Unified Approach." *Quality and Quantity* 48:457–74.

O'Brien, Robert M. 2015a. *Age-period-cohort Models: Approaches and Analyses with Aggregate Data*. New York: Chapman and Hall.

O'Brien, Robert M., Kenneth Hudson, and Jean Stockard. 2008. "A Mixed Model Estimation of Age, Period, and Cohort Effects." *Sociological Methods and Research* 36:402–28.

Phillips, Julie A. 2014. "A Changing Epidemiology of Suicide? The Influence of Birth Cohorts on Suicide Rates in the United States." *Social Science and Medicine* 114: 151–60.

Rodgers, Willard L. 1982. "Estimable Functions of Age, Period, and Cohort Effects." *American Sociological Review* 47:774–87.

Schaie, Klaus Warner. 1965. "A General Model for the Study of Developmental Problems." *Psychological Bulletin* 64:92–107.

Shahpar, Cyrus, and Guohoa Li. 1999. "Homicide Mortality in the United States, 1935–1994: Age, Period, and Cohort Effects." *American Journal of Epidemiology* 150: 1213–22.

Smith, Herbert L. 2004. "Response: Cohort Analysis Redux." Pp. 111–19 in *Sociological Methodology*. Vol. 34, edited by R. M. Stolzenberg. Oxford, UK: Basil Blackwell.

Yang, Yang, Wenjiang J. Fu, and Kenneth C. Land. 2004. "A Methodological Comparison of Age-period-cohort Models: Intrinsic Estimator and Conventional Generalized Linear Models." Pp. 75–110 in *Sociological Methodology*. Vol. 34, edited by R. M. Stolzenberg. Oxford, UK: Basil Blackwell.

Yang, Yang, Sam Schulhofer-Wohl, Wenjiang J. Fu, and Kenneth C. Land. 2008. "The Intrinsic Estimator for Age-period-cohort Analysis: What It Is and How to Use It." *American Journal of Sociology* 113:1697–736.

Yang, Yang, and Kenneth C. Land. 2006. "A Mixed Models Approach to the Age-period-cohort Analysis of Repeated Cross-Section Surveys: Trends in Verbal Test Scores." Pp. 75–97 in *Sociological Methodology*. Vol. 36, edited by R. M. Stolzenberg. Oxford, UK: Basil Blackwell.

Yang, Yang, and Kenneth C. Land. 2013. *Age-period-cohort Analysis: New Models, Methods, and Empirical Applications*. New York: Chapman and Hall.

Author Biography

Robert M. O'Brien is a professor emeritus of sociology at the University of Oregon, where he taught for over 30 years. He specializes in criminology and quantitative methods and has published extensively in both areas. His interest in age-period-cohort models was kindled by a talk given by Bill Mason at the University of Oregon in the late 1980s. He published his first article using age-period-cohort models in *Criminology* in 1989. Recently he published a book titled *Age-Period-Cohort Models: Approaches with Aggregate Data* (Chapman and Hall, 2015). Other recent publications include "Dropping Highly Collinear Variables from a Model: Why It Typically Is Not a Good Idea" (*Social Science Quarterly* 2016), "Age-Period-Cohort Models and the Perpendicular Solution" (*Epidemiologic Methods* 2015), and "Estimable Functions of Age-Period-Cohort Models: A Unified Approach" (*Quality and Quantity* 2014).

Submission Information for Authors

Sociological Methodology (SM) is the only American Sociological Association periodical publication devoted entirely to research methods. It is a compendium of new and sometimes controversial advances in social science methodology. Contributions come from diverse areas and have something new and useful—and sometimes surprising—to say about a wide range of methodological topics. *SM* seeks qualitative and quantitative contributions that address the full range of methodological problems confronted by empirical research in the social sciences, including conceptualization, data analysis, data collection, measurement, modeling, and research design. The journal provides a forum for engaging the philosophical issues that underpin sociological research. Papers published in *SM* are original methodological contributions including new methodological developments, reviews or illustrations of recent developments that provide new methodological insights, and critical evaluative discussions of research practices and traditions. *SM* encourages the inclusion of applications to real-world sociological data. *SM* is published annually as an edited, hardbound book.

The content of each annual volume of the journal is driven by submissions initiated by authors; the volumes do not have specific themes. Editorial decisions about manuscripts submitted are based on the advice of expert referees. Criteria include originality, breadth of interest and applicability, and expository clarity. Discussions of implications for research practice are vital, and authors are urged to include empirical illustrations of the methods they discuss.

Manuscripts should be submitted electronically to http://mc .manuscriptcentral.com/smx. Submitting authors are required to set up an online account on the SAGE Track system, powered by Scholar One. The submission fee of $25 is payable through SAGE Track. Submission of a manuscript for review by *Sociological Methodology* implies that the article has not been previously published and that it is not under review elsewhere.

For full manuscript submission guidelines see sm.sagepub.com. For further information about the journal, visit the ASA journal page at http://www .asanet.org/research-and-publications/journals/sociological-methodology. Inquiries concerning the appropriateness of material are welcome. Prospective authors should send inquiries to soc-methodology@psu.edu.